SEMIOTICS IN POLAND
1894–1969

SYNTHESE LIBRARY

STUDIES IN EPISTEMOLOGY,

LOGIC, METHODOLOGY, AND PHILOSOPHY OF SCIENCE

VOLUME 119

SEMIOTICS IN POLAND

1894–1969

Selected and edited
with an Introduction by

JERZY PELC
University of Warsaw

D. REIDEL PUBLISHING COMPANY
DORDRECHT: HOLLAND / BOSTON: U.S.A.

PWN—POLISH SCIENTIFIC PUBLISHERS
WARSAW: POLAND

Library of Congress Cataloging in Publication Data

Pelc, Jerzy, comp.
Semiotics in Poland, 1894–1969.

(Synthese library; 119)
Translation of Semiotyka polska, 1894–1969.
Includes bibliographical references and index.
1. Semantics (Philosophy). 2. Philosophy, Polish.
I. Title.
B4690.S4P4413 149′.94 77-22405
ISBN 90-277-0811-8

This translation has been made by Olgierd Wojtasiewicz from
Semiotyka polska 1894–1969
ed. by Jerzy Pelc
published in 1971 by Państwowe Wydawnictwo Naukowe, Warszawa

Distributors for Albania, Bulgaria, Chinese People's Republic, Czechoslovakia, Cuba, German Democratic Republic, Hungary, Korean People's Democratic Republic, Mongolia, Poland, Rumania, Vietnam, the U.S.S.R. and Yugoslavia
ARS POLONA
Krakowskie Przedmieście 7, 00-068 Warszawa 1, Poland

Distributors for the U.S.A., Canada and Mexico
D. REIDEL PUBLISHING COMPANY, INC.
Lincoln Building, 160 Old Derby Street, Hingham, Mass. 02043, U.S.A.

Distributors for all other countries
D. REIDEL PUBLISHING COMPANY
P.O. Box 17, Dordrecht, Holland

Printed in Poland by Z. G. im. M. Kasprzaka

CONTENTS

PREFACE TO THE ENGLISH EDITION

In the Introduction to the Polish-language version of the present book I expressed the hope that Polish studies in semiotics would before long be numerous enough to make possible another anthology on semiotics in Poland containing material published since 1970. That hope has in fact come true.

The fact that semiotic research has been gaining momentum in this country is reflected in the growing interest in the discipline, in expanding international contacts, and in the steady increase in the number of publications.

Thus, 1972 saw the setting up of the Department of Logical Semiotics, headed by the present writer, at Warsaw University Institute of Philosophy. The seminar on semiotics, which I started in 1961, had met more than two hundred times by the end of 1976; since 1968, meetings have been held jointly with the Polish Semiotic Society. Another seminar, confined to university staff and concerned with logical semiotics, which was initiated in 1970, had met more than fifty times by the end of 1976. The former seminar often plays host to foreign visiting professors; so far scholars from Australia, Belgium, Britain, Canada, Czechoslovakia, France, the German Democratic Republic, Italy, the Netherlands, the Soviet Union, and the United States have attended. It is regularly attended by Polish scholars from all over the country and is interdisciplinary in nature; among the speakers and participants are logicians, philosophers, linguists, information scientists, praxiologists, psychologists, theorists of art, literature, theatre and film, architects, zoosemioticians. The other is attended by postgraduates and junior research workers, mostly Poles who specialize in logic, philosophy, linguistics, psychology, sociology, and history of art. It is also worth noting that the university curriculum for linguists (some 1500 students annually) now includes a course in logical semiotics.

The Yale University Department of Linguistics and the Indiana University Research Center for Language and Semiotic Studies have suggested to Warsaw University that they would co-operate on a regular basis with the latter's Department of Logical Semiotics; Professor

Thomas A. Sebeok, chairman of the Indiana Center, held lectures in Warsaw in 1976, and a staff member of the Department of Logical Semiotics attended a semestral postgraduate course in Indiana in the same year.

The Polish Semiotic Society, founded in 1968 and affiliated with the Polish Academy of Sciences, has over one hundred members specializing in various fields. By the end of 1976 it had held over one hundred meetings, attended by speakers from all over Poland and from several foreign universities. It sponsors the publication of *Studia Semiotyczne*,[1] the first eight volumes of which contain 98 papers, about one half of them being texts of papers previously read at meetings of the Polish Semiotic Society, held jointly with the Warsaw University Department of Logical Semiotics. Every volume of *Studia Semiotyczne* is accompanied by two separate booklets containing English-language and Russian-language summaries respectively. Beginning with Vol. VI, some papers in *Studia Semiotyczne* appear in English or in French; in the future, *Studia Semiotyczne* will carry papers in congress languages only. The Polish Semiotic Society also sponsors a series of monographs, two of which have already appeared, while another two are ready for the press.[2]

The Polish Semiotic Society promotes contacts between Polish scholars and semioticians abroad.

International conferences on semiotics, apart from those listed in the *Introduction* to the Polish edition, were held in Warsaw in 1965 and in 1976, and 1973 saw the appearance of the materials from the symposium on semiotics held in Warsaw in 1968.[3]

Polish semioticians are active in international bodies concerned with semiotic research, namely in the International Association for Semiotic Studies and several are on the editorial board of *Semiotica*, the IASS quarterly.[4] They also participated in the IASS meeting in Parma 1971, and in the First Congress of the IASS, Milan 1974.[5]

Since the *Introduction* to the Polish edition of this book was written Polish semioticians have taken part in the following international meetings which were mainly or to a large extent concerned with semiotics: the semiotic seminars held by the Centro Internazionale di Semiotica e Linguistica at Urbino, Italy, in 1968, 1969, and 1971 through 1974; the semiotic conference organized by the Hungarian Academy of Sciences at Tihany, Hungary, in 1974; the symposium on philosophy and language, held at the Institut des Hautes Études de Belgique (Brussels 1975); the 1975 Linguistic Institute, organized at Tampa, Fla., by the Linguistic Society of America jointly with the State University System of Florida (this meeting included Polish Seminars in Semiotics and the North American Semiotics Colloquium, at which the Semiotic Society of America was founded, with Henry Hiż of the University of Pennsylvania, a native Pole, as its president); the Semiotisches Colloquium, held by the Technische Universität and Staatliches Institut für Musikforschung Preussischer Kulturbesitz (West Berlin, 1975); the colloquium on semiotics and pragmatism, held at Stuttgart University in 1976; the Symposium on Semiotics and Theories of Symbolic Behavior in Eastern Europe and the West, organized by Brown University in Providence, R.I., in 1976; the II Wiener Symposium über Semiotik, organized by the Arbeitskreis "Semiotik" des Oesterreichischen Linguistischen Programms and Wiener Sprachgesellschaft (Vienna, 1976).[6]

The above list, given by way of example, disregards such international events as the meetings of members of the Institut International de Philosophie at Cambridge (1972), at Varna (1973), at Dubrovnik (1974), in Iran (1975), at Bern (1976), and the philosophical and logical congresses in Bucharest (1971), Varna (1973), New Delhi (1975), and congresses and symposia on literary history and linguistics, e.g., Bellagio (1973) and Bielefeld (1975), which were partly concerned with semiotics and at which Polish scholars were also present.[7]

When we discuss the activity of Polish scholars in semiotics, mention is also due to those who permanently reside abroad, namely Henry Hiż (University of Pennsylvania, Philadelphia), Jerzy Kalinowski (CRNS, France), Włodzimierz Krysiński (Carleton University, Ottawa), Eryk Landowski (Paris), Edward Stankiewicz (Yale University, N.H.), Jan Srzednicki (Melbourne University), Ryszard Zuber (Paris University), and those who are temporarily working outside Poland, such as Irena Bellert and Anna Wierzbicka (in Canada and Australia, respectively).

A dictionary of semiotic terminology, initiated by Irena Bellert, is an example of an international project in which Poles living on both sides of the Atlantic are collaborating.[8]

Another instance of the expansion of Polish semiotic research are the lectures given abroad by members of the Polish Semiotic Society. They have so far covered three continents, including such centres as Oxford, Exeter, Regensburg, Bielefeld, Venice, Urbino, Helsinki, Paris, Brussels, West Berlin, Amsterdam, Leyden, Utrecht, Groningen, Wassenaar, Cambridge (Mass.), Ottawa and Tokyo.[9]

In publications which, in all probability, best reflect advances in any discipline, and this applies to semiotics as well, a brief preface cannot do justice to the full bibliographical data, and obviously is not intended to do so. That is why I confine myself to listing in a footnote, a few items by way of example only, from among the dozens of books and papers written by Polish authors on this subject in recent years. Some of the items listed there have the character of reports and directories[10] which will supply the interested reader with further information. I believe that anyone who adds the nearly one hundred papers which have so far appeared in *Studia Semiotyczne* to the items listed in the footnote will agree that the publications which have appeared since 1969, the *terminus ad quem* of the present book, would, in fact suffice to make not one new *Semiotics in Poland*, but several anthologies of this kind.

Warsaw, March 1977 *Jerzy Pelc*

NOTES

[1] *Studia Semiotyczne* (Semiotic Studies), edited and prefaced by Jerzy Pelc; appears annually, Vol. I in 1970, Vol. VIII in 1977.

[2] Jerzy Pelc, *O użyciu wyrażeń* (*On Uses of Expressions*), Wrocław-Warszawa 1971; Barbara Stanosz, Adam Nowaczyk, *Logiczne podstawy języka* (*Logical Foundations of Language*), Wrocław–Warszawa 1976; Jerzy Pelc, *Semiotics: an Overview* (to appear); *Lingwistyka i język. Studia z semiotyki lingwistycznej 1660–1969* (*Linguistics and Language. Studies in Linguistic Semiotics* 1660–1969), (ed.) Jerzy Pelc (to appear).

[3] *Recherches sur les systèmes signifiants: Symposium de Varsovie* 1968, (ed.) Josette Rey-Debove, *Approaches to Semiotics*, 18, The Hague 1973.

[4] Poland is represented on the IASS Executive Board by Jerzy Pelc and Stefan Żółkiewski, and on the Editorial Committee of *Semiotica* by Jerzy Pelc. By the end of 1976 *Semiotica* had fourteen Polish contributors.

[5] The IASS meeting at Parma was attended, on behalf of the Polish Semiotic Society, by J. Pelc and S. Żółkiewski, and the congress in Milan, by Andrzej Bogusławski, Janusz Chmielewski, Leon Koj, Witold Marciszewski, Maria R. Mayenowa, Jerzy Pelc, Barbara Stanosz, Jan Sulowski, and Stefan Żółkiewski, while Marian Mazur and Maria Nowakowska sent in papers.

[6] The following Polish scholars took part in the international conferences mentioned: J. Pelc and S. Żółkiewski at Urbino (S. Żółkiewski and J. Kuryłowicz are members of the Scientific Committee of the Centro Internazionale di Semiotica e Linguistica at Urbino); A. Bogusławski and J. Pelc at Tihany; J. Pelc in Brussels; G. Dydel-Wróblewska, J. Jadacki, J. Pelc, Z. Saloni, B. Stanosz, M. Świdziński at Tampa, Fla.; W. Marciszewski, J. Wierzchowski in West Berlin; H. Buczyńska-Garewicz in Stuttgart; L. Koj, M. R. Mayenowa, J. Pelc at Providence, R. I.; M. Nowakowska and A. Schaff in Vienna.

[7] Meetings organized by the Institut International de Philosophie were attended by the following Polish members of the Institute who are interested in semiotics: I. Dąmbska, J. Pelc, and A. Schaff. In Bucharest and in Varna fairly large groups of Polish scholars were present, and many of them attended discussions on the philosophy of language. In New Delhi and in Bellagio the only representative of Polish semioticians was J. Pelc, who was also the only one to be present in person in Bielefeld as M. Nowakowska, whose paper was read during the proceedings, was unable to attend.

[8] In connection with this project, J. Pelc visited McGill University in Montreal in 1976 as scientific adviser, and was followed, a few months later, by Władysław Stróżewski, of the Jagiellonian University in Cracow. Several Polish scholars have prepared entries for the dictionary.

[9] Regensburg: A. Bogusławski; Tokyo: W. Kotański; Oxford and Cambridge (Mass.): J. Chmielewski; Venice: S. Żółkiewski; Bielefeld: M. Nowakowska and J. Pelc; the remaining places: J. Pelc.

[10] For instance: Witold Doroszewski, *Elements of Lexicology and Semiotics*, in: *Approaches to Semiotics*, The Hague–Paris–Warszawa 1974; Tadeusz Kowzan, *Littérature et spectacle*, in: *Approaches to Semiotics*, The Hague–Paris–Warszawa 1974; Mieczysław Wallis, *Arts and Signs*, in: *Studies in Semiotics*, Lisse, Holland, 1975. A fragmentary bibliography of Polish publications on semiotics is to be found in *Studia z historii semiotyki* (*Studies in the History of Semiotics*), (ed.) Jan Sulowski, Wrocław–Warszawa, Vol. I, 1971, Vol. II, 1973, Vol. III, 1976. Bibliographical data and a kind of chronicle of semiotic research in Poland are given by Jerzy Pelc, 'The Development of Polish Semiotics in the Post-War Years', *Dialectics and Humanism*, Vol. 1, Warsaw 1973 (reprinted in *Semiotics*, X, 4, 1974); and Jerzy Pelc, 'Introduction' to '*Semiotics: an Overview*' (to appear).

ACKNOWLEDGEMENTS

My thanks are due to Professor Olgierd Wojtasiewicz of Warsaw University for the translation of the book, and to Polish Scientific Publishers in Warsaw and D. Reidel Publishing Company in Dordrecht, Holland, who have undertaken its publication.

Jerzy Pelc

Warsaw, March 1977

FROM THE INTRODUCTION TO THE POLISH EDITION

I

In my Introduction to *Logic and Language*,[1] written in 1965, I expressed the hope that that anthology of texts on semiotics, the first to be published in Poland, would not be the only book of its kind to appear in this country. The publication of the present book is a partial realization of that hope. The period separating the appearance of these two books saw the birth of the Polish Semiotic Society, which subsequently prepared and sponsored *Studia semiotyczne* (Studies in Semiotics),[2] a collection of papers intended to originate a new periodical. Moreover, the International Association for Semiotic Studies was formed in Paris to unite the various national semiotic societies, including the Polish one, and Poles have served on its board. This Association has its own periodical. At first, it was a section, entitled "Studies in Semiotics — Recherches Sémiotiques", in the internationally sponsored bimonthly *Social Science Information — Information sur les Sciences Sociales*, but by 1969 it had evolved into a separate quarterly *Semiotica*. Both periodicals have had Polish representatives on their editorial boards, and both have included papers by Polish researchers. The Polish Scientific Publishers have published *Prakseosemiotyka* (Praxiosemiotics) by Tadeusz Wójcik,[3] *Studia z historii semiotyki* (Studies in the History of Semiotics),[4] *Semantyka a pragmatyka* (Semantics Versus Pragmatics) by L. Koj,[5] and *Wstęp do logicznej teorii pytań* (An Introduction to the Logical Theory of Questions) by T. Kubiński.[6] A volume on the semiotics of art is being prepared by the Polish Academy of Sciences, Research Unit for Contemporary Culture. *Dociekania semantyczne* (Sematic Investigations) by Anna Wierzbicka[7] and *Studies in Functional Logical Semiotics of Natural Language* by J. Pelc[8] have also appeared. The seminar on the logical semiotics of natural language, which was first attached to the Department of Logic of Warsaw University and later to the Department of Logical Semiotics (the setting up of the last-named unit being in itself a sign of the progress made in the emancipation of semiotics as a separate discipline), has continued to hold regular meetings, of which many have been in conjunction with the Polish Semiotic Society. These have been attended by scholars from all over Poland and occasionally by foreign visitors as well. The two departments in turn have sponsored a research group concerned with the logical semiotics of natural language, which in addition to logicians includes representatives of other disciplines from all over the country. Poland has played host to two international conferences on semiotics: the first at Kazimierz-on the Vistula in 1966, and the second in Warsaw in 1968. The papers read at Kazimierz were subsequently published in *Sign, Language, Culture*,[9] while those read in Warsaw appeared in *Recherches sur les systèmes signifiants, symposium de Varsovie 1968*.[10] Both include many contributions by Polish authors. Three other international symposia on semiotics in which Polish scholars actively participated were held in Italy in 1968, 1969 and 1971. Finally, *The Proceedings of the 14th International Philosophical Congress*, held in Vienna in 1968, include papers and communiqués on semiotics, mainly logical semiotics, submitted by Polish authors.

These are encouraging facts and worthy of mention. They testify to the advances made in the science of signs, both in Poland and abroad. But these advances did not take place by chance. The emergence of semiotics as an interdisciplinary science that brings together certain areas of interest of logicians and linguists, philosophers and praxiologists, psychologists and sociologists, information theorists and cyberneticists, theorists of art and literature, technologists, neurophysiologists, bibliographers, library scientists and representatives of other disciplines, reflects the present-day need to counteract the rapid and excessive specialization that might result in an undesirable departmentalization of science and a dissipation of research efforts.

This development of semiotics, in order to prove useful, must be controlled, and not spontaneous. It is important that semiotics should not become a science of everything, and hence of nothing. Nor should it become a hunting ground for dilettanti who lack competence to work within their own respective fields. It is clear that interdisciplinary sciences, especially those which expand rapidly and are fashionable at a given time, have to face that danger. Countermeasures may be of various kinds. One urgent and important — if difficult — task would seem to be the precise definition of the subject matter, methods and scope of semiotics. A first step in that direction might consist in realizing clearly the differences between logical, or — more broadly — philosophical, semiotics and semiotics as practised in other disciplines, for instance, in linguistics. Such differences will be noticed easily by anyone who takes the trouble to read, carefully and thoroughly, a number of logical or philosophical texts on issues in the science of signs and then compares them with texts written by linguists on the same subject. *Semiotics in Poland* offers a sample of Polish achievements in logic of language and philosophy of language between 1894 and 1969. A study of the evolution of problems, interests and trends in semiotic research over several decades in various countries and academic circles will help logicians to acquire a better understanding of the subject matter, scope and methods of logical semiotics, and will provide those who are not logicians with an awareness of past and present activities in that discipline.

I will repeat here, as a *ceterum censeo*, the appeal which I formulated in my introduction to *Logic and Language*, an appeal which calls for co-operation among representatives of various disciplines who — each in his respective field — are engaged in semiotic research. This applies above all to co-operation between logicians and linguists. This is, of course, not to say that I would dismiss co-operation within a much broader group of specialists. It merely reflects my belief that it is more reasonable and more realistic to set ourselves a succession of partial tasks. The publication of *Semiotics in Poland* represents the fulfillment of one of these partial tasks and, I believe, a response to my appeal, or at least it is so intended. Our next partial task — modest too, but realistic — should, in my opinion, be a selection of texts on linguistic semiotics. I address this appeal to my linguist colleagues.

II

The title *Semiotics in Poland, 1894–1969* is far from being precise, but if it were to be precise it would have to be awkwardly long. Let us begin with the dates, which are the most precise — though not totally precise — element. The year 1894 saw the appearance of Kazimierz Twardowski's book *Zur Lehre vom Inhalt und Gegenstand der Vorstellungen, Eine psychologische Untersuchung*. Excerpts from it, entitled

On the Content and Object of Representations, are the earliest texts included in the present anthology. But of course Twardowski's book was written before 1894, presumably between 1891 and 1893, so that the *terminus a quo* might be shifted back a bit. Fewer qualifications need to be made regarding the *terminus ad quem*, i.e., the year 1969. Ryszard Wójcicki's paper *The Semantic Concept of Truth in the Methodology of Empirical Sciences*, the latest of the texts included in this book, was written, or at least given its final shape, in 1969 precisely, when it was also published. Those papers which have not been published previously — i.e., *Names and Predicables* by Peter Thomas Geach, *The Semantic Functions of Oblique Speech* by Witold Marciszewski, and *Analyticity and Apriority* by Adam Nowaczyk, were all written before 1969, but in any case Marciszewski's paper was revised by him as recently as in 1969. Thus, *Semiotics in Poland* mirrors the development of semiotic thought over at least seventy-five years, which is not a brief period. In the present book, history meets with contemporaneity.

In continuing our analysis of the title of this book we encounter the word 'Poland', which also requires certain clarification. Let us disregard the fact that the aforementioned paper by Twardowski, *On the Content and Object of Representations* was written in German, since the nationality of the author, and not the language of the original, was adopted as the criterion of Polishness. But in that case why have we included a paper by an author called Peter Thomas Geach, who was born in London, is the son of an Englishman who was a professor in an Indian university, and who himself studied at Oxford and later became a professor in an English university and a member of the British Academy? This is a somewhat romantic case which, I think, deserves to be explained, and I hope Professor Geach will not take it amiss if I relate here what he himself once told me. He has almost no recollection of his mother and was brought up by his father. When he was a small boy, he came to blows with one of his schoolmates, and his adversary, in order to wound him to the quick, called out: "You are not a true Englishman, but a damned Prussian!" Following this incident Geach asked his father if it was true that his mother had been a German. His father then explained that "Your grandfather was a German subject, but he was a Pole." In fact, Geach's maternal grandfather, whose surname was Sgonina, had left Pomerania, his homeland, several years before

World War I, and settled in Britain together with his family. It was only as a result of his scuffle with a schoolmate that Geach learned of his origins. Towards the end of his studies at Oxford he came across *Harvest in Poland* by Geoffrey Dennis. The author had been employed as a private tutor to a young Polish aristocrat and had spent many years in country residences in Poland. In his memoirs he mentions people and landscapes, customs and traditions, Polish hospitality, and Polish food. And since he experienced his first love in Poland and spent his youth there, his recollections were warm and heartfelt. This book impressed Geach very strongly. He suddenly recalled his schoolmate's words, a few vague scenes from his early childhood, and even a few Polish words which he had heard from his mother. Quite unexpectedly, he began to develop an emotional interest in his mother's country. He started learning Polish; after his father's death, he found in his papers a Polish grammar, which his mother had used. He then started buying handbooks of the Polish language and books in Polish. Today, he speaks and writes a faultless Polish, which is quite elaborate and sometimes even flavoured with appropriately inserted archaisms. He visits Poland almost every year, and is proud of his Polish ancestry on the distaff side. He manifests an obvious patriotism, and in his heart he is as Polish as he is English. When offered a book on the clandestine teaching performed by the staff of the University of Warsaw during the Nazi occupation[10] he replied in a moving letter in which he writes that, when he saw in Cracow how Polish children played near the Royal Castle, he could see by himself that Poland's World War II heroes had not died in vain. I felt that Geach's paper on *Names and Predicables* which he wrote in Polish and intended for this anthology, not only could, but should be included in a book which has Poland in its title. Another paper of his, *Russell's Theory of Descriptions*, has been published by me in *Logic and Language*, where it is included among texts by British authors. I think that both these decisions can be factually justified.

The first word in the title of the present book is 'semiotics.' It cannot be denied that this is the most troublesome word of the lot, and that it requires the most explanation. To formulate an exhaustive and satisfactory explanation I would have to write a separate theoretical study, preceded by thorough research, or perhaps even a separate book. Since that is impossible now, I must necessarily confine myself to a few general comments. At the end of the last century, from which date the first papers in this book derive, no Polish university had a separate chair of either logic or psychology. A professor of philosophy, besides covering such philosophical disciplines as the history of philosophy, epistemology, and ethics, had to lecture on logic and psychology as well. Lectures on the philosophy of culture, and sometimes those dealing with the history of philosophy, covered certain concepts and problems of sociology, while other concepts and problems of sociology were taken up by historians and jurists. Studies in experimental psychology were initiated in Poland by Kazimierz Twardowski at the University of Lwów in 1901. He also organized the first psychological laboratory (1907), which he transformed into a research centre in 1920. The first separate chairs or departments of logic in the philosophical faculties of Polish universities were not set up until several years after World War II. This was also the case with the first periodical concerned with logic, *Studia Logica*, which was founded by Kazimierz Ajdukiewicz in 1953. Thus, even though problems of logic and psychology had emerged as separate wholes at a much earlier date (in the case of logic the distance in time is two thousand five hundred years), and even though they later gave rise to specialized fields of research, the organization of research and teaching lagged far behind. Moreover, the organizational position could not fail to be reflected in the character of research and teaching at the university level. The division of tasks and professions remained obscure. This was particularly true in the period in which the opening texts in this anthology were written, namely the papers by Kazimierz Twardowski. Within one and the same paper he appears on one occasion as a logician, on another, as a psychologist, and on still other, as an epistemologist. That is why it would be not only difficult, but simply erroneous, to classify a given paper of his as purely psychological, or purely philosophical, or purely logical. To do so would be to ignore the facts. *Mutatis mutandis*, the same situation arises when one tries to determine whether a given text written by him is, or is not, semiotic in nature. All his writings are dominated by a striving to remove misunderstandings resulting from the vagueness and ambiguity of expressions. In that sense they are written from a semiotic standpoint and imbued with elements of semiotic analysis; they express specified semiotic tendencies and are an implementation of a consistent semiotic programme, even if their subject matter does not lie within the sphere of semiotics. On the other hand, these very same writings of his often reveal a striving to base concepts on the foundations of data drawn from inner experience, and from that point of view they are partly psychological and partly epistemological in nature, even if their subject matter does not lie within any of these fields. In view of this I have decided to include in this anthology not only those of Twardowski's papers (e.g., *Issues in the Logic of Adjectives*) whose very title shows that they lie within the sphere of the logical theory of natural language, or its logical semiotics, but also those whose title — and, in part, subject matter — would place them in the border area between psychology and epistemology (e.g., *On the Content and Object of Representations*). In the latter case, however, I have selected only those fragments of a given paper which include either semiotic analyses or ideas outside the field of semiotics which later proved to be of immense significance for the development of that discipline. I mention here all the practical decisions which I have had to make, because they were a result of a state of things which might be of theoretical interest, too. The facts discussed above shed light on the meaning of the term 'semiotics', which occurs in the title of this book. Now, in its initial period semiotics did not appear as a distinct discipline, even though problems of the science of signs and language recurred in many papers. Nor can we, for the reasons mentioned above, speak about any clear difference between philosophical semiotics and logical semiotics when referring to that period; it would even be difficult to speak about differences between semiotics as pursued by philosophers and semiotics as pursued by logicians. Yet gradually, in the writings of Twardowski's immediate disciples, the distinct nature of semiotics research was already becoming more clearly marked, as can be seen by the subject matter and titles of their various papers. While Twardowski's dissertations on the whole were outside the sphere of semiotics, and were philosophical, or psychological, or philosophico-psychological in nature and merely included concepts and issues from the sphere of the logical theory of natural language, i.e., certain elements of the logical semiotics of that language or applications of that discipline, the situation changes when we come to his immediate successors:

for their papers belong to semiotics. Dominated by the logical trend, they include certain elements of epistemology (or, speaking generally, philosophy), or of psychology. But these extralogical admixtures gradually disappeared, and in the works of the younger generation of researchers, published after World War II, they are almost entirely absent. At the same time, following the setting up of research and teaching centres concerned with logic and attached to the universities and the Polish Academy of Sciences, and following the appearance of a periodical devoted especially to logic and the emergence of the academic profession of logician (as distinct from that of philosopher), writings about the logical theory of natural language, i.e., logical semiotics, have come to possess a distinct character of their own. Their authors use the concepts of logic, especially — and on an increasing scale — those which have been developed in formal logic and mathematics. Artificial languages, as well as natural language, are more and more often the direct subject matter of analysis. Terminological distinctions are introduced from time to time: research on artificial languages is called pure or logical semantics, while research on natural language is called logical semiotics. But the former deals with natural language, too, in so far as the artificial languages it investigates share common characteristics with natural language. For instance, Ryszard Wójcicki's paper on *The Semantic Concept of Truth in Empirical Sciences*, included in this anthology, falls within the scope of logical semantics thus understood.

Semiotics in Poland shows the development of semiotic thought in this country over seventy-five years. That is why the opening items in this volume are so different in character to those which come later. We thus have to do with an evolution of the concept of semiotics as well. Hence the difficulty when we have to explain briefly what the first word in the title really means. We can arrive at a proper answer only after having read the book. By a proper answer, we mean one which will take into account the historical perspective and evolutionary changes, as well as certain permanent characteristics that are common to most, if not all, the papers included in the present anthology. But what are these common characteristics?

We are talking about semiotics as pursued by philosophers or logicians and as such it differs from semiotics as pursued by representatives of other disciplines, e.g., linguists. It is a semiotics which is mainly concerned with the study of natural language, and hence the subject matter of its investigations partly coincides with that of linguistic semiotics. It is a semiotics which in its initial stages was developed by philosophers interested in epistemological and psychological issues, and that is why — because of its origin and traditions — it pays a great deal of attention to concepts and problems from the sphere of pragmatics, and not only semantics and syntax; accordingly it embraces spheres of interest which are close to psychology (particularly psychosemiotics), and praxiology (particularly praxiosemiotics). It is a semiotics which since its very inception has been marked by a strong logical trend that has intensified with the passage of time; and as such — even in its initial period, when it went beyond the bounds of logic and was pursued by philosophers who did not confine themselves to logical research — it has been mainly, owing to this dominant trend, a logical theory of language, and largely natural language at that. It is a semiotics which does not confine itself solely to analysing concepts of the logical theory of language, but also applies its own conceptual apparatus to analyses of concepts, mainly semiotic in nature, which are to be found in the various specialized disciplines. As such it is linked to the methodologies of the various disciplines. Its connections with these specialized disciplines are especially strong in view of the fact that most researchers concerned with semiotics, and hence most of the authors included in this book, besides possessing philosophical or logical training, have been trained in some other discipline as well.

What other links connect the papers to be found in this anthology? Their style of *work*. I have deliberately used a word that evokes associations of effort, the goal of which is the production of something both practical and useful. I have also deliberately distorted the time order somewhat by placing the following two texts at the very beginning of the book: *On Clear and Obscure Styles of Philosophical Writing* and *Symbolomania and Pragmatophobia*. The opinions formulated therein may serve as a motto or manifesto: it is not "... [our] duty to rack our brains over what the author of a philosophical work, had in mind when writing in an obscure style ..."; "... an author who does not know how to express his thoughts clearly does not know how to think clearly either ...";[11] let us not forget that "symbols do symbolize something" and let us prevent "symbols and operations performed on them", which originally served "as means to an end," from becoming an end in themselves; let us not place symbols above things.[12] Such recommendations, whether formulated explicitly or implicitly, have, I feel, been observed by the authors of the papers included in this book, and this is what links all these papers together, regardless of their time of writing, divergences of theoretical standpoints, or their focus of interest. Our thoughts turn to the man who, "having found Poland a fallow field, overgrown with weeds, rolled up his sleeves and started pulling out the weeds and planting nutritious vegetables."[13] So many scholars, both his contemporaries and those who came later — not to mention their writings — owe so much to Kazimierz Twardowski's eminent mind, as well as to his personality, which commanded universal respect. These scholars took up his ideas and continued and expanded them in their research. That is why Twardowski's influence as a teacher and an educator is still being felt and will continue to be felt in coming years, too.

His personal authority has become legendary. I will add here a story I heard from my parents, who at one time attended his lectures and lived in the same apartment house in Lwów. During World War I, when Lwów was occupied by foreign troops, Twardowski, walking down the street, saw an armed and mounted soldier riding on the pavement instead of in the roadway. He firmly reproached the soldier for breaking the traffic rules, snatched his horse by the bridle and led it to the roadway. His conduct on that occasion revealed something that was so characteristic of Twardowski as a teacher: his tirelessness in preserving a sense of order and his uncompromising insistence on discipline. But to him, order and discipline were not what a pedant or a sergeant on the training ground might have in mind. He meant by them much deeper and much more important values of human conduct, and hence of intellectual activity as well. That is why I have mentioned the above story. The effect he had on his disciples (which in certain Polish academic circles has survived to the present day) consisted in making them think and speak in an orderly manner; and it is this which they have transmitted to their own students, and which has found reflection in the style of work shown by the papers included in this book, or at least in the overwhelming majority of them.

III

Semiotics in Poland is not intended as an anthology of papers by writers of the Twardowski school. Such a school does not even exist, if by that term we mean the same philosophical opinions, the same research problems and similar specializations. Twardowski himself was not an adherent of any single philosophical system, and in his research and teaching work he did not confine himself to any single field. He worked in epistemology and psychology, general and formal logic, the history of philosophy, ethics, and the methodology of teaching. His disciples and their successors include representatives of various disciplines (logic, epistemology, ethics, praxiology, psychology) and advocates of various philosophical doctrines (radical conventionalism, reism, etc.). What binds them together is the style of work described above. But that was not the criterion of selection when this book was being compiled. Nevertheless if it consists almost exclusively of papers by Twardowski, or by his disciples and the disciples of his disciples, that is because such is the state of things with regard to semiotic research in Poland. On the whole other Polish logicians and philosophers have simply not concerned themselves with these issues. [14] But they have been taken up by four generations of researchers who can be traced back to Twardowski: his own students, who included Ajdukiewicz, Kotarbiński, Czeżowski, Dąmbska and Romahnowa; students and disciples of Kotarbiński and/or Ajdukiewicz, and students and disciples of these students and disciples. It is no accident that the ancestral line of Polish students of semiotics originates with a man who attached so much importance to a clear, explicit, and orderly formulation of thoughts. He has left his mark on his successors, because his own example encouraged them to investigate the problems of language and has thus guided them in their research.

In order to show the development of semiotic thought in Poland, the texts included in the present anthology have been arranged chronologically, but not by the dates of their writing or publication, rather by the generations to which their authors belong.

Hence the first group consists of papers by Kazimierz Twardowski. The second is made up of papers by his direct disciples, i.e., Kotarbiński, Czeżowski, Ajdukiewicz, Dąmbska and Łuszczewska-Romahnowa, in the order of their dates of birth. The third group is by the first generation of Kotarbiński's disciples, namely Ossowska, Ossowski, and Kotarbińska. Here the principle of seniority, which is otherwise observed throughout the book, breaks down: if it were to be followed strictly, then the papers by Dąmbska and Romahnowa would have to follow Kotarbińska's paper, and not precede Ossowska's. I think, however, that the fact that Dąmbska and Romahnowa were among Twardowski's disciples, while Ossowska and Kotarbińska were students of Kotarbiński, who himself was a direct disciple of Twardowski, is more important than the dates of birth. Next follow two names which are not to be found on the above ancestral line: neither Geach nor Schaff is a disciple of Twardowski or of his students, and the position of their texts has been determined by their dates of birth. The next group is made up of papers by Suszko, Przełęcki, Stonert and Pelc, who belong to the middle generation of disciples of Ajdukiewicz and/or Kotarbiński. As can be seen from the above, the division into generations is not free from overlapping, which is due to the fact that at certain times Twardowski lectured simultaneously with his direct disciples and with disciples of his disciples; later the same applied in the case of Kotarbiński, Czeżowski and Ajdukiewicz; today, Twardowski's own students are still active, and three successive generations of university teachers are at work alongside them. Two papers written jointly by Suszko and Kraszewski are placed immediately following Suszko's text because of the continuity of authorship. Koj and Marciszewski, whose papers come next, did not study under Twardowski's disciples, but Koj during his doctoral studies worked under Łubnicki, who had been a student of Twardowski, and Marciszewski worked after he had already received his doctoral degree under Ajdukiewicz for a spell. The concluding papers are by Wójcicki, Stanosz, and Nowaczyk, members of that generation which not only attended lectures and seminars held by Ajdukiewicz and Kotarbiński, but also studied under their disciples, and under disciples of their disciples. Wójcicki, Stanosz, Nowaczyk, and their coevals are not the youngest generation of Polish logicians, since they now have their own post-graduate students, who — let us hope — will creditably continue the fine traditions of the seventy — five years here under review, and whose papers will be included in a new *Semiotics in Poland*, to cover the period from 1970 onwards

The above comments on the arrangement of the material may be rounded off by adding that where more than one paper by the same author is included, they are placed in order of publication (the only exceptions being, as has been mentioned, the two opening papers by Twardowski).

It is debatable whether the principle of the sequence of scholarly generations adopted here mirrors the development of semiotic thought and the evolution of the very concept of semiotics better than the usual arrangement of papers according to their date of writing or publication. A shortcoming of the system adopted in this book is that the chronological order of the appearance or writing of the papers is broken. For instance, Ajdukiewicz's papers, which were written in 1958 or 1959 and first printed in 1969, precede Ossowski's paper, which was published in 1926. On the other hand, an arrangement based on dates of publication would result in our having papers by the same authors scattered throughout the book, which would shatter the totality that is formed naturally by papers of one and the same author. The underlying idea of the system adopted in this anthology is the conviction that the essential theoretical views of a researcher are usually shaped during the early stages of his development *qua* researcher, and that this process is closely linked to the particular generation of scholars to which he belongs, and to the school which moulds him intellectually. This is not to say that we may ignore a person's individual evolution or disregard far-reaching changes in a scholar's theoretical approach or even striking incompatibilities between views held by him at an early age and those expounded much later (for instance, Wittgenstein's opinions as reflected by *Tractatus Logico-Philosophicus*, on the one hand, and *Philosophical Investigations*, on the other). Yet, if we do not confine ourselves to exceptional cases, but look for what is typical, we will find more essential links between papers by one and the same author, even if distant from one another in time, than between contemporary papers written by various authors belonging to different scholarly generations. That is why the genetic arrangement based on the sequence of generations seems more natural to me than an arrangement determined by dates of publication, since the former system better reflects the organic evolution of opinions. Should, however, any reader think otherwise, he can look up the relevant data in the appendix entitled 'Sources of the Texts'.

There he will find the date of first publication of the texts included in this book; he can then read them in any order he finds suitable and thus form his own picture of the evolution of semiotic thought in Poland.

This picture will become richer in detail if the reader acquaints himself with the data pertaining to other papers on semiotics by each of the authors represented in this anthology. These data are to be found in the appendix entitled 'Biographical and Bibliographical Notes'. The authors' names are arranged alphabetically, and each entry briefly describes a given author's academic career and lists his publications on semiotics, in chronological order. These lists are intended to be complete; moreover, they include not only titles of papers on semiotics, in the broadest sense of the term, but also of papers which in some way refer to semiotics, even if semiotics is not their subject. This, too, has been done intentionally: the idea is to provide information for those who also want to know where to look for those papers which provide a secondary or indirect contribution to semiotics.

These biographical and bibliographical data are the stuff for a future handbook of the history of semiotics in Poland. They are also intended for those who will use *Semiotics in Poland* as a substitute for a badly needed but as yet unwritten handbook of semiotics. I hope that the present anthology may serve not only as a scholarly publication to be used mainly by logicians, philosophers, linguists and others interested in the problems of signs and languages, but also — in view of the clarity of the papers it includes — by students and by all those who want to broaden their general knowledge.

***

I am indebted to the authors of the texts included in this anthology for their kind permission to have their papers reprinted here, and also for their advice and co-operation.

Warsaw, 1971 *Jerzy Pelc*

Notes

[1] *Logika i język. Studia z semiotyki logicznej* (Logic and Language. Studies in Logical Semiotics). Selected, translated, introduced and annotated by Jerzy Pelc. Warsaw 1967, XXIX+559 pp.

[2] *Studia Semiotyczne* (Studies in Semiotics), (ed.) Jerzy Pelc, Ossolineum 1970 - 1977. (Eight volumes have appeared so far. — Tr.).

[3] Tadeusz Wójcik, *Prakseosemiotyka. Zarys teorii optymalnego znaku* (Praxiosemiotics. An Outline of the Theory of Optimized Sign). Warsaw 1969, 289 pp.

[4] Studia z historii semiotyki (Studies in the History of Semiotics), (ed.) J. Sulowski, Ossolineum 1971 - 1976 (Three volumes have appeared so far—Tr.).

[5] Leon Koj, *Semantyka a pragmatyka.* Stosunek językoznawstwa i psychologii do semantyki (Semantics Versus Pragmatics. The Relation of Linguistics and Psychology to Semantics), PWN 1971, 132 pp.

[6] Tadeusz Kubiński, *Wstęp do logicznej teorii pytań* (An Introduction to the Logical Theory of Questions), PWN 1971, 116 pp.

[7] Anna Wierzbicka, *Dociekania semantyczne* (Semantic Investigations), Wrocław 1969, 201 pp.

[8] Jerzy Pelc, *Studies in Functional Logical Semiotics of Natural Language*, Mouton 1971, Janua Linguarum, Series Minor, No. 90, 238 pp.

[9] Sign, Language, Culture, (eds.) A. J. Greimas, R. Jakobson, M. R. Mayenowa, S. K. Shaumian, W. Steinitz, S. Żółkiewski, Mouton 1970, Janua Linguarum, Series Maior, No. 1, XX+723 pp.

[10] *Z dziejów podziemnego Uniwersytetu Warszawskiego* (From the History of Underground Warsaw University), Warsaw 1961, 316 pp.

[11] Cf. Kazimierz Twardowski, 'On Clear and Obscure Styles of Philosophical Writing' in the present book.

[12] Cf. Kazimierz Twardowski, 'Symbolomania and Pragmatophobia' in the present book.

[13] Tadeusz Kotarbiński on Kazimierz Twardowski in an article in *Pion*, 21 (138), 1936.

[14] An exception is Roman Ingarden, especially as the author of Chaps. IV to VII of *Das literarische Kunstwerk*, Halle 1931 (later also published in a Polish-language version). Ingarden, too, had attended Twardowski's lectures before moving to Göttingen to study under Husserl.

ON CLEAR AND OBSCURE STYLES OF PHILOSOPHICAL WRITING

by

Kazimierz Twardowski

(Fragments)

We often hear complaints about the obscure and intricate style which philosophers sometimes use to express their thoughts (...).

The question then arises (...) whether style of philosophical writing and/or speaking must in certain cases be obscure. (...) Many people think (...) that sometimes even a clear-thinking philosopher who wishes to express his thoughts as clearly as possible is unable to do so because of the intricacy of the matters and issues he is discussing. (...)

It is not clear, however, what are the grounds for the opinion that certain philosophical matters and issues cannot be presented in a clear manner. It is difficult to imagine that someone could succeed in demonstrating that all the writings concerned with a given philosophical subject matter are marked by an obscure style. On the other hand, it would be much easier to demonstrate that certain philosophers can quite clearly discuss subjects which are commonly held to be difficult and intricate. This gives rise to the supposition that the obscurity of the style in which some philosophers write is not an inevitable consequence of the factors inherent in the subject matter of their analyses, but has its source in the vagueness and obscurity of the way they think. (...)

This opinion finds support in the extremely intimate connection that exists between thought and speech: the more abstract the thought expressed by speech, the more intimate the connection. (...) Human speech is not only an outward manifestation of thought, but also its instrument, which makes our abstract thinking possible. When thinking, we think in words, and hence in speech.

Accordingly, it is not the case that we are able to think first, and only later to array our thoughts in words, nor that we are initially able to think about a philosophical problem in a clear manner, but that subsequently — when we come to express these clear thoughts in words — we must content ourselves with their obscure presentation simply because of the intricacies of that problem. This is not the case because our thoughts, especially when concerned with abstract subjects, immediately present themselves to us in verbal attire, in the most intimate possible connection with the words of speech. And if, when expressing our thoughts aloud or writing them down, we encounter doubts and difficulties, if we select words and change the original order in which our thoughts have developed, we do so precisely because such an outward manifestation of our thoughts in speech or writing reveals certain obscurities which we failed to notice when those thoughts were first developing in our minds. Such an outward manifestation of thoughts slows down their course, presents them to us a second time at a slower pace, and thus helps us to discover shortcomings which we did not notice at first.

It may be that some authors even then cannot perceive the obscurity of their thoughts (...), and hence the obscurity of their own style.

Now if the above remarks are correct, then they largely free us from the duty to rack our brains over what the author of a philosophical work had in mind when writing in an obscure style. This is worth the effort only if we already have grounds for believing that he thought clearly. (...) But if we do not have such a belief, then we may quietly assume that an author who does not know how to express his thoughts clearly does not know how to think clearly either, and therefore his thoughts do not deserve our efforts to guess them.

SYMBOLOMANIA AND PRAGMATOPHOBIA

by

Kazimierz Twardowski

(Fragments)

Symbols, in the sense of conventional signs used in certain disciplines instead of words, have rendered incomparable services to those disciplines; what is more, some disciplines would have been unable to make even a single step forward had they not been using a system of symbols from the very beginning. (...) When working with (...) symbols we abstract from the concepts and objects which they symbolize (...). We combine and rearrange these symbols in various ways, performing a number of operations on them, and in this way we obtain certain results. But these results, when still in symbolic form, require interpretation; that is why, after having performed our operations (...), we have to move again from the land of symbols to the world of the concepts and objects which they symbolize (...). It is only when we do this that we reach the goal which symbols and operations performed on them make it easier or even possible to reach.

From this it follows that, when using and working with symbols we must very scrupulously take into account the fact that they are the means to a particular end. That is why a given symbolism — for brevity we will use this word to cover both the symbols themselves and the operations performed on them — must be very precisely adjusted to the concepts and objects to be symbolized and must be subjected to a rigorous inspection again and again in order to avoid difficulties in the final interpretation of the results obtained by means of that symbolism and expressed in its language.

Such difficulties emerge if the results are at variance with those convictions of ours which are independent of any symbolism (...). We must then ask ourselves whether we have to reject these convictions which are independent of symbolism and to accept the results obtained contrary to those convictions as a result of using a given symbolism, or whether, vice versa, we have to accept those convictions which are independent of symbolism as correct, and the results obtained by the use of symbolism as erroneous. (...)

There are, however – or so it would seem – persons for whom the question formulated above does not exist, or who have a ready-made answer to it, for they have an unfaltering belief in the infallibility of the symbolism they use (...). Symbols and operations performed on them, originally means to an end, become for them an end in themselves, an object of ardent love and a source of great intellectual delight.

Such an exuberant passion, which dominates a person's way of thinking and acting, is — as we know — termed a 'mania'. We speak of 'graphomania', 'erotomania', etc. Why then should we not speak of 'symbolomania' as well? (...)

Symbolomania is associated with, and always accompanied by, what we might term 'pragmatophobia', which is, as it were, the negative complement of the former. The latter term is used here to refer to an aversion for things, i.e., those things which symbols, being names of things in the most general sense of the word, symbolize. (...) A symbolomaniac shuns things and prefers neither to think nor to hear about them.

Algebraic, or symbolic, logic likewise exposes its exponents to the risk of developing symbolomania and pragmatophobia, for in that discipline every theorem "must be proved in a strictly deductive manner by means of formal rules of calculus, without reference to the meaning of the symbols with which we work".[1] The principle of not referring to meanings of symbols, while quite correct when it comes to carrying out logistic operations, must not be understood in such a way that symbols have no meaning at all. In that case symbolic logic would degenerate into a purely formal theory of combinations, devoid of any content. Nor must this principle be understood in such a way that, in the event of a divergence between theorems which we owe to the symbolism of symbolic logic and convictions which are independent of that symbolism, we should always refer from the meaning of the symbols and from the things which they denote to the symbols themselves, and never from the symbols to their meanings and to the things they denote. In that case symbolism would triumph over what it symbolizes, and any person who was interested exclusively in such a triumph would be a model symbolomaniac and pragmatophobe. (...)

(...) There are two factors which must not be disregarded when we speak about circumstances that favour the expansion of symbolomania and pragmatophobia in symbolic logic. One factor is the newness of the symbolism of symbolic logic, and of symbolic logic itself, as opposed to traditional, or classical, logic, which is sometimes looked down upon by the former. (...) Advocates of symbolic logic sometimes credit themselves with the discovery of truths and the formulation of concepts which others have attained without resorting to symbolic logic. Hence, if in predisposed minds the enthusiasm that accompanies symbolic logic as the newest way of treating certain logical problems concentrates on the symbolism of symbolic logic above all else, then the distance which separates them from symbolomania and pragmatophobia will prove short indeed.

The second factor is the mathematicoidal nature of symbolic logic and its symbolism. Mathematicoidal, and not mathematical (...). But it is easy to forget this distinction and to ascribe the character of mathematical symbolism to the symbolism of symbolic logic as well, and hence also all the advantages and privileges of the former. This may lead us into errors of which the origin is none other than mathematical symbolism. (...)

The tendency to place symbols above things may result in bending things to comply with symbols, that is, making statements about things according to what follows from symbol-based assumptions and operations, regardless of what things tell us about themselves, or even contrary to what they tell us about themselves. (...) But symbolomania and pragmatophobia do their worst damage when they affect the training of young scientists and scholars. Young minds easily fall prey to the charms of various symbolisms, even the less perfect ones; they are attracted by the relative ease with which symbols can be handled, whereas they encounter various difficulties when they come face to face with things themselves. They are also attracted by a certain charm of mystery which surrounds all symbolisms and which is attainable only to those who have learned its language.[2] It is therefore easy to understand that some people, who are dazzled — and to some extent, justly — by symbolism, lose their coolly critical attitude when assessing its role and significance, and develop a way of thinking which, even if subsequently liberated from the onesideness born of symbolism, affects the way in which they approach problems, both theoretical and practical, by giving priority to form before content, appearances before reality.[3] (...) Whatever else could be said in a derogatory sense about symbolomania and pragmatophobia, does not apply to symbolism itself, which is a magnificent and indispensable instrument of certain types of research. (...) But, nevertheless, whoever confines himself, as a symbolomaniac and pragmatophobe does, to a *mechanical* handling of symbols and fails to see any other operations or goals to which those operations can lead him, becomes like a man who wishes to confine himself solely to mechanical movements, renouncing all action guided by conscious reflection. And such a man would merely be an *homme machine*, and could hardly be classed as a *homo sapiens*.

Notes

[1] Cf. Z. Janiszewski, 'Logistyka' (Symbolic Logic), in: *Poradnik dla samouków* (A Guide for the Self-taught), 1915 edition, Vol. I, p. 453.

[2] Cf. H. Poincaré's formulation, "Ce language n'est compris que de quelques initiés, de sorte que les profanes sont disposés à s'incliner devant les affirmations tranchantes des adeptes." (*Science et mèthode*, Paris 1908, p. 152).

[3] It is not hard to see that there is an analogy between a symbolomanic pragmatophobe and a bureaucrat, if we accept the definition of bureaucracy as formulated by the late Professor Klein of Vienna, who calls bureaucracy "*formelle Rechthaberei in Verbindung mit sachlicher Gleichgültigkeit*."

ON THE CONTENT AND OBJECT OF REPRESENTATIONS

by

Kazimierz Twardowski

(Fragments)

3. *Names and representations*

(...) Mill, when writing about names, asks whether it is more proper to consider names as names of things or as names of our representations of things.[1] (...) The word 'Sun' — according to Mill — is a name of the Sun, and not the name of representation of the Sun; he does not deny, however, that a name evokes in, or conveys to, a listener only a representation, and not the thing itself. The function of a name thus appears to be twofold: a name conveys to a listener the content of a representation, and at the same time names an object. But we have been persuaded that we must distinguish three, and not two, elements of every representation: the act, the content, and the object. If a name really gives an exact linguistic image (*Bild*) of the corresponding mental relations, then it must also have a correlate of the act of representation. This does in fact happen, and the three elements of a representation: the act, the content, and the object, have counterparts in the triple task which every name has to perform.

By 'name' we are to understand everything which logicians of old schools called a 'categorematic sign'. But all linguistic means of denoting which are not merely syncategorematic (such as 'father's', 'around', 'nevertheless', etc.), but also do not themselves form a complete expression of a proposition, emotion, or act of will (requests, questions, commands, etc.), but are merely an expression of a representation, are categorematic. The expressions 'the founder of ethics', 'the son who offended his father' are names.[2]

What then is the task which names have to perform? Obviously, to evoke in a listener the specified content of representation.[3] In uttering a name, one intends to evoke in one's listener the same mental content with which one is filled oneself. If a person says: the 'Sun', the 'Moon', the 'stars', he wants those who hear him also to think about the Sun, the Moon, and the stars. But when the speaker wants, by means of those names, to evoke a specified mental content in his listener, he is simultaneously betraying to him the fact that he, the speaker, finds that content in himself, and hence he wants the listener to represent to himself what he (i.e., the speaker) represents to himself.[4] In this way a name performs two functions. First, it informs that the person who uses it has a representation of something; it indicates that a mental act is taking place in the speaker. Secondly, it evokes a specified mental content in the listener. This content is what is meant by the *meaning* of a name.[5]

That, however, does not exhaust all the functions of a name. A name also performs a third function, i.e., the function of naming objects. Mill says that names are names of things, and to substantiate his statement he correctly refers to the fact that we use names to convey something about things, etc. Thus the naming of objects turns out to be the third task which a name has to perform. A name accordingly has three functions. First, to inform about an act of representation taking place in the speaker. Secondly, to evoke a mental content, i.e., the meaning of a given name, in the listener. Thirdly, to name the object represented by the idea which is the meaning of that name. (...)

5. *So-called 'objectless' representations*

(...) Those representations to which allegedly no object corresponds are of three kinds. First, there are representations which simply include a negation of every object, i.e., a representation of nothing. Secondly, there are representations to which by assumption no object corresponds, since their content includes mutually contradictory formulations, such as a 'round quadrangle'. Thirdly, there are representations to which no object corresponds, because up to now no such object can be indicated in accordance with our experience. We shall consider these three kinds of *objectless* representations in order to analyse the arguments adduced in favour of their existence.

1. As for the representation denoted by the word 'nothing' (...) it seems doubtful to me whether the word 'nothing' is categorematic, i.e., whether it denotes any representation at all, in the way that such words as 'father', 'court', 'leaf' do. The meanings of the words '*nihil*' and '*non-ens*' generally used to be equated, and today 'nothing' is also believed to be simply a replacement for the expression 'non-entity' ('*nicht-etwas*'). But if this is so, then it seems necessary to ask what such words as '*non-ens*', i.e. 'non-entity', mean.

What the schoolmen called infinitation, that is, the composition of a word with 'non', usually yields a new word with a well-defined meaning. A word combined with 'non' is used to make a dichotomous division of a representation.

But it is not the representation whose name is preceded by 'non' which thus becomes divided in a dichotomous manner. When we say 'non-Greeks', we do not thereby divide Greeks into those who are Greeks, and those who are not. What has been divided is a superordinated concept, 'human beings' let us say. (...) Only a disregard of the fact that infinitation yields a dichotomous division of a superordinated concept could have led to the astonishing opinion that 'non-human' may mean — without considering any superordinated concept applying to both humans and non-humans — everything which is not a human being, and hence an angel, a house, passion, the voice of a trumpet, etc. (...)

If infinitation actually yields a dichotomous division of a given superordinated concept, then it is evident that such expressions as 'non-Greeks', 'non-smokers', taken in the sense indicated above, are far from being devoid of meaning, and it is quite correct to classify them as categorematic. Accordingly infinitation as such does not deprive an expression of its categorematicity. It is, however, also evident that this dichotomous effect of infinitation depends on a certain condition: there must be a representation which is superordinated to that which is the meaning of the infinited name. If no such representation exists, then the infinited name has no meaning. It is obvious that the word 'something' denotes a representation to which no other representation is superordinated, because that which is superordinated to something must itself be something, and hence in the case now under consideration one and the same thing would at the same time be superordinated to something and equi-ordinated to it. But the infinitation of the word 'something' presupposes something superordinated to 'something', and hence it presupposes a nonsense. In any case, this kind of infinitation is not possible in the same sense as is the infinitation of a name such as 'Greeks' etc. Avicenna already draws attention to this fact and, on the basis of the arguments adduced above, he concludes that such infinitation as '*non-res*', '*non-aliquid*', '*non-ens*' are inadmissible.[6] And if we subject to closer scrutiny the function which the word 'nothing' performs in language, then we find that this word is in fact syncategorematic and that it is not a name, but rather a component of negative sentences. "Nothing is eternal" means "there is not anything that would be eternal"; "I see nothing" means "there is not anything that I see", etc.

If what has been said above is correct, then the argument in favour of the existence of objectless representations, based on the word 'nothing', collapses by itself, since the word 'nothing' does not denote any representation. (...)

2 and 3. The second group of allegedly objectless representations consists of those whose content includes mutually exclusive characteristics. The representation of an acute-angled square belongs to this type. But a closer examination of this class of representations shows that those who claim that no object comes under such a representation are confusing certain concepts. This confusion is easy to discern if we consider the three functions which are the attributes of a name. For here, too, we find all three functions: informing, denoting, and naming. In uttering the name 'acute-angled square', one is informing one's listener that one is experiencing an act of representation. The content assigned to that act of representation forms the meaning of that name. But that name not only means something, but also names something, viz., something which combines in itself contradictory properties and whose existence is immediately denied if we have to pass judgement on what has been named. Yet something has undoubtedly been named by that name, even though this something does not exist. And that which has been named differs from the content of the name, because, first, that content exists whereas that something does not, and, second, we ascribe to that which has been named properties which certainly are not attributes of the content of the representation. For were that content to have those contradictory attributes, it would not exist, and yet it does exist. The content of the representation is not that to which

we ascribe the property of being acute-angled and at the same time the property of being square; what the name 'acute-angled square' names is the carrier of these properties, i.e., something which does not exist but is nevertheless represented. Thus an acute-angled square is something represented not in the sense in which the content of the representation is represented, because the content of the representation exists; an acute-angled square is something represented in the sense of an object of representation, which in this case is rejected, but is nevertheless represented as an object. (...)

The confusion caused by those who claim that objectless representations exist consists in their taking the non-existence of the object of representation as its not-being-represented. But every representation represents an object, which may or may not exist, just as every name names an object regardless of whether that object exists or not. (...)

(...) If something 'exists' as represented in the sense of an object of representation, then its existence is no existence in the proper sense of the term. By adding the formulation "as an object of representation" we modify the meaning of the word 'existence'; something which exists as the object of a representation does not in fact exist at all, but is merely represented. The phenomenal, intentional existence of an object is opposed to the real existence of that object, an existence which forms the content of an acknowledging judgement.[7] (...)

Notes

[1] Cf. J. S. Mill, *System der induktiven und deduktiven Logik*, von Th. Gomperz, Leipzig 1884, Part I, Chap. 2, Sec. 1.

[2] Marty, 'Ueber subiektlose Sätze', in: *Vierteljahrschrift für wissenschaftliche Philosophie*, VIII, p. 293.

[3] Cf. F. Brentano, *Psychologie vom empirischen Standpunkte*, Leipzig 1874, Part II, Chap. 6, Sec. 3; see also Marty, op. cit., p. 300, and Mill, loc. cit.

[4] Sounds and other objects whose representations are used to evoke in another intelligent being certain representations connected with them, are, for that other being, usually – though not always – the sign (*Kennzeichen*) of the fact that the said representations take place in the consciousness of that being who produces those sounds or other objects. (Cf. B. Bolzano, *Wissenschaftslehre*, Sulzbach 1837, Sec. 285).

[5] "Etymologically the meaning of a name is that which we are caused to think of when the name is used." (Cf. W. S. Jevons, *Principles of Science*, p. 25). In any case we define as the meaning of a word that spiritual content which it is the task

and the ultimate goal of that word to evoke in a listener (either naturally or on the strength of usage), if that word can as a rule attain that goal. A name is a sign of the representation which a listener is supposed to develop in himself, and at the same time it is a sign of the representation which has developed in the speaker. A name expresses that representation only so far as it makes it possible to realize this fact. (Cf. Marty, loc. cit.)

[6] Cf. K. Prantl, *Geschichte der Logik im Abendlande*, Vol. II, p. 356.

[7] Cf. F. Brentano, op. cit., Part II, Chap. 1, Sec. 7.

ACTIONS AND PRODUCTS. COMMENTS ON THE BORDER AREA OF PSYCHOLOGY, GRAMMAR, AND LOGIC

by

Kazimierz Twardowski

(Fragments)

Contents: 1. The purpose of this paper. — 2. The action and phenomenon aspects. — 3. Examples. — 4. The various degrees of difference between the meaning of a verb and that of its corresponding noun. — 5. The etymological figure. The inner complement (inner object). — 6. The criterion of the inner nature of a complement (an object). — 7. Phenomena which result from actions. — 8. The relations of these phenomena to the inner complement (object). — 9. The concept of a product. — 10. Kinds of actions and products. — 11. Words which refer to physical and psychophysical actions and products. — 12. Words with refer to psychophysical and mental actions and products. —13. The double ambiguity of such words. — 14. Nouns which denote actions as well as products. — 15. Nouns which denote dispositions as well as products and actions. — 16. The function of verbal nouns. — 17. Analogies in other languages. — 18. The procedure by which separate words denoting actions and products are obtained. — 19. This procedure does not apply to the dispositional meaning of such words. — 20. The product of an action and the object of the product. — 21. Is the distinction between action and product justified? — 22. A fact which argues in favour of such a distinction. — 23. Products which vanish together with actions; their actual and potential existence. — 24. Products which last longer than actions. — 25. The classification of products into non-durable and durable. — 26. What makes the durability of products possible? — 27. The real nature of durable products. — 28. Mental products are not durable. — 29. Both physical and psychophysical products may be durable. — 30. Mental products find expression in psychophysical ones. — 31. Moreover, psychophysical products sometimes express mental products. — 32. Mental products as meanings. — 33. Mental products which find expression in durable psychophysical products. — 34. The potential existence of mental products in psychophysical products. — 35. The fixation of non-durable products. — 36. Various ways of fixing non-durable products. — 37. Writing as a method of making mental products durable. — 38. Products which have been made durable considered as apparently independent of the actions which produce them. — 39. The identification of products. — 40. Such identification is also one of the causes of the apparent independence of non-durable products. — 41. "Quasi-products" are another such cause. — 42. Substitutive products. — 43. Psychophysical substitutive products. — — 44. The function of substitutive products in logic. — 45. Conclusion.

1. In such pairs as 'to race' — '(the) race', 'to jump' — '(the) jump', 'to cry' — '(the) cry', 'to speak' — '(the) speech', 'to think' — '(the) thought', 'to err' — '(the) error', 'to judge' — '(the) judgement' the first word denotes an action; the meaning of the second word, as related to the first, will be analysed in this paper.*

2. It might at first be supposed that the difference between two such words is only one of grammar, and not of logic, i.e., that they differ only as to form, and not as to meaning. Thus the words '(the) race', '(the) jump', etc., would denote the same actions as the words 'to race', 'to jump', etc. In fact '(the) race', '(the) jump', etc., might be said to denote actions, but at the same time it would be impossible to deny that those nouns — precisely because they are nouns — do not convey the action aspect so strongly as the verbs 'to race', 'to jump', etc., do, but do bear out another aspect, which might be termed the phenomenon, or event aspect. When we speak about a race, a jump, etc., we are probably thinking of a fact, an occurrence, something which happens or takes place, rather than an action which someone performs. We say, for instance, that a given race is for home-bred horses, and this formulation clearly brings out the phenomenon, or event aspect; if, on the other hand, we wish to emphasize the action aspect we instead use verbal nouns (*substantiva verbalia*); we say, for instance, that we enjoy walking in the forest or climbing mountains. (...)

5. Grammarians long ago drew attention to the mutual relationship between such words and spoke of "etymological figures", by which they meant constructions in which a noun, formed from the same stem as a given verb, functions as its complement, or object, which in such cases is termed an inner, or cognate, complement, or object. (...) Of course, such a relationship between a verb and a noun is not restricted to etymological figures. (...) **

7. What we have already said about the relationship between the meanings of verbs and nouns combined in pairs like those quoted above, applies therefore to verbs with inner complements, or objects. This also enables us to understand the meaning of an inner object.

8. It may be said generally (...) that the relationship between a verb

and its corresponding inner object expresses a relationship between an action and that which results from that action. When we fight, a fight results; when we think, thoughts result; when we order, an order results; when we sing, a song results, etc.

9. That which results from an action may be termed the product of that action. Hence we may say that a jump is a product of jumping, a song is a product of singing, an error is a product of erring, etc., with the proviso that there is a gradation from those cases in which a product almost merges with the action of which it is a result to those in which a product is clearly distinct from the action.

10. The examples which we used above to explain the concept of a product referred to actions and products of various kinds, which can be reduced to two basic kinds, namely physical actions and products, and mental actions and products. The former category covers 'to race — (the) race', 'to jump — (the) jump'; the latter 'to think — (the) thought', 'to judge — (the) judgement', 'to intend — (the) intention'.[1] Among physical actions and products we should single out those that are psychophysical, namely those in which a physical action is accompanied by a mental one which somehow affects that physical action and, accordingly, its product; a product obtained in this way is termed psychophysical, too. This applies to the actions and products denoted by: 'to cry — (the) cry', 'to sing — (the) song', 'to speak — (the) speech', 'to lie — (the) lie', etc. (...)

12. There are words which may denote psychophysical (or even physical) actions or products, as well as those which are mental in nature. The word 'approach' has such a triple meaning (a road in the physical sense of the term, a method manifested in a book on a scholarly subject, an attitude toward a problem — Tr.), the word 'opinion' has a double meaning: either that of a mental product, also called 'a conviction' 'a view', 'a sentiment', or that of a psychophysical product, as in the case of the term 'sentence' in its grammatical sense. Likewise, the words 'state' and 'negate' either denote a physical action of judging, determined as to its quality, or a psychophysical action of uttering an affirmative or negative judgment.

14. There is no doubt (...) that we often use a noun to indicate an action, with the result that such nouns become ambiguous, since they may alternately denote actions and their products. This is the case, for instance,

in the formulations 'to submit a request' and 'to begin with a request' (products and action, respectively — Tr). Likewise, the noun 'judgement'. is used in both meanings, i.e., as a product and as an action from which that product results. In the former sense we say that certain judgements are logical consequences of certain other judgements; in the latter we speak about judgements as mental functions and say, e.g., that judgement is a cognitive action. To distinguish these two meanings we sometimes say 'judgement in the psychological sense of the term', i.e., action, and 'judgement in the logical sense of the term', i.e., product. (...)

15. But the ambiguity of such words as 'judgement' does not stop here for (...) we sometimes use 'judgement' (...) in the sense of a disposition[2] to pass judgements, e.g., when we say that 'he has a sound judgement of the facts'. Here, of course, we mean an ability to pass a pertinent judgement, and the dispositional nature of a judgement in this sense consists in the fact that 'sound judgement' is a relatively permanent characteristic of a given person, whereas the action of passing a pertinent judgement is something momentary. It is also in the dispositional sense of the word that we say that, e.g., education is meant to train not only the memory, but one's judgement as well.[3] (...)

16. The risk of confusing an action and its product is naturally greater than that of confusing a product or action with the corresponding disposition. (...) Verbal nouns usually retain their action meaning: 'to speak — — speaking — speech', 'to question — questioning — (the) question', 'to reason — reasoning — (the) reason', etc. (...) (In other cases, a verbal noun may denote both action and product, as in the case of 'painting' in English; differences between the various languages, even within the Indo-European family, make cross-language comparisons difficult — Tr.)

21. Probably none would deny that we have to distinguish those meanings in which the words under consideration refer to dispositions from those in which they refer to actions and products. Doubts may, however, arise as to whether it is really necessary to make a distinction between those meanings in which the words refer to products of actions and those in which they refer to actions. (...) These doubts will, however, be dispelled by a certain fact which makes such a distinction necessary. (...)

22. The fact is that we predicate many things about products which we do not predicate about actions. We say that certain questions are

incomprehensible, which is not to say that posing questions is an incomprehensible action, etc. (...)

23. While in the cases discussed above the distinction between an action and its product must be substantiated, there are other cases, in which that distinction is self-evident. The products mentioned so far might be termed non-durable, that is, products which exist only as long as the action which produces them. A cry exists as long as the action of crying continues (...), a thought, as long as someone is thinking, (...). It is true that we often (...) say that the thoughts or ideas of a sage may outlive him, but in this case we do not mean the continuing existence of certain products — an existence which would be independent of any action; what we do mean is a repetition, from generation to generation, of actions and products similar to those produced by earlier generations or by that sage. Likewise, we say that ideas, beliefs and desires are inherent in us even though at a given moment we do not perform the appropriate actions. It is common knowledge that such a formulation means merely this: that we have certain dispositions which in the future may make us produce the same products we produced in the past. Hence when we speak of a continuing existence of products of this kind we mean either repetition of the same actions and products, or their potential existence. That is why these products may be termed non-durable in the sense of their not actually existing longer than the actions which produce them.

24. There are, however, products which can — and usually do — last longer than the actions which produce them. Examples of such actions and products include: 'to draw — a drawing', 'to write — the writing', (...) etc. In each pair, the noun, coupled with a corresponding verb, may function grammatically as its inner object. We employ etymological figures and speak of (...) 'printing prints', 'building buildings', and so on; and when we abstain from using a cognate object (...) we speak of 'erecting buildings', etc. That we are dealing here with an inner object is proved not only by the fact that we can use the etymological figures, but also by the fact that each noun denotes something which only results from a corresponding action, as was the case in the previous examples: 'to jump — the jump', 'to believe — the belief', 'to judge — the judgement' etc. Nor may we speak here about an outer object, since the action indicated by the verb is not transferred to what the noun

denotes, although these verbs may also have outer objects (...).[4]

26. Certain products may continue to exist after the action from which they result has been completed because these actions are transferred to something, or affect something, which existed before the particular action began, and which continues to exist after that action has been completed; speaking very generally, this something can be called the material of the action. When making a plait we make a plait of something (we plait something). (...) The action itself, from which a durable product results, consists in transforming, or modifying the material; this action changes the configuration of its particles or changes it in some other way. (...) Hence, strictly speaking, the product of that action is merely a new configuration, modification, or transformation[5] of the material, since that material had existed before the action began. This is why, when calling a 'drawing' the product of (the action of) drawing we do not mean that the particles of graphite and paper are products of the action of drawing since the product is only a given configuration of particles of graphite on paper. (...) But since a configuration, an arrangement, a form, etc., exist only in material of one kind or another, hence, loosely speaking, we refer to a given whole — a particular configuration of particles of graphite, i.e., lines — as a drawing. (...)

29. As far as actions and durable products are concerned we may (...) speak only of physical products, while singling out psychophysical products as a separate subgroup. A human footprint in sand is a durable physical product of a physical action of a man walking on the sand if his consciousness is not involved in that action; a human footprint in sand, if made consciously and intentionally, is a durable psychophysical product of a psychophysical action. Paintings, sculptures, etc., are psychophysical products because they result from psychophysical actions, that is physical actions accompanied by mental actions which affect the course of these physical actions and hence also the products that result from them.

30. Because of the relationship that arises in such cases between a given psychophysical product which is perceptible to the senses and a corresponding mental product which is not subject to sensory perception, the psychophysical product becomes the external expression of the mental product. This relationship arises both in the case of non-durable

and durable psychophysical products. Thus a cry may be an expression of pain; a movement of the head, an expression of an affirmative judgement; a jump, an expression of fear. Likewise, a footprint may be an expression of a desire to find support for one's body; a drawing, an expression of an artist's conception; a stab, an expression of the stabber's anger toward the person stabbed.[6] In all these cases the mental product is expressed externally in the corresponding psychophysical product because that psychophysical product results from a psychophysical action, and not from a purely physical one.[7] Thus the statement that a mental product is expressed by a psychophysical one — in other words, that a mental product finds expression in a psychophysical one—can be reduced to two elements: first, that the mental product (together with the corresponding mental action) is in part the cause of the emergence of the psychophysical product, and second, that this mental product, and the corresponding mental action, are not subject to sensory perception, whereas the psychophysical product is. (...)

31. A mental product which is manifested in a psychophysical product (...) is sometimes said (...) to be expressed by that psychophysical product. But we say this only under a certain condition, namely if the psychophysical product in which a mental product is manifested can itself become a partial cause of the emergence of an identical or similar mental product by evoking a mental action which is identical with, or at least similar to, the action whence that product has resulted. Thus, for instance, (...) a drawing in which the artist's conception is manifested, truly expresses that conception only if a person who views the drawing develops a conception analogous to that which the artist had. (...) If the drawing is such that another person — or perhaps even the artist himself, some time later — does not, on viewing it, develop a conception analogous to that which the artist had when drawing it, then the drawing is *incomprehensible* and does not express that conception, even though that conception has found in it an expression — albeit a very inadequate one in this particular case.[8]

32. Psychophysical products which express certain mental products are also termed 'signs' of those mental products, and the mental products themselves are termed their respective 'meanings'.[9] Thus any mental product which bears to a psychophysical product the relation of being expressed by the latter is a meaning. We accordingly speak of the mean-

ing of a cry, the meaning of a drawing, the meaning of a gesture, the meaning of a blush, etc. Linguistic "expressions" are also those psychophysical products in which certain mental products — i.e., thoughts, judgements, ideas, etc. — find their expressions. At the same time, linguistic expressions usually express those mental products which are their meanings and of which they are signs, with greater precision than other psychophysical products do. (...) A psychophysical product which is endowed with too many meanings actually becomes a psychophysical product without meaning.[10]

33. At the moment when a psychophysical action takes place — owing to which a mental product is expressed — i.e., finds expression in the corresponding psychophysical product (for instance, fear finds expression in a cry while a person is crying, an idea or a thought finds expression in a drawing while a person is drawing) — both the mental product and the psychophysical product exist simultaneously. (...)

In the case of non-durable psychophysical products, the psychophysical product vanishes at more or less the same time as the mental product, and sometimes the mental product may even outlast the corresponding psychophysical product, but not vice versa. Hence, at the moment when the psychophysical product ceases to exist, the mental product ceases to be expressed. When a moan ends, suffering ceases to be expressed, even though it may continue to exist. The opposite is the case when it comes to durable psychophysical products: even though a given conception has vanished from the artist's mind, the drawing in which he expressed it continues to exist, and the conception will continue to be expressed by the drawing as long as the latter exists. Hence a mental product which no longer exists still finds expression in a psychophysical product which continues to exist. By finding its expression in a durable product a non-durable product "survives in it" and thus takes on the appearance of durability, *non omnis mortuus est*, since a psychophysical product to whose emergence it had contributed continues to exist.

34. Something similar also occurs when a mental product is the meaning of a psychophysical one, that is, when the latter expresses the former — in other words, when it is a sign of the former. (...) Now, if a sign is a durable product, then a durable partial cause of the emergence of a non-durable mental product exists as long as the sign exists. In the example above, a drawing is that durable partial cause of a thought

which, when the cause is completed, develops as a mental product in the person viewing the drawing. That mental product is not durable in itself: it exists only as long as the mental action which produces it. Such a mental product may come into being a number of times, but it will always be non-durable. But even when this mental product does not exist, i.e., when the mental action that produces it does not take place in anyone, one of the partial causes which at any moment may give rise to that non-durable product still continues to exist: it is that durable psychophysical product which is a sign of the mental product under consideration. Now, just as we say that the cause continues to *exist* in the effect, so we also say that the effect exists potentially in the cause, be it even a partial cause. We say accordingly that a mental product which is a meaning of the corresponding psychophysical product, that is, of the corresponding sign, exists potentially in that psychophysical product, i.e., in that sign. That mental product, that meaning, that content, etc., which bears a specified relation to the psychophysical product, takes on the appearance of being inherent in that psychophysical product, of being included or embodied in it; all these formulations obviously mean nothing other than that psychophysical product is one of the partial causes of the emergence of the said mental product, so that that mental product exists potentially (but neither actually nor in fact) in the said psychophysical product. [11] (...)

35. Durable psychophysical products thus impart to non-durable mental products the appearance of durability by being durable effects and durable partial causes of the latter. This is why mental products which bear that relation to durable psychophysical products may be called fixed, and we may accordingly speak of the fixing of non-durable products. Such a fixing is not restricted to mental products, but may apply to non-durable physical and psychophysical products as well. It always consists in establishing such a relation between a non-durable and a durable product that the latter, while being an effect of the former, becomes a partial cause which, when combined with the remaining partial causes, gives rise to an identical or similar non-durable product. For instance, by recording phonographically a cry we make the action which has produced the cry, i.e., a non-durable product, indirectly produce marks on the record, and hence, durable products. (...)

37. The process of fixing mental products, such as thoughts, ideas,

emotions, desires, decisions, etc., in writing, in print, etc., is even more complex. This is so because in such cases we do not directly fix a non-durable mental product by means of a durable psychophysical product (...); instead we fix non-durable psychophysical products which express non-durable mental products. Besides a single series of non-durable mental products we have two series of psychophysical products, one of which consists of non-durable products, and the other of durable ones. When we think, we are performing mental actions of which our thoughts are the products; these are non-durable mental products.

At the same time we are performing the action of speaking which yields non-durable psychophysical products, such as the words, sentences, etc., we utter. It is only these non-durable psychophysical products which we fix by actions that yield durable psychophysical products, namely written signs in the most general sense of the term.

38. As a result of being fixed, non-durable products assume not only the appearance of being durable, but also the appearance of products that are to some extent independent of the actions which yield them. (...)

39. If a durable psychophysical product evokes the mental product expressed in it, whether successively in one and the same person or either successively simultaneously in a number of different persons, then it obviously evokes not one product only, but as many products as there are actions yielding them. Such products are not identical with one another, but differ from one another to a greater or lesser degree. It is enough to recall how widely different mental products develop in different persons who are influenced, for instance, by one and the same picture, or by one and the same statement (...). Yet as long as we consider the said psychophysical product to be a product which expresses a mental product, the differences among the mental products it evokes may not be too great, and all those various mental products must reveal a number of common characteristics. It is precisely those common characteristics, those elements which the individual mental products sharing, which are usually considered to be the meaning or the content of a given psychophysical product, on the obvious condition that these common characteristics correspond to the intention in which the psychophysical product has been used as a sign. That is why we also say that a given statement evokes in various persons *the same* thought, whereas in fact it evokes as many thoughts as there are persons in-

volved, and, moreover, these thoughts are not identical. Yet we disregard the differences among them and consider only those elements which match one another and which match the corresponding elements in the thought of the author of the statement, to be that thought which is the element of that statement. We accordingly speak of only one meaning of a sign — leaving cases of ambiguity aside — and not about as many meanings as there are mental products which that sign evokes or may evoke in the persons whom it influences. Now, meaning conceived in this sense is no longer a specific mental product, but something we attain by the operation of abstraction performed on given products.[12] (...)

42. (...) There are products which result from a certain action in such a way that they imitate or replace other products which result from other actions. They might be called artificial or substitutive products. A human footprint in clay may be made artificially, which is to say that instead of being the result of a person making an imprint of his foot in clay, it may be modelled in clay by hand. (...) If non-durable products fixed by means of durable products may be called 'petrifacts', then such substitutive, artificial products may be called 'artefacts'.[13]

43. Such artefacts can often be found within the sphere of psychophysical products. Ample use is made of them, e.g., by an actor who assumes postures intended to express certain emotions. In fact, however, the actor usually merely imagines those emotions, so that his posture, in itself a psychophysical product, is not a result of genuine emotion, which usually finds expression in such a posture, but a result of an imagined emotion, i.e., an inner representation of such emotion. Thus, such an imagined emotion is a product which is substitute for a genuine emotion, and the posture referred to above is likewise an artificial product, since it is not a real expression of emotion, but merely its assumed, pretended image.

44. Another example for the use of artefacts is provided by logic. A proposition as a product of the action of judging, i.e., of making judgements, is expressed in propositions, i.e., psychophysical products which result from the psychophysical action of uttering or making statements. Such statements thus express propositions, so that propositions are meanings of such propositions.[14] We can, however, produce artificial, substitutive statements which are not expressions of propositions

actually made, but expressions of artificial products which are substitutes for propositions actually made, namely propositions which are merely imagined. Hence, the respective meanings of these artificial statements will not be propositions which have actually been made, but propositions which have merely been imagined, i.e., inner representations of judgements.[15] Artificial statements include not only the symbols used in logic, such as *SaP*, $a < b$, but also the statements used in it and consisting of words of a given language. This is so because a logician (or a grammarian, etc.), when making such statements, whether orally or in writing, by way of example, usually does not make propositions which constitute the meanings of those statements. Sometimes he could not make such propositions, even if he wished to do so; for if he wishes to give an example of inference which is formally correct, although it exclusively employs propositions materially false, and if he utters or writes the following statements:

All triangles are squares,
All squares are circles,
―――――――――――――
All triangles are circles,

then his utterances are not psychophysical products which express propositions actually made. Instead they express propositions which are merely conceived, and these are mere substitutes for real propositions; likewise, those utterances are substitutes for real ones, that is, for those which express real propositions. In this case, a psychophysical artefact expresses a mental artefact.[16]

45. (...) Within the whole of research, it is possible to single out a group of disciplines of which mental products are the only or main object of study. These disciplines might perhaps best be called "the humanities" (...). By moving further in the same direction we could try to define psychology, which after all is a fundamental discipline within the humanities. The distinction between mental products and mental actions, as well as the distinction between the various kinds of mental products, may prove quite useful. Similar considerations may contribute to an explanation of the mutual relationships between the various disciplines within the humanities in general (...).

Notes

* The differences between Polish and English make some of K. Twardowski's comments pointless in an English-language version; in many other cases, what is essential in Polish is of marginal importance in English. (Tr.)

** This refers to what in traditional English grammars is termed 'verbs taking cognate objects' (Tr.)

[1] Since human speech in many cases has separate words for actions and for products (not only physical, but mental as well), logicians have long referred to mental products as something different from actions, even though they may not always have realized the difference clearly. (Cf. B. Bolzano: "Bei den Worten: ein Urteil (...) eine Behauptung stellen wir uns sicher nichts anders vor, als etwas das durch Urteilen (...) und Behaupten hervorgebracht ist." *Wissenschaftslehre*, I, 1837, p. 82.

[2] I am using the term 'disposition' in the sense formulated by Höfler (*Psychologie*, 1897, Sec. 12), who also (op. cit., Sec. 6) draws attention to words which promiscuously denote actions (products) and dispositions.

[3] Some authors use the term 'judgement' in a fourth meaning, namely to denote what is usually termed '*enuntiatio*', '*Aussage*'. (...) In this fourth meaning the term 'judgement' accordingly denotes a certain psychophysical product. See also footnote[14].

[4] The same applies to (...) the verbs discussed earlier. A judgement is an inner object of judging, but that to which it refers is an outer object of judging. (...) For a distinction between the inner and the outer object of a representation see my paper *Zur Lehre vom Inhalt und Gegenstand der Vorstellungen*, Vienna 1894.

[5] These words are of course used here in the sense of a product, and not in that of an action of modifying or transforming.

[6] A relationship analogous to that which occurs here between actions also occurs between products. A corresponding mental action is expressed in a psychophysical action: a sensation of pain in shrieking, thinking about a drawing in (the act of) drawing, etc.

[7] Since a psychophysical product may result from several psychophysical actions, and not from a single action alone, various mental products can be expressed in it. For instance, what is expressed in a drawing may be the idea of a drawing which a person has when drawing, the idea which we want to convey through the intermediary of his drawing, his intention to convey that idea, etc. Thus, a psychophysical product expresses certain mental products directly and others indirectly with varying degrees of indirectness. (...)

[8] This is why a rigorous distinction must be made between the formulation "a psychophysical product (subject) expresses a mental product (object)", on the one hand, and the following formulations which are synonymous with one another: "a mental product is expressed, or finds expression, in a psychophysical product", "a psychophysical product is an expression of a mental product". This distinction enables us to eliminate various misunderstandings. For instance, it can help to

explain the very controversial issue of the relationship between music and emotions. For, while everyone agrees that emotions (and thoughts) experienced by the composer at the time of composing can be expressed in a musical work, it does not in the least follow that that work expresses those emotions. See also Hanslick, *Vom Musikalisch-Schönen* and Hausegger, *Die Musik als Ausdruck*, Wien 1885.

[9] For a detailed theory of signs and meanings see Martinak's *Psychologische Untersuchungen zur Bedeutungslehre*, Leipzig 1901.

[10] Regarding terminology it should be noted that the verb 'to mean' is itself ambiguous. When we say that a word means something, we intend to convey that it has a meaning. The verb 'to mean' interpreted in this sense corresponds to the Latin verb '*significare*' and to the German verb '*bedeuten*'. Instead of saying that a word or, more generally, a psychophysical product, means something, we may say that it has a meaning, and also that it comprises a meaning, that a meaning is connected with it, that a meaning is inherent in it, that it expresses a meaning. (...) We also say that psychophysical products denote something; this is to say they denote the outer objects of those actions from which those psychophysical objects result. Thus, we say, with respect to words, that they not only mean something, but also denote something, for some words come into existence as a result of names being given to objects. Hence the psychophysical product of this action is a name, i.e., a word whose meaning is the representation of a given object. At the same time, a name or a word denotes (mentions) that object. Thus, 'Sophroniscus' son' expresses a concept which is its meaning, but which also mentions a certain person. Likewise, the word 'triangle' expresses a concept, which is its meaning, and denotes (mentions) all the objects that come within that concept. Cf. my paper *Zur Lehre vom Inhalt und Gegenstand der Vorstellungen*, Sec. 3.

[11] As a result, the word 'meaning' takes on another meaning besides those specified in footnote[10], namely the meaning of being able (of course, together with other partial causes) to evoke a mental product in a person whom a psychophysical product, as a sign of that mental product, affects, i.e., the meaning of the ability of making a person realize an appropriate mental product. (...)

[12] For details see E. Husserl, *Logische Untersuchungen*, Vol. II, Halle 1901, especially pp. 97ff, where he discusses *ideale Bedeutung*.

[13] 'Artefact' is a standard English term, hence Twardowski's suggestion to introduce it as a neologism is lost in the English-language version, even though the meaning Twardowski would like to impart to it differs from the standard lexical meaning of that word in English. 'Petrifact', intended by Twardowski to be the antonym of the former and coined by him in Polish, is etymologically almost self-explanatory in English — much more so than in Polish. (Tr.)

[14] We have mentioned above that some people call a proposition that which occurs here as a statement. This is done, for instance, by J. Łukasiewicz, who defines a judgement as "a series of words or other signs which state that an object has, or has not, a certain property"; cf. his *O zasadzie sprzeczności u Arystotelesa* (On the Principle of Contradiction in Aristotle), Cracow 1912, p. 12. But when treating a proposition as a series of words or other signs he is obliged to make

a distinction between this series of words or other signs and that which is the meaning of the series. He does so, in fact, when he refers to 'synonymous propositions', which he defines as those which "express the same idea in different words" (op. cit., p. 25). Now this idea expressed in words is, of course, nothing other than a judgement in the sense of a product of the action of judging. Hence if the word 'proposition' were to be used to denote "a series of words or other signs" which express such an idea, there could be no word with which to denote the idea itself.

[15] The significance in our mental life of propositions which are merely conceived is due, *inter alia*, to the fact that they are components of all our concepts; cf. my paper *Wyobrażenia i pojęcia* (Ideas and Concepts), Lwów 1898, esp. Sec. 11. See also K. Twardowski's paper *O istocie pojęć* (On the Essence of Concepts) – Ed.

[16] Bernard Bolzano was the first to substantiate in detail this view of the subject matter of logic. Those judgements which are made independent, in the manner described above, of the action of judging he terms '*Sätze an sich*'. Besides them he also mentions '*Vorstellungen an sich*', i.e., representations which are similarly made independent of the action of representing. Cf. his *Wissenschaftslehre*, 1837, Vol. I, Secs. 19–23 and 48–53. (...)

ISSUES IN THE LOGIC OF ADJECTIVES

by

Kazimierz Twardowski

(Fragments)

The classification of adjectives into determining (...) and modifying (...) adjectives is well known. The former, when joined to a noun, add either a positive or a negative characteristic to its meaning (a 'learned man', an 'inexperienced person'); the latter, when joined to a noun, take away its original meaning. But, while losing its original meaning, a noun, when combined with a modifying adjective, becomes a name of an object to which the noun, taken in its original meaning, may no longer be applied (an 'artificial limb', a 'forged banknote', a 'former minister', etc.). Furthermore, like adjectives, adverbs and adverbial phrases may also be classified into those that determine and those that modify.

This classification does, however, give rise to certain doubts, primarily because the concepts of determination and modification are not parallel: determination may be interpreted as a simple function, whereas modification is a complex function. (...) The function of modifying (...) includes, first, the function of a partial removal of the content of the idea expressed by a given noun, and second, the function of replacing that removed part of the content — which is a result of combining a given adjective with a given noun — by other positive or negative characteristics. Thus, for instance, the adjective 'forged', when joined to the noun 'banknote', removes from the content of one's representation of a banknote those characteristics which are attributes only of non-forged banknotes, and replaces them by those characteristics which are attributes only of forged banknotes. On the other hand, certain characteristics are common to both forged and non-forged banknotes, e.g., the fact that they are pieces of paper with an appropriate text printed on them. The modifying adjective in no sense displaces the latter characteristics from the representation expressed by the noun 'banknote'. Now the second-named function of modifying adjectives does not apparently differ in nature from the proper function of determining adjectives. On the other hand, as regards the first function of modifying adjectives, this appears rather as the opposite of the determining function, since it removes part of the content of the representation expressed by a given noun, instead of enriching that content.

In view of the above it would be difficult to consider determining and modifying adjectives as two parallel types of adjectives. But the classification of adjectives into these two types is imprecise as it does not cover all types of adjectives. For instance, there are adjectives which perform only the first function of modifying adjectives; these might be called abolishing adjectives. An example is the adjective 'alleged', at least as used in certain cases. When joined to the noun 'shape' it removes from the representation of shape all those characteristics which combine to yield the idea of shape; that is why an alleged shape is no shape at all.

But there is yet another type of adjective, which adds no characteristics to the content of the representation expressed by the noun to which it is joined (...); adjectives of this type neither add nor remove anything, but as it were underscore, emphasize, confirm or preserve the characteristics included in the content of the representation expressed by the noun; in some cases they restore those characteristics if they have been removed by an abolishing adjective or by the abolishing function of a modifying adjective. They might accordingly be called preserving, confirming, or restitutive adjectives.[1] This group includes such adjectives as 'true' or' real', when used in such formulations as 'true friendship', a 'real fact'.

If we accordingly analyse adjectives from the point of view of the function they perform with respect to the content of the representation expressed by the noun to which they are added, then we obtain four types of adjectives (...). The first major type includes those adjectives which perform a simple function, which in turn appears in two variations. The first variation consists in modifying the content of the representation, either by enriching it or reducing it, and covers both determining and abolishing adjectives. The second variation does not change the content, but confirms, strengthens, or restores it, and covers confirming adjectives. The second major type covers those adjectives which perform a complex function, i.e., both removing and determining. This includes the modifying adjectives, which might also be called abolishing-and-determining adjectives. The above classification may be depicted as follows:

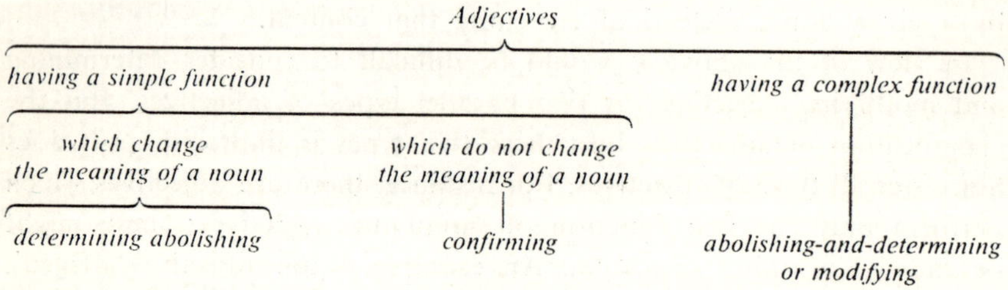

NOTE

[1] Attention has been drawn to them by F. Brentano. See Häfler, *Logik* 2nd. ed., 1922, footnote on p. 186

A SURVEY OF LOGICAL AND SEMANTIC PROBLEMS

by

Tadeusz Kotarbiński

With regard to the semantic aspect of language, the division of logical research into three branches (Morris, Carnap) is beginning to be fairly well established. Distinction is made between syntactics, semantics and pragmatics. Syntactics covers the issues resulting from the relationships between expressions; semantics covers the issues resulting from the relationships between expressions and the objects to which they refer; pragmatics covers the issues resulting from the relationship between the users of expressions and the expressions they use. The whole forms semantics in the broader sense of the term, in this case often called semiotics. We shall be interested in the first two branches. We want to show with what they are concerned, in order to promote intellectual contacts between logicians and linguists.

Logicians distinguish expressions which are meaningful with respect to a given language (this reference to a chosen language is indispensable in view of the fact that strings of language signs, having the same structure, may happen to belong to two or more language systems) and such as are not. The latter may either totally lack any semantic aspect from the point of view of the language in question, or may be endowed with indefinite meaning, or may be semantically faulty. These are called nonsensical, and among them are statements which are evidently false, absurdities or contradictory statements, and syntactically unconnected expressions, whose component parts are knit together in a way which does not agree with their semantic categories. For every meaningful expression belongs to a definite semantic category (Husserl) or to a definite logical type. The basic semantic categories are those of sentence and term, and the other logical types can be formed by sentence-forming or term-forming functors of the various kinds (Leśniewski). The kind of a given functor is determined by whether it combines terms with terms into a certain whole, or sentences with sentences, or terms with sentences, by the number of such elements, and by whether the whole thus formed is a sentence or a term. Thus we have sentence-forming functors of two sentential arguments — for example, the sentential connective 'if...then...' (rendered by one symbol in the artificial logical notation); sentence-forming functors of two term arguments — for example, the copula 'is' in an elementary sentence (that is, a singular sentence of the 'subject+copula+predicative word' type — a sentence of the '*A* is *B*' type with a singular term as the subject); term-forming functors of one term argument — that is, term negation; term-forming functors of two term arguments — for example, 'and' connecting two terms; and many others. It may be added that logicians are also interested in the ambiguity of connectives in ordinary usage — for instance, the ambiguity of the connective 'and' in English, where it may be conjunctive (*A* is *B* and *C*), or enumerative (*A* and *B* are *C*), or synthesizing (*A* and *B* jointly form *C*). Logical types only roughly coincide with parts of the sentence as distinguished by traditional grammar in syntactical analysis. The study of semantic categories is useful in eliminating hidden nonsense, and in building a rational hierarchy of the branches of formal logic, a hierarchy based on the logical type of the variables occurring in the formula used in a given branch.

These distinctions do not relieve logicians of the need for an analytic definition of 'meaning'. It is now common knowledge that it does not suffice to accept the meaning of a word or an expression to be what is associated with it when it is experienced or comprehended (Ossowska). When it comes to terms, it seems correct to seek the essence of their meaning, or connotation, or content, in that set of properties which we want to ascribe to a given object when we predicate a given term about it. On a closer investigation it is necessary to distinguish a number of variations of the comprehension of the meaning of a term (Ajdukiewicz). The search for a general analytic definition of the meaning of expressions ultimately leads to an attempt to define analytically the phrase 'means that...', as applied to sentences. Such tentative definitions so far formulated are, as the present author knows them, all invalidated by the error of the vicious circle type, when endeavours are made to define this phrase by the phrase 'means that...', which is characteristic of analytic definitions, or when attempts are made to explain the meaning of that phrase by the word 'expresses', since the latter must in turn be referred to some language, and hence to a system of meanings attached to definite signs (Ossowska). Another difficulty is the handling of the functor 'means that...' as forming sentences which have the form of an intensional function. Every sentence can be analysed as a result of a substitution of values in a sentential function (that is, in a string of signs which differs from a sentence in that instead of certain constant elements of a given sentence it includes variables of the corresponding semantic categories). Now sentential functions are classified into extensional and intensional (Russell). The truth-value, that is, truth or falsehood, of a substitution for an extensional function depends solely on the extensions of the values substituted for the variables, whereas the truth-value of a substitution for an intensional function depends on the meaning of the values substituted for the variables. Extensional functions with sentential variables are what are called truth functions, characterized by the fact that the truth-value of a result of a substitution of values for the variables of such a function is determined exclusively by the truth-values of the sentences substituted for the sentential variables, regardless of the various meanings of those sentences; it is as if the extension of all true sentences were the same, and the extension of all false sentences were also the same (Frege, Schröder). Such are all the formulas of the sentential calculus. On the other hand, the intensional function 'The sentence *S* means that *p*', where *S* is a term variable, and *p* a sentential variable, does not satisfy this condition. Hence it is an intensional function, as is also the function '*A* knows that *p*' (for instance, Copernicus knew that the earth revolves around the sun, but did not know that a world war broke out in 1914, although both *that*-clauses have the same logical value since both are true). The problem of eliminating the intensional functions, which depart from the rules of inference through the substitution for the variables of definite constant values of sentences, is still vital. The suspicion arises that intensional functions are faulty forms of expression, and that every correct sentential form should be extensional. Hence the tendency to abandon attempts at defining 'meaning' and to use only the concept of synonymity (Ajdukiewicz, Kreczmar), since the function '*A* is synonymous with *B*' is extensional. But there is great difficulty in analytically defining synonymity without entering a vicious circle, if every analytic definition has the form '*A* is synonymous with *B*'.

Synonymity is most closely connected with the search for the general conditions of mutual translatability of two linguistic systems and the general conditions of definability of a given term by means of a given system of terms. Certainly, any object which satisfies one of two synonymous sentential functions must also satisfy the other. And a given sentential function, which includes only term variables, is satisfied by all and only those objects, the names of which, when substituted for the variables, change that function into a true sentence. The object of which a given term is a name in a given language is a designatum of that term with respect to that language. This reveals the mutual relationships between the concepts of satisfaction, denotation, designatum, term and truth. Logic tries to elucidate these relationships, which have given rise to what is known as the semantic concept of truth (Tarski), which is a modern continuation, freed from common objections, of the classical interpretation of truth as agreement with reality. By that conception, the truth of a sentence consists in that it is satisfied by all objects, the concept of satisfaction being not defined by reference to truth, but introduced in a different way.

We have been using the concept of term in general and the concept of the name of a particular object. Whoever might demand in relation to those expressions analytic definitions, which would reduce them to some more comprehensible terminology, would thereby raise a problem whose solution also involves the risk of a vicious circle. For if we do not confine ourselves to the statement that a given expression is the name of the object it refers to, and ask further, what is meant by the fact that a given object is referred to by a given expression, the answer seems obvious that this is the object about which the name is predicated in a true sentence. And when we in turn try to explain which object is called that which is predicated in a sentence, we obtain the answer that such an object is that whose name is the subject of that sentence.

Other difficulties are raised by the problem of classification of terms. Let us first recall the classifications of terms into categories in the sense that comes from Aristotle — that is, classifications which distinguish names of things, properties, relations, etc. There are various such classifications, which on the one hand are considered to be lists of the kinds of terms, drawn with respect to the ontological kinds of the designata of those terms, and on the other, to be lists of the ontological kinds of objects. The category of events is usually added to those specified above. This grammatical *a priori* ontology gave rise to the controversy about reism or concretism. This is an endeavour to verify the hypothesis that wrongly formulated ontological issues will vanish if we observe the principle that in the ultimate formulation we do not use terms other than names of concrete objects (Leibniz). The concretists think that there are genuine terms and apparent terms (onomatoids), and that the latter are certain nouns, adjectives, noun phrases or adjectival phrases which because of their appearance are taken to be terms, but are not terms; they can be used meaningfully only in substitutive formulations, and can always be eliminated in favour of genuine terms alone. The latter are always names of things, so that there are no other objects than objects from the category of things (Brentano). This results in the necessity of making a distinction as between empty genuine terms (such as 'a mountain higher than Mt. Everest') and onomatoids (such as 'seniority'), since genuine terms are general, singular and empty, according to whether a given term has more than one designatum, or one designatum only, or no designatum at all. The distinction consists in that a genuine empty term is defined analytically by linking terms, which have designata, by such a functor as links certain other terms, having designata, into a compound term which also has designata; on the contrary, an onomatoid cannot be defined in this way. For instance, 'a mountain higher than Mt. Everest' is a genuine term, although it is empty, because for instance the expression 'a fir higher than this apple-tree' is a genuine term that has designata. This classification of genuine terms competes with another, by which the singularity and the generality of a term are distinguished not by the number of designata, but by the place which is proper to a given term in the structure of elementary sentences. A singular term would be that which may occur as the subject and may not occur as the predicative word in such a sentence. Such a distinction would result in a duality of the semantic category of terms. There is also a controversy as to whether certain objects are general objects. A general object would be such as has only those properties which are common to all the designata of a given general name. Both this definition and the arguments for (Ajdukiewicz) and against (Leśniewski) the existence of general objects are semantic and logical in nature. The striving of the reists, like the striving of the nominalists who deny the existence of general objects, are manifestations of the campaign of logical criticism against hypostases, understood as an illusory guessing at the existence of certain entities because of the existence of certain words and nominal and adjectival phrases, as if each such word or phrase were bound to have some designatum or designata.

Logical semantics also wages a campaign against equivocations and vagueness of expressions. Distinction is made between terms which are unclear, indistinct, and vague as to their meanings. We comprehend a given term clearly if we correctly apply it to its designata, and to them only; we comprehend it distinctly if we are able to specify, in the form of an analytic definition, the set of the properties combining to form its content (Jevons). These are contrasted with unclear and indistinct comprehension of terms; vagueness is unclearness in marginal cases (for example, when we do not know whether a person aged forty is still 'young' or not). Analytic definitions are the best remedies against unclearness and indistinctness, and moreover one may renounce the use of an unclear or indistinct term and use another one instead — the meaning of which would come closer to the former but would be established by a synthetic definition. Especially in the case of embarrassing vagueness, either the limit of applicability of a given term is set arbitrarily, or the formulation of statements using a vague adjective (such as 'young') is replaced by the formulation of statements using a corresponding but much less vague adjective in the comparative degree (for example, 'younger'). Attention has been drawn to the fact that it is hardly possible to quote a term which would be totally clear and devoid of vagueness (Chwistek), so that all the improvements can only reduce those defects, but cannot eliminate them completely.

All this shows the great importance of definitions for the improvement of the semantic functions of language. That is why logicians spare no effort to develop the theory of the technique of building definitions, both analytic and synthetic. That theory describes and compares the following methods of forming analytic definitions, which are important especially in bringing out the meanings of expressions in natural (ethnic) languages: the Socratic deductive method, the Socratic inductive method, the etymological method, the philological method, and the phenomenological method. Further, admissible forms of definitions are worked out in the analysis of the artificial symbolisms used in the deductive systems of mathematics and formal logic, especially in connection with synthetic definitions (Dubislav). In addition to the classical form of defining *per genus et differentiam*, which is particularly useful in defining terms, there are definitions by abstraction, recursive definitions, useful especially in defining functors, definitions in use, and other forms of definitions.

Various issues emerge when efforts are made to build a deductive system of semantics analogous to the deductive systems of propositional calculus, the calculus of terms, set theory and geometry. In general, a large number of semantic and logical problems accompanies the construction of deductive systems in any field. By way of example, let us quote one of the methods of demonstrating the agreement and the mutual independence of the axioms of a system. This is achieved by assuming (Hilbert) that the constant symbols of that system have no meanings and by imparting to them, by way of trial, such meanings for which all the axioms would prove true (agreement), and then some other meanings for which some axioms would prove true, and the others would prove false (independence). Another example is the effort to formalize deductive systems — that is, to pursue the procedure of deducing secondary statements of a deductive system (theorems) from their primary statements (axioms) independent of the meanings of the symbols used in the system, and to deduce them only from the shape and position of symbols (Hilbert). As a technical ideal of semantics, this seems paradoxical. As regards the deductive systematization of semantics, the first is issue whether in a rational sequence of exposition of deductive systems semantics should be discussed first, or whether it should follow formal logic and mathematics. Opinion is in favour of the latter interpretation, for although semantic issues must be raised when discussing any of these two disciplines — for example, when explaining the uses of symbols and formulating the conditions of meaningfulness of strings of symbols belonging to a given system — yet it must not be forgotten that the axioms and theorems of formal logic and mathematics are free from names of symbols, and that names of symbols are joined to the terms of those systems only in semantics. The problem arises, in connection with attempts to build a deductive system of semantics, whether that would be an *a priori* discipline. Logical semantics is also deeply interested in the problem of the essence of *a priori*, and as such it analyses the supposition that all *a priori* knowledge is nothing but self-evident knowledge that can be derived without recourse to empirical data, exclusively from the meanings of the symbols used in a given statement. Hence the criticism of the Kantian distinction between analytic and synthetic judgements, and the efforts to demonstrate the analytic nature of the mathematical statements (Couturat), in connection with the reduction of the mathematical statements to statements of formal logic (Frege, Russell). Finally, the possibility of a full formalization of the deductive system of semantics has become a topical issue (Chwistek).

To illustrate the difficulties resulting from certain problems of semantics, we shall quote an antinomy in that field, one not yet solved in a way accepted by all the experts. Let us consider that class of words of a given ethnic language — for example, the English language — every element of which has the property which constitutes its meaning (for instance, the word 'word' is itself a word, the word 'English' is itself English), and compare it with that class in the same language every element of which does not have the property which constitutes its meaning (the word 'house' is not a house, the word 'branch' is not a branch). Let the words of the former class be called autological, and the words of the latter, heterological, and let us now consider whether the word 'heterological' is autological or heterological. In answering that question how can we avoid a contradiction, since if the word 'heterological' is autological it follows that it is heterological, and vice versa (Grelling)? Another antinomy arises when we ask whether the description, 'The least number which cannot be univocally defined in less than a hundred words of the English language', defines any number univocally (as its only designatum), or not (Richard). The efforts intended to free semantics from such complications make it imperative to take into account the difference between the ordinary and the material supposition of terms (in the latter case, a given word being used as a name of words built like itself), and also to adopt a stand with respect to the rule which forbids, under the penalty of obtaining nonsense, the use of a given term as its own designatum. Hence follows the distinction between a language and its metalanguage, which includes names of symbols used in the former, and hence in turn the theory of metalanguages of increasingly high orders, and the study of logical and semantic relations between them (Russell, Leśniewski, Tarski). All these theoretical procedures are intended above all to free reasoning from the nonsense concealed under the misleading veil of words, and in particular to free reasoning from nonsensical questions clad in the appearance of meaningfulness. The study of the concept of question and the structure of questions (Ajdukiewicz) serves the same purpose.

The survey made above shows that the efforts of logicians in the study of semantics do not in the least contradict the work of linguists in the same field. The linguist distinguishes the component forms of an ethnic language and describes their semantic functions, also from the genetic point of view. Further, it is the linguists who have to see to it that pupils at school speak and write correctly — for example, from the point of view of standard English. This task cannot be separated by the teacher from his concern that the pupils should speak and write logically — that is, clearly and consistently as to meaning, and if necessary also with comprehension of the meanings of the words they use. Here is the meeting ground of the linguist, the logician concerned principally with the building of ideal artificial languages, and the teacher of logic who has to correct the formulations of his pupils as to clearness and consistency. It would be most desirable that logicians should acquire linguistic knowledge and that linguists should cover with their researches languages of deductive systems, and also study how ethnic languages perform those semantic functions in which formal logic and logical semantics are interested.

THE REISTIC OR CONCRETISTIC APPROACH

by

Tadeusz Kotarbiński

The responsibilities of the secondary school teacher undoubtedly include the striving to make the pupils understand meanings of words as clearly and as distinctly as possible. This task acquires particular importance in the teaching of those disciplines which in library catalogues are called philosophical. This is so because the principal shortcomings of those disciplines, which also account for protracted controversies — for example, in epistemology, ontology, general theory of value, etc. – do not consist in defective observation or experimentation, or in using wrong forms of inference, but are mainly reducible to the habit of thinking, and correspondingly speaking, vaguely. Hence, whether he so desires or not, the teacher must build a system of verbal explanations and, as it were, compile a dictionary of those terms which sow confusion. And since every subject taught leads to some philosophical issues, every teacher must try to contribute to such a philosophical dictionary, and the professional teacher of philosophical subjects must help him in that, working out the dictionary not only for his own use, but also for use by teachers of other disciplines. For instance, the controversy is revived from time to time as to whether mathematics is an empirical or a purely deductive science. I have no intention of solving that problem here. My point is only that the controversy would not be chronic if the participants would distinctly realize the ambiguity of the term 'empirical'. In the genetic sense, a statement can be understood as empirical only if the person concerned has ever previously observed something, or if he has observed at least one of the objects denoted by one of the terms involved in that statement. In the methodological sense, only such a statement which for its founding requires at least one observation statement as a premiss is empirical.

We are thus struggling with confusion due to unnoticed ambiguity, which becomes particularly vicious in the case of similar meanings, as in the example above. In other cases, trouble arises from the vagueness of a term about which we do not know whether it covers by its extension a given border-line object. Very often there are differences in the appraisal of the scientific character of a given theorem or its foundation; for instance — are observations on animal behaviour, made in the natural environment without laboratory control, scientific if they are fairly (?) careful and made by an expert (?), or are they not scientific even if both these conditions are satisfied? In other cases, we correctly apply a given term to appropriate objects, but we cannot specify the properties which combine to form its meaning and are helpless in the face of obscure theories in which that term plays an important role. By way of example we may mention here the term 'time' in the general theory of events or the term 'truth' in the foundations of epistemology. In still other cases, we witness delusions concerning the existence of alleged entities, delusions which are largely explained by the presence in our language of nouns and adjectives of various kinds — for example, general and abstract, such as 'equality', 'law', etc.

The best method of eradicating such defects in thinking, which originate from wrong suggestions of language, would be to avoid all such stigmatized words. But such a principle, not qualified by additional explanations and limitations, would result in a most detrimental sterility of research. We must know how to render innocuous the various words, while preserving the wealth of valuable content which they carry. In some cases, we can save the full content, essential in given investigations, by discarding an unnecessary auxiliary problem which requires the use of a hopelessly vague term. Thus, in order to group all the elements of a given set into two classes according to size, it is not necessary to consider which are 'large' and which are 'small' (these being extremely vague terms), but it suffices to see which in a given pair is larger than the other, and to group in one class all those which are larger than those grouped in the other class. We use here the term 'greater than', which is incomparably less vague than are the terms 'great' and 'small'. An example of a similar procedure can be found in making the analysis of a deductive system independent of the terms 'truth', 'true', etc. These are not, we admit, classical examples of vagueness, but they are controversial enough to make their elimination, whenever this can be done without losing the sense of a given inquiry, a justified endeavour. In the theory of deductive systems this is achieved by interpreting axioms as the primitive statements of a given system, and interpreting theorems as theses which are deducible from axioms in conformity with rules of inference, neither the interpretation of the primitive character of the axioms nor the interpretation of the rules having any reference to 'truth'. Sometimes in the case of a confusing word, it suffices to accept a thesis or theses containing that word, and the controversy is resolved. For instance, when we witness a controversy as to whether man is a machine, it suffices, in order to take a stand in the controversy, to accept the statement that machines do not move of their own will and do not perforof actions. But in most cases we have to find definitional equivalents m given terms. We then proceed either synthetically, by imparting to a given word some arbitrary meaning (as was done by logicians with the sentential connective 'if... then...', whose function was radically modified as compared with the colloquial usage), or we proceed analytically, by trying to grasp the meaning of a given word in a definition which specifies those properties which combine to form its intuitive meaning. The controversy about the existence of universals revives from time to time like a many-headed hydra. The philosopher who has to take sides in that controversy will first define the term '*universale*' by stating, for instance, that a '*universale*' is any such object, associated with a given general term, as has only those properties which are common to all the designata of that term — that is, common to all the individuals referred to by that general term. When this definition is adopted, the thesis about the existence of universals can easily be reduced *ad absurdum* if it is assumed that: (1) '*A* exists' means the same as 'some object is *A*'; (2) every object has either a certain property or its negation; (3) every designatum of a general term has a specific property.

This concludes the lengthy introduction, the task of which was to persuade the reader that it would be well to think of some general recommendations outlining the methods of working out all the explanations to be used in a philosophical dictionary and serving to eliminate all that Francis Bacon called *idola fori*.

The reader's attention will now be drawn to only one of such recommendations, to which the present author attaches special importance — namely, to the principle of reism. What is the requirement formulated by that doctrine? This only — that all ultimate formulations, and hence also all ultimate explanations of words, should include no nouns and adjectives other than concrete nouns and concrete adjectives. The point is not, of course, that such ultimate explanations should consist of concrete nouns and/or concrete adjectives alone, to the exclusion of connectives, negation particles, copulas, etc. The essential point is that these ultimate explanations should include no nouns and adjectives other than concrete ones. Instead of 'noun or adjective' let us say briefly 'term'. The principle of reism can then be formulated as follows: let us see to it that we know how to reduce every statement to a form which contains no other terms than concrete ones.

The question now arises: which terms are to be accepted as concrete. These are terms of three kinds. The first kind consists of singular names of persons or things, terms which we use as grammatical subjects in true singular sentences about individual persons or things (of the type '*A* is *B*') — that is, proper names such as 'Plato', 'Rome', etc. The second kind includes general names of persons or things — that is, terms which we use as grammatical subjects in true general sentences (of the type 'every *A* is *B*') about persons or things, such terms as 'man', 'city', etc. As for the terms of the second kind there is no doubt that they can be meaningfully used as predicative words both in general and in singular sentences. Some people doubt, on the other hand, whether the terms of the first kind can be used meaningfully as predicative words. Our opinion is that these two kinds must be treated on an equal footing in that respect. But this controversy is not essential for the principal issue here under consideration. What is important is the discussion of the third kind of terms — that is, concrete empty terms. These are terms which cannot be subjects of any true singular or general sentence about persons or things, but are by definition reducible to certain combinations of singular or general terms which are names of persons or things; these combinations are such that if some other singular or general names of persons or things are combined is this way, then the whole becomes a singular or general name of persons or things. This clumsy formula becomes simpler if we introduce abbreviations: 'denoting term' for 'singular or general name of persons or things', and 'elementary sentence' for 'true singular or general sentence about persons or things'. This yields the following formulation: an empty term is a term which cannot be the subject of an elementary sentence, but is by definition reducible to such a combination of denoting terms that other terms, combined in this way, yield a whole which is a denoting term. Let the word 'chimera' serve as an example. It is defined in ancient mythology as a lion's head with a goat's body and serpent's tail. No such monster has ever existed, and hence it is an empty term. But if its denoting elements (that is, 'lion's head', 'goat's body', 'serpent's tail') are replaced by other denoting terms, 'head', 'thorax', 'trunk', we obtain the whole 'head with thorax and trunk', which is a term and denotes any (mature) insect. A similar operation cannot be performed on those terms which are not empty concrete terms, although they also do not denote persons or things. They are such words as 'smoothness', 'relationship', 'tune', 'shift', and, generally, what are called names of properties, relations, contents, events, etc. All these nouns and adjectives which have the appearances of concrete terms but are not concrete terms, we take the liberty of calling apparent names or onomatoids. Hence the world of terms is divided into concrete terms (or genuine terms) and apparent terms (or onomatoids), and the principle of reism requires elimination from all ultimate formulations of all onomatoids, so that only concrete terms — singular, general and empty — are left.

The reader will certainly ask for the foundation of this principle. First of all, we shall refer to the teacher's experience. Is this not the way to explain words in a natural manner, natural at least from the didactic point of view? Should we wish to explain to a child what the word 'similarity' means, should we not show him in turn several pairs of objects which look alike, and say: 'Look, here are two sparrows: this one is grey and that one is grey, this one is hopping and that one is hopping, this one has a short beak and that one has a short beak. These two birds are similar one to the other. And here are two windows: both are rectangular and both have rectangular panes, separated by thin pieces of wood. These two things are similar one to the other. Do you understand now what is similarity?' Or suppose that in the class we encounter in the text the word 'recovery', which the pupils do not understand. We shall tell them: 'Whenever a person has been ill, and later was better, and now is still better, we speak of recovery'.

Suppose then that the principle of concretism, understood as a guiding line for teachers, has been founded sufficiently by reference to the psychological naturality of such a guiding line. But the reader will ask in turn — How can that psychological naturality be explained? Reference, quite correct, to the course of development of language in the life of both individuals and ethnic groups does not provide an exhaustive answer. It is a fact that names of persons and things appear earlier in a language than do other nouns, but this fact in turn calls for an explanation. Here the reist will risk a thesis which is no longer founded on didactic considerations, but is of an ontological nature. That thesis states that every object is a thing or a person, and that this justifies everything which we have tried to explain above. Let us subject that ontological statement to closer scrutiny.

Since every object is a thing or a person, and 'entity' and 'individual' are extensional equivalents of 'object', then to put it briefly, there exists nothing else but things and persons — of course, if the word 'exists' is understood in its fundamental sense. In that fundamental sense: *A* exists, is the same as: a certain object is *A*, which is equivalent to the formulations: some entity is *A*, some (at least one) individuals are *A*, etc. Each of these formulations can either be shortened into: something is *A*, or expanded pedantically: for some *x*, *x* is *A*. Now the reist deems that only things and persons exist, since it is true only of things and persons that certain objects are things or persons. Should a person further ask about the definition of the term 'object', we should have to refer to the meaning of the copula 'is' in singular empirical statements (such as 'this is green', with an indication of a leaf; or 'the Earth is spherical'; or 'I am gay'; or 'Peter is a carpenter') and say that only that, and all that, is an object about which we may meaningfully formulate a singular sentence (of the type '*A* is *B*') with the copula so understood.

From that point of view it is true that Vesuvius exists, and that the dinosaur exists (of course not necessarily now, but among past objects); it will be false that a centaur or a cyclops exists; but it will be nonsensical to say 'the eruption of Vesuvius exists' or 'the simultaneity of action and reaction exists'. Consequently, we may say correctly that centaurs and cyclops do not exist, but we may not say correctly 'the eruption of Vesuvius does not exist' or 'the simultaneity of action and reaction does not exist', for the formulations which are here rejected are not subject to a meaningful negation, but must be eliminated as nonsenses. The negation of a false statement is a true statement, the negation of nonsense is itself nonsense. It would be nonsense to say 'over whereby exists', but to say 'over whereby does not exist' would be nonsense, too. The reservation must be made that we do not mean the words 'over whereby' in their possible material supposition, but we take them here in their usual role. The other reservation may also prove advisable. It can meaningfully, and even correctly, be said that — for example — simultaneity of action and reaction exists, if the word 'exists' is comprehended in its secondary sense, for instance, in a sense in which that statement would mean the same as 'a body presses another body when it is itself pressed by that other body'.

To recapitulate. The point is that the world of objects is identical with the world of things and persons, and therefore only things and persons exist in the essential sense of that word, and hence the conclusion that only names of things and persons may occur in such a meaningful, and correspondingly in such a true, statement, from which results the assertion of existence of designata of such terms. The concretist has nothing against metaphorical formulations, against using words in their secondary senses. Without sentences with nouns and adjectives other than the names of things and persons we could not communicate briefly and quickly. In lectures, let us not renounce apparent names of properties, relations, events, etc. The point is only that we should know in each case how to eliminate any such apparent term, for only then shall we interpret correctly the reality, which consists exclusively of things and persons.

This view evokes appraisals which differ greatly from one another. Some people see in it an excessive and over-bold simplification of the picture of reality, while others think that these are self-evident consequences of conventions about meanings of words. It would be a hopeless task to try to bring reism into agreement with all the appraisals by its opponents who themselves differ so much in their opinions. But we make it a point to be on good intellectual terms with logicians who are past masters in exact notation. What appraisal of concretism can we expect from them? I think rather a favourable one. Using his own language, a semantician would classify what are called the names of properties, relations, contents, events, etc., in the various semantic categories, and say that these are semantic categories other than either the category of subjects or the category of predicates of inherential and subsumptive statements. An inherential statement says that a given individual belongs to a given class, while a subsumptive statement says that the totality of individuals of a given class belongs to some other class. The scheme of an inherential statement may be symbolized $x \in M$ (x is an element of the class of M's, or x is one of the M's or, x is an M), and the scheme of a subsumptive statement, $M \subset N$ (the class of M's is included in the class of N's, or whatever is an M, is an N, or any M is N). Now a logician will simply never think of substituting for x anything but a singular name of an individual, and for M or N, a general name of the individuals belonging to a given class, in doing which he will never identify any individual with any property, relation or event. Hence, if he wants to write down the statement saying that the relation of equality holds between x and y, he will resort to another scheme and write $x = y$, which is a special case of the general scheme xRy, which can be included neither in the scheme of inherential statements nor in the scheme of subsumptive statements. A logician will moreover conclude that the reists identify the totality of individuals with the totality of things and persons, against which he will not raise any objections, but will only say that deciding this issue is not a task of formal logic. It is perhaps superfluous to add that the objection against reism that in logic there are statements which do not include any terms — for instance, any theorem of the sentential calculus, be it the principle of transposition: $\Pi\, p, q\, [(p \rightarrow q) \rightarrow (\sim q \rightarrow \sim p)]$, where p and q are sentential variables, would be out of place. The reists do not assert that sentences without terms are impossible. They only claim that if a sentence in its ultimate formulation, without any abbreviations and substitutions, includes some nouns or adjectives, then these are names of concrete terms, whether singular or general or empty.

On the other hand, the reists will find it more difficult to reach an agreement with those ontologists who accuse them of an excessive simplification of reality. We mean here those who accept the ontological interpretation of Aristotelian categories or of any similar list. All entities are there classified as substances, properties, relations, events, etc. Now the reists accept only one ontological category — namely, that which has traditionally been called the category of substances. And please note, even within that category they eliminate Aristotelian substances in the secondary sense — that is, general substances, in other words the universals. What remain are only the Aristotelian substances in the primary sense: Socrateses, horses, stones — in a word, individual things or persons; the term 'thing' becomes modernized and covers everything which is temporal and spatial and physically defined — for instance, physically influencing something else. Guessing at the existence of anything else is considered by the concretists as committing a hypostasis — that is, imagining the existence of some entities — caused by a delusion due to the existence of certain nouns. This gives rise to a number of special problems — for instance, whether the term 'class' is a genuine term or an onomatoid, that is, whether certain objects are classes or not. The answer will depend on which of the two uses of that term, distributive or collective, is involved. If it is said that the class of *M*'s is included in the class of *N*'s, whereby it is meant that whatever is an *M* is an *N*, then the term 'class' is used distributively, and the term itself is an onomatoid which vanishes in the ultimate formulation. If, however, the term 'class' is used to denote a group of school pupils, the term 'class' occurs as a name of a team, of an object of which the individual pupils are component parts.

Some of these hypostases enjoy support not from physicists themselves, but from many philosophical interpreters of physics. In that field, any discussion is extremely difficult, since one of the most topical but unperformed tasks of concretism is to work out a dictionary of mathematics and physics in the reistic interpretation. In this connection the present author will raise only one point, where the issue seems to him clear enough. Now it is not possible to accept such an interpretation of experimental data in which a particle of a physical body, and hence a small object, would have, under certain conditions, to be identical with a wave. The noun 'wave' is a special case of the general term 'process' or 'event', and as such, from the reistic point of view, yields nonsense when substituted for a name of a thing in an ultimate formulation. To say about a thing that it is a wave is equally nonsensical as to say about a thing that it is a mode or a quality. A similar attitude must be adopted by the concretist with reference to the opinion that reality is a combination of changes which are not changes of anything. The term 'change', being also a special case of the term 'event', is an onomatoid. To think that the world consists of changes is to build reality out of hypostases, and if it is moreover thought that those changes are, as it were, subject-less, this is as if a person were to think that a hospital consists of diseases, recoveries, and deaths, without patients. For the reist, reality is a fabric composed of changing things. Stress is laid on 'fabric' and 'changing', in order to avert a possible objection that for the reist the world is a 'static conglomerate' ('a mere sum') of 'rigid and changeless solids'.

After all that has been said above, it would not be astonishing if the reader wondered at a certain dualism to be seen in the formulations used by the concretist. Why is reference always made to things and persons? Does he accept the Cartesian distinction between two kinds of substances, not reducible to one another, called *res extensae* and *res cogitantes*? Can they not all be reduced to one of the two kinds, either the extensive entities, or physical objects, or the sentient entities, that is, psychic beings? In the present author's opinion, this can be done, and for that purpose he chooses the first solution: for him, any psychic being is a physical object, and hence John who experiences emotions, hears, thinks, and makes decisions, is the same John who talks, writes, and in general moves in a purposive manner. This is not only reism, but also somatism, which is a special case of the former. In principle, however, there might be reists who would reject somatist monism, and accept dualism or spiritualist monism. Somatism is certainly a form of materialism, but it does not imply mechanism. The assumption that every object is a physical object does not in the least mean that the laws of mechanics sufficiently explain everything that happens to objects, for it seems that the laws of mechanics do not even explain all that happens to objects in those respects in which physics is concerned with objects. *A fortiori*, they do not explain those changes with which psychology is concerned. Thus, somatism faces the immense task of formulating the conceptual structure of psychology from the reistic point of view. That standpoint excludes the description of basic psychological statements such as statements about smells, tunes, tastes, immanent coloured patches, senses of thoughts, and the like contents of perceptions, images and concepts, or acts of perception, of imagining, or of conceptual comprehension, etc. All these nouns are onomatoids. Elsewhere, the present author has taken the liberty of calling radical realism that view which accepts what are called names of contents to be onomatoids and hence prohibits their use as subjects in elementary sentences. From the standpoint of reistic somatism, efforts are made to interpret basic psychological statements as statements about persons (the latter being, of course, identical with certain physical objects). There would be, for instance, such statements as: '*X* experiences as follows: *A* is *B*'. As applied to a given case, this would yield: 'John sees so: this is black'. Of course, the sentence of the type '*A* is *B*' is a special form of sentence, and in the general formula of the psychological statement its place may be occupied by a sentence of any form. Ir we take *p* to stand for an arbitrary sentence, we obtain the following general formula of the psychological statement: '*x* experiences as follows: *p*', or '*x* is experiencing as follows: *p*'. Thus the only specific name element of the psychological statements would be the term 'experiencing' or one of its special forms: 'seeing', 'wishing', etc. All these terms are predicable about persons, are names of persons (the latter being, from the somatist standpoint, identical with certain physical objects). And here are some other examples of psychological statements: 'John thinks so: $2+2 = 4$', 'John feels so: they are playing sadly', 'John doubts so: do angels exist?', 'John desires so: be happy', Perhaps it would even not be too bold to generalize this formula so that not only a sentence, but any phrase referring to how a given person experiences, could be substituted for '*p*'. It might even be an inarticulate exclamation, so that a given psychological statement would be: 'John experiences so: Oh!'

Having worked out the principles of reism in its somatist and radical realist form, the present author for some time had an illusion of having invented something new. But soon his attention was drawn to the fact that such a view was represented by Franz Brentano in the last years of his life. In the writings dictated by Brentano, who was then blind, and appended to a new edition of his *Psychologie vom empirischen Standpunkt* in Meiner's Philosophical Library (cf. No. 193 of that publication, Chapter entitled 'Von den Gegenständen des Denkens', and *passim*, Leipzig 1915), we find explicit warnings against hypostatizing those nouns which are not names of things and persons, and an explicit tendency to accept things and persons as the only entities. Brentano was thus undoubtedly a reist, a concretist, although he did not use such terms to describe his standpoint. As a former priest, he stopped at the threshold of somatism and never crossed it. To the end of his life he remained a dualist reist. He was not the first reist: he himself found that even in Leibniz's works appears what is probably the first formulation of that principle. In *Nouveaux essais snr l'entendement humain* Leibniz said that all thorny metaphysical issues would vanish immediately if we made sure that in formulating theorems we should use only names of concrete objects. With this, for him somewhat unexpected, formulation Leibniz initiated the brief series of the forerunners of reism, which in a very vague form may be traced back even to the Stoics. He was, however, a spiritualist reist. The present author knows of no consistent and conscious somatist reists. Is that so because the doctrine is erroneous? This is not known, but the claim may be hazarded that it is a true doctrine, and that is why the reader's attention was occupied with its analysis.

COMMENTS ON THE MEANING OF WORDS

by

Tadeusz Kotarbiński

Handbooks of logic make a distinction between the denotation and connotation of names. Every name (in other words: every potential predicative term) designates every object and only such objects about which it (i.e. the name) may be predicated in a true statement. There are, however, polysemic names, where a given name designates certain objects with respect to a given meaning, and certain other objects with respect to another meaning. Now, the denotation of a name with respect to a given meaning is the totality of objects which it designates with respect to that meaning; while the meaning of a name — in other words, its sense — may be called its connotation. Hence polysemic names have more than one connotation each, even though whenever a given name is used as a predicative term only one of its connotations, as it were, manifests itself; this is so because in each case when a given name is used as a predicative term it is used in a specified meaning.

In answer to the question, 'What is a connotation?', something resembling the following is usually suggested: the connotation of a given name is the set of those properties which are ascribed to the objects to which it refers when used as a predicative term. To put it more briefly, the connotation of a predicative term is the set of those properties which it predicates about something. Thus the term 'connotation' is explained by reference to the formulation 'a set of properties', that is, by reference to an onomatoid, since the term 'property' is an onomatoid. This is so because it is a name, if by a name we mean any word or nominal phrase which can be used as a predicative term in specified sentences, i.e., which, when so used, yields a meaningful whole. And the formulations 'redness is a property', 'length is a property', 'malice is a property' are meaningful. Yet among names we can differentiate names of concrete objects from all remaining names. The names

'Africa', 'Polyphemus', 'man', 'gnome', 'molecule' are names of concrete objects, whereas the names 'colour', 'relationship', 'acceleration', 'contradiction' are not. The present author takes the liberty of calling the names of concrete objects — genuine names, and other names — apparent names or onomatoids. He also adopts the concretist approach, which makes us strive in all explanations to arrive at genuine names and to do without onomatoids. Hence in the present instance, too, his intention is to explain the term 'connotation' without resorting to the term 'property'.

Let us do this by reference to examples. We shall consider the term 'vigorous', as used to describe a given person. To the question, 'What is the connotation of this term?', we might receive the answer: 'vitality', 'enterprise', 'persistence'. We have just used three notorious onomatoids. How are we to get rid of them? How are we to answer the question without resorting to them? Now let us try the following answer: when we say that a person is vigorous we want the addressee to think that person is lively, enterprising, and persistent. The onomatoids 'vitality', 'enterprise', 'persistence' have disappeared and have come to be replaced by genuine names: 'lively', 'enterprising', 'persistent'. And this is always the case: we can always get rid of onomatoids in this way. Consider another example: give the connotation of the adjective 'ramshackle', as used to describe a house. The standard traditional answer might be: the connotation of this term consists of the following properties: being a building; being old; being in a state of disrepair or dilapidation. But we can avoid onomatoids if we say that a person who uses the word 'ramshackle' with reference to a house wants his addressee to think that that object is an old and dilapidated building.

Thus, in general terms, instead of saying: 'the connotation of the term *T* used in a given utterance consists of the properties *a*, *b*, *c*, ...', it suffices to say: 'when predicating *T* about a given object, one wants to make one's addressee think that that object is *a*-y, *b*-y, *c*-y...' (where '*x*-y' bears the same relation to *x* as 'white' does to 'whiteness', 'strong' to 'strength', etc.).

The above holds true for the explanation of genuine names, such as 'vigorous' when used with reference to persons, i.e., concrete objects, and 'ramshackle' when used with reference to houses, which are also concrete objects. In these cases they can be eliminated and replaced by genuine names, such as 'lively', 'old', 'dilapidated', etc. There are, however, cases in which we have to explain the meaning of an onomatoid by giving its connotation. Suppose that someone calls a football player's movements on the field 'sluggish'. The word 'sluggish' is used here as an onomatoid, since it is a name of a person's movements, and hence it is not the name of a concrete object, since movements are not concrete objects (the moving player is a concrete object, but in the present instance the word 'sluggish' is used with reference to a person's movements, and not to the person himself). The standard traditional answer to a question concerning the connotation of the word 'sluggish' as used in this way would probably be as follows: the connotation of the word 'sluggish', as used in this way, consists jointly of the properties: slowness, inertness, and indolence. But we can avoid the use of onomatoids if we formulate the answer like this: when we say a person's movements are 'sluggish', we want the addressee to think that that person moves slowly, inertly, and indolently. This eliminates onomatoids even though no genuine names are used to replace them. The new formulation includes adverbs, but their occurrence in an utterance in no way contradicts the principle of concretism.

The idea outlined above is criticized first and foremost by those who disagree that when we say that John is such and such we want our addressee to think that John is precisely such and such. Communication of content is not identical with propaganda, and we often say something just to communicate an idea to someone, and without any intention of inducing him to accede to the idea. This, however, is a basic misunderstanding. To 'think' that something is thus and thus and to 'believe' that something is thus and thus are two different things. If a person believes that Mars is inhabited, he not only thinks so, but also holds, or is at least of the opinion, that it is so. But even a person who holds otherwise, but understands what the other person has said, must have thought the same thing, even though without any conviction, without sharing that other person's opinion. To put it briefly, what is meant here is a process of thinking that is necessary for an understanding of another person's utterance.

Thus the first objection can easily be refuted. The second is more substantial. Here opponents criticize the use of the sentential connectives 'that' and 'to' (in the formulation 'we want the addressee to think that ...'). They do so because, in logical terminology, these are — unlike such conjunctions as, 'and', 'or', 'if... then...' — intensional functors, and not extensional ones. A conjunction is extensional, which means that if a statement *A* — which consists of statements *B* and *C* linked by that conjunction — is true, then any other statement consisting of two statements linked by the conjunction will remain true, if any of these two statements (say, *B*) is replaced by any true statement, if *B* is true, or by any false statement, if *B* is false. Likewise, if *A* is false, then it will remain false if any of these two statements (say, *B*) is replaced by any true statement, if *B* is true, and by any false statement, if *B* is false. The same applies to the conjunctions 'or' and 'if... then...' (which is used to form conditional statements), as interpreted, for example, mathematics. Thus, in the case of to the conjunction 'and', the compound statement '$2+3 = 5$ and $2\times2 = 4$' is true, since a statement formed of two statements linked by 'and' is true if and only if both component statements are true. Hence the compound statement quoted above remains true if one of its component statements, e.g., '$2+3 = 5$', is replaced by any true statement, even if it be taken from a thematically remote field, e.g., 'a magnet attracts iron'. In that case, we obtain a statement which looks strange (for we are not inclined to use 'and' to link statements which differ widely from a thematic point of view) but nevertheless remains a true statement, namely 'a magnet attracts iron and $2\times2 = 4$'. On the other hand, since the compound statement '$2+3 = 5$ and $2\times2 = 6$' is false because its second component is false, hence we obtain a false compound statement if the second component is replaced by any false statement, be it even if it be taken from a remote sphere of research, for instance, if the new compound statement thus obtained is '$2+3 = 5$ and the bubonic plague is not a contagious disease'.

Now, unlike the conjunctions 'and', 'or', 'if... then...', etc., the conjuctions 'to' and 'that' are not extensional, which makes it impossible to carry out many reasonings of a specified kind about the compound statements obtained by means of these conjunctions. For instance, we are not free to draw new conclusions from them by replacing true component statements by just any other true component statements of our choice. For instance, from the true statement that Pasteur endeavoured to discover a method of treating rabies we cannot draw the conclusion that Pasteur endeavoured to have poison gases used in war, even though it is true that a method of treating rabies has been invented and that poison gases came to be used in war. Likewise, we cannot conclude from the statement 'The ancients knew that the square of the hypotenuse equals the sum of the squares of the other two sides' that the ancients knew that radioactive substances exist. We are not justified in concluding this even though the geometrical statement after 'that' in the original compound statement and the physical statement after 'that' in the alleged conclusion are both true.

Why do we tire the reader with this lengthy and elementary explanation? We do so in order to emphasize the concerns of logicians regarding certain problems which cannot fail to arouse the interest of linguists as well. Logicians usually endeavour to eliminate all intensional, i.e., non-extensional, functors from scientific statements; they do so not only because the compound statements constructed by means of such intensional functors cannot be subjected to the operation of replacing true component statements by other true statements, but for another reason as well. They suspect that the occurrence in a given statement of an intensional conjunction always proves the superficiality of the analysis of what has been stated; they also suspect that a sufficiently penetrating analysis should always result in the formulation of a statement that is free from such conjunctions. Hence, whoever remains satisfied with formulations which include such conjunctions as 'to' or 'that' lays himself open to the accusation of being stuck midway in his semantic analysis. This is precisely the objection that can be raised against the above attempt to establish a method of answering the question as to what is the connotation of a given predicative term, in other words, what a given predicative term connotes, or — to put it more correctly from the concretist point of view — how a given predicative term connotes (in contrast to what it denotes, i.e., what objects combine to form its denotation).

We prefer to say 'how it connotes' rather than 'what it connotes', because the pronoun 'what' has the nature of a name, and suggests fallaciously that predicative terms somehow refer to certain objects which are their connotations, whereas — in accordance with the ideas outlined above — the term 'connotation' is not a genuine name, but an onomatoid, and hence a word that may be tolerated in substitutive phrases only, but must be eliminated from both questions and answers formulated in a definitive manner.

We can try to generalize what has been said above with reference to the connotations of predicative terms. Note that 'connotation' (or 'sense', or 'meaning') is a special case of 'meaning' in general. (If we use the term 'significance', we must, of course, bear in mind that whenever reference is made to linguistic entities 'significance' is used in the sense of 'meaning', and not of 'importance'). Names alone connote, but they are not alone in having meanings, since all expressions and words, and even certain component parts of words, have meanings. Now let us try to construct schemata of answers to questions about the meaning of other words or expressions, according to the pattern of the schemata of answers to questions about the connotation of a given predicative term. Consider an example, such as the adverb 'reluctantly', as used in a specified context. In trying to describe its meaning we explain that we say that a person acted reluctantly if we want the addressee to think that that person acted slowly and unwillingly. In general terms, whenever we are asked about the meaning of a given adverb, we ought first of all to complete the question, as it were, by formulating a sentence in which that adverb occurs. In the case discussed above, we formulate the sentence 'a person acted reluctantly'. Next we should explain what the speaker wants to achieve by making such a statement. *Mutatis mutandis*, we should handle questions about the meanings of conjunctions in a similar way. If someone is interested in the meaning of the conjunction 'although', we may, by availing ourselves of sentential variables, answer as follows: for all sentences p, q: if we say that p although q, we want the addressee to think that q, and that p, and that it seemed that once q is the case, then p should not be the case (e.g., 'he is weak although he is healthy' is the same as 'he is healthy, and he is weak, and it seemed that once he is healthy, he should not be weak').

It is easy to see that in developing the last example we have left the safety of the shore and come out onto the open sea. By the safety of the shore we mean confining ourselves to an interpretation of actual meanings which occur in a specified utterance made by someone, while the open sea is the sphere of potential meanings which are involved when we ask not about the meaning of a given word when uttered by a specified person *hic et nunc*, but about its meaning "in a given language", e.g., in English. We feel, however, that in view of our tasks there is no essential difference between sailing close to shore and on the open sea. What does the word 'tasty' mean in ordinary English? The question is not about what John wanted to evoke when he used that word today to describe a portion of food, but about what anyone wants to evoke when he uses the term 'tasty' as an ordinary Englishman might do to describe something. One more thing. Objections are sometimes raised against the above schema of answers of the type: what does a given word mean? Some people claim that this schema does not provide appropriate answers to such questions. If a question refers to a word W, then the appropriate answer should begin with the formulation: 'The word W means...', or: 'In language L, the word W means...'. The present writer does not feel that such a structural adaptation of answers to questions is always obligatory. A person who is asked 'How do you feel?' may quite rightly answer 'I am in high spirits', and the advice 'Turn to the right' may be a reasonable answer to the question 'Which street will take me to the market?'. The same applies to the case under consideration in this paper. But if a person prefers to begin his answers in the manner indicated above, we cannot object. The schema of answers may then become more intricate, and with reference to the conjunction 'although' it would be: "The conjunction 'although' means that, for all sentences p, q, ...," and so on, as above.

THE CONTROVERSY OVER DESIGNATA

by

Tadeusz Kotarbiński

Language is a common object of research for both linguists and logicians, but linguists *qua* linguists and logicians *qua* logicians are interested in different things. The vocabulary and expressions of a given language, its structure, the factual connections between its elements and forms and its meanings, its history and developmental trends, as well as the differences and similarities between the various languages, the genetic relationships between them, and everything that characterizes human speech as a unique fact in human history — these are the considerations which linguistics takes into account in studying the shaping of languages during the long process whereby culture is formed. In the field of logic, however, language is interesting only as a more or less suitable instrument for a pertinent and possibly definitive formulation of judgements and their substantiations. This difference in basic interests has led to differences in mental habits and conceptual apparatus. Linguistics has given rise to the development of the concepts of vowel and consonant, declension, suffix, parts of speech, and later also phoneme and morpheme, etc., while to logic we owe the concepts of material supposition, connotation, semantic category, functor, metalanguage, etc. It is high time to establish closer contacts, so that every modern linguist will be thoroughly acquainted with what logicians have arrived at in the course of their reflections as to the conditions of logically correct speech, and that every modern logician should be able to use concepts developed by linguistic researchers, who describe existing languages as they really are and likewise are concerned with correctness of speech, but correctness of a different kind. For a linguist who specializes in, say, English linguistics, is competent to provide information as to how we should, and how we should not, speak if we want to speak standard English.

Since, however, every person with even as little as an elementary education ought to be able to speak his own language well, and also to use language correctly from a logical point of view, it follows that linguists and logicians meet within the field of secondary education. This makes it imperative for them to reach an understanding on matters of common interest, to avoid confusion in pupils' minds.

On the basis of numerous conversations the present author has to conclude that the term 'designatum' gives rise to misunderstandings. He would be glad if this brief paper might contribute to eliminating such misunderstandings or at least to reducing their harmful effects in the field of public instruction.

How is this term now used by logicians? It seems correct to state that they speak mainly, and probably most of them speak exclusively, about designata of names, that is, of predicative terms in sentences which have a subject-copula-predicative term structure. For them, the designatum of a given name is that object and only that object about which that name may be predicated in a true statement. Thus, for instance, New York City is the only designatum of the name 'the greatest American town', since it is only about New York City that we may truthfully say that it is the greatest American town, because the statement 'New York City is the greatest American town' is true, whereas the statements 'Rhode Island is the greatest American town', 'Nevada is the greatest American town', 'Oklahoma is the greatest American town', are not true, and that applies to any other statement in which something else than the New York City would be indicated as its grammatical subject. Likewise, Henry Palmerston, William Gladstone, Winston Churchill are designata of the name 'Englishman', because that name may be predicated about each of them in a true statement. On the other hand, neither Johann Sebastian Bach, nor Aristotle, nor many others, can be designata of that name, since any statement about any of them to the effect that he was an Englishman would be false.

So far so good. Everything is clear and indisputable. But now someone asks whether Ben Salem is the designatum of the name 'the capital of New Atlantis', or whether Horatio is the designatum of the name 'Hamlet's friend', or whether Phoenix is the designatum of the name 'the bird that rises from its own ashes'. When it comes to such and similar questions, which, speaking loosely, refer to fictive entities, wide differences of opinion immediately arise.

Many of those who take part in such discussions answer in the negative, and in doing so they cite the fact that neither the city of Ben Salem, nor Hamlet's friend Horatio, nor the Phoenix exist (or existed or will exist), and hence nothing can be predicated about them in any true statement. For a sentence based on the subject-copula-predicative term structure, that is, for a sentence of the type '*M* is *N*', or 'every *M* is *N*', or 'some *M*'s are *N*', to be true it is necessary that the object *M* or some objects *M* exist. If an object does not exist, then nothing can truthfully be predicated about it, just as no action can be performed on it since there is nothing to be acted on. This opinion prevails among logicians, since they stress the great difference between 'to mean' and 'to denote' and between their analogues 'connotation' and 'denotation', and between 'meaning' ('sense') and 'designatum'. In all these pairs, the first elements characterize linguistic units, i.e., words and expressions, regardless of the relations they bear to extra-linguistic reality; some say that these terms are 'semiotic', but not 'semantic', in nature. On the other hand, the second elements in the above pairs are not merely 'semiotic', but 'semantic' as well, since they characterize linguistic units (words and expressions) with respect to the relations they bear to extra-linguistic reality, according as they reflect an element of that reality, or miss the mark entirely. What a given name means in a given language — in other words, its connotation or the set of properties which combine to form its sense, or meaning — has to do solely with the intentions of the persons who use that language. On the other hand, what, if anything, makes up the denotation of a given name, as used in a given language — that is, what it denotes in that language, and what, if anything, is the object designated by that name, i.e., what is its designatum — is determined not only by the intentions of the users of that language, but also by extra-linguistic reality, i.e., by the fact whether there are, or are not, objects existing in the extra-linguistic world which comply with those intentions.

Other parties to the dispute, mainly linguists, when hearing such or similar arguments protest vigorously and demand that Ben Salem be accepted as the designatum of the name 'the capital of New Atlantis', Horatio, as the designatum of the name 'Hamlet's friend', Phoenix, as the designatum of 'the name of the bird' described above, and generally, that various fictive entities be accepted as designata of their names. They argue that we speak about Ben Salem, Horatio, the Phoenix, even though these are fictive entities, even though no such city, person, or bird exists. Yet in our minds we have representations of such objects, and this is not a question of senses, or meanings, or connotations, of those names, since meanings, connotations, etc., are sets of properties, and we do not think of a set of properties when we have representations of Ben Salem, Horatio, the Phoenix, etc. In such a case we have in mind a city, a person, a bird, etc., which are endowed with sets of properties, but which are not sets of properties. They are substances rather, fictive substances, to be sure, but substances nevertheless, and we need a term which will describe the speaker's attitude toward them precisely as that of a person who speaks about them regardless of whether these substances in fact exist or whether he merely imagines them to exist. Would not the term 'designatum' serve this purpose best? Let us accordingly use it this way, and let us say boldly that 'Cerberus' is the designatum of the name 'the many-headed hound who guards the entrance to Hades', that Rosinante is the designatum of the name 'Don Quixote's horse', etc., in the same way as we do not hesitate to say that Virgil is the designatum of the name 'the author of the *Aeneid*'.

It is reasonable to satisfy real, and not just imaginary, needs, and hence our duty to reach an agreement on the choice of a term that will satisfy the needs indicated by the linguists who take part in the discussion. Some say that such a term is already in circulation and has a fairly long tradition behind it. The term in question is an 'intentional object'. If we have a representation of something, then that something becomes the object of our attention, we direct our thoughts to it (in a sense) and we think about it (in a sense); that something accordingly becomes an object of our intention and is for us an intentional object. When we speak about that something by using a certain name, then it is, so to say, the intentional counterpart of that name, or its intentional object. All this is true, but we have to say that such a mode of speaking is rather lengthy and clumsy. Moreover, the term 'intentional object' is customarily used only with reference to our representation of fictive objects, and not on all occasions, regardless of whether we have a representation of an existing or an imaginary thing.

If we look for a more suitable word we come across the term 'denominatum'. And hence our proposal that we should say that Mephistopheles is the denominatum of the name 'the devil to whom Doctor Faustus sold his soul', and that we should not say that Mephistopheles is the designatum of that name — unless we believe that devils exist. Let us say that an object *P* is the denominatum of a name *N* as used in a given expression if that name *N* makes us form a representation of *P* regardless of whether *P* does, or does not, exist, that is, regardless of whether a fragment of the world is *P* or whether, in a free formulation, *P* is a fictive object. Furthermore, let us say that a name makes us form a representation of a given denominatum. In such a case nothing prevents us from saying, in a rational manner, if we so wish, not only that the denominata of the name 'Muse' include Clio, Euterpe, Thalia, Melpomene, and the other members of the Parnassus group, but that a general fictive object, a 'Muse in general', a 'universal', is also the denominatum of that name — in this case a general denominatum (we would use this formulation if a person, when speaking about a Muse, has that in mind). On the other hand, let the words 'to designate' and 'designatum' remain in their "semantic" role as connected with the existence of the object of a given name. Let a name *N* in a given language designate only its designata, and let the designatum of *N* be only that object about which that name may be truthfully predicated — hence an existing object only, and not a fictive one.

This would also apply to names which do have designata. Thus, for instance, when considering the word 'razor' and treating it as synonymous with the expression 'shaving knife', we may say that a razor is the denominatum of the name 'shaving knife' not only with reference to any real razors, and not only with reference to any razor from the world of fiction, if we have any such a razor in mind, but also with reference to a universal 'razor in general', if we have in mind anything like that; in such a case we would have to do with a general denominatum. All this freedom of formulation can be explained by the fact that an expression of the type '*P* is a denominatum of a name *N*' is a substitutive abbreviation. Were it not that — were it to be taken literally, its acceptance would result in the acceptance of the existence of an object *P* concerning which it could be claimed that it is a denominatum of *N*. For that expression is a special case of the schema '*A* is *B*', and for a statement constructed according to that schema to be true, the existence of an object *A* would be required. But those expressions which are substitutive abbreviations do not comply with that rule—if we bear in mind what they are substitutive abbreviations of. If we say, e.g., that "a shaving knife is a denominatum of the name 'razor'" we merely claim that "whoever understands the name 'razor' has a shaving knife in mind", and if we say that a person has a shaving knife in mind we do not claim indirectly that any shaving knife exists; likewise, when we say that a person has a millegenarian in mind we do not claim that any such old man exists. To have something definite in mind is the same as to imagine how it would be if it were so. It is obvious that the real nature of our acts of imagination does not in the least guarantee the real nature of what we imagine.

It is to be hoped that these remarks may contribute to the adoption of a common terminology in a field which is of common interest to linguists and logicians and which, while being also the common field of their teaching activity, is at the same time a field of their discordant verbal habits.

TOKEN-REFLEXIVE WORDS VERSUS PROPER NAMES

by

Tadeusz Czeżowski

Token-reflexive words (the term originates with H. Reichenbach, *Elements of Symbolic Logic*, 1947; B. Russell's term was 'egocentric particulars', cf. his *Human Knowledge, Its Scope and Limits*, 1948; the problem seems to have been first taken up by E. Husserl in his *Logische Untersuchungen*, Vol. II, 1901, where the term used is '*okkasionelle Ausdrücke*'), like the personal pronouns 'I', 'you', the demonstrative pronouns 'this', 'that', the time and place indicators 'now', 'today', 'here', etc., are names which change their respective meanings according to who utters them and under what circumstances. Hence they are mainly used to denote certain objects temporarily, during the utterance of a given sentence, even though they may be used repeatedly and even regularly to denote one and the same object; for example, a person regularly uses the pronoun 'I' whenever he refers to himself. We link them with the denoted object by indicating that object in various ways (linguists call them ostensive or deictic names), and hence they are singular names. But if we denote a certain object at a certain moment by a given token-reflexive name, this does not preclude our using the same name, at some other moment, to denote some other object, not identical with the former. It is as if a token-reflexive name, lacking a constant meaning of its own, acquires one only temporarily and as it were derivatively as a result of the indication, by its user, of its designatum.

Proper names resemble token-reflexive names in that they also denote individual objects which must be indicated when their respective proper names are assigned to them; moreover, one and the same proper name may be assigned to various persons and/or things without any inconsistency between the various uses of the name (e.g., the name of a country such as 'France' used as the name of a passenger ship; the personal names 'Raffaello' and 'Michelangelo' used in analogous functions; the anglicized name 'Rome', which was used initially to denote the famous Italian city, and later also came to denote at least two towns in the United States (this applies to many other names of European towns which later came to be used as names of American towns); various proper names used as brand names of cigarettes, etc.). Proper names, too, lack any meaning that could be attached to them permanently, and acquire such meanings, as it were, derivatively by being attached to certain designata. This is what distinguishes proper names from singular terms such as 'the least common multiple of the numbers 2, 4, and 7', each of which has a constant meaning of its own, so that each is the name of a single specified object which is defined by that meaning. The relationship between token-reflexive words (names), on the one hand, and proper names, on the other, consists in the fact that when we wish to assign a proper name to a person or thing we utter a statement which establishes the relation of identity between the object indicated by the token-reflexive name in question and the object to which we assign that proper name, that statement being 'this object is called so and so' or 'this is *N N*' or some other equivalent formula. In accordance with the laws of syntax a proper name may occur in the predicate of a sentence only in those cases where we identify the person or thing named by that proper name with the object indicated by a token-reflexive name or to which some other proper name has been assigned.

Token-reflexive names may be compared to certain singular terms, namely those which denote an individual object defined in relation to another individual object, e.g., 'the author of *Hamlet*', 'the highest mountain in the world', etc. The similarity consists in the fact that expressions of both kinds denote individual objects, and they do so by indicating the relation which each such object bears to another individual object.[1] In the case of token-reflexive names the second member of the relation is the sentence I utter when using a token-reflexive name: 'I am hungry' indicates the (grammatical) subject denoted by the pronoun 'I' as the person who utters that sentence; 'this is black' indicates the object denoted by the pronoun 'this' as the object to which the sentence refers; 'yesterday was Monday' indicates the day which preceded the day when the statement was made.

When analysing token-reflexive expressions we have to distinguish individual signs of a word or a sentence from a word or a sentence in the lexical or grammatical sense. In the following two sentences: 'Cracow lies on the Vistula' and 'the oldest Polish university is in Cracow' we encounter one and the same lexical word 'Cracow', but two different individual signs.[2] To make a distinction we introduce the term 'token' to denote an individual sign (which may be an inscription or an utterance), and the term 'type' to denote a word or sentence in the grammatical sense (or an expression in any symbolic language), i.e., a set of equiform tokens. Thus, instead of saying that the word 'Cracow' occurs in the two sentences quoted above, we shall say that those sentences include two tokens of the type 'Cracow'.[3]

By referring to our previous description of token-reflexive expressions we shall say that:

(A) Every token-reflexive name used individually may be replaced equivalently by a phrase which incorporates the token-reflexive expression 'this token'.

Thus, for instance, 'I' is the same as 'the person to which this token refers' (i.e., the person who has just said or written this; 'now' is the same as 'the moment to which this token refers' (i.e., the moment in which this sentence, which incorporates the word 'now', was uttered); 'this table' is the same as 'the table to which this token refers' (i.e., the table indicated in the words just uttered). The relation 'refers to' will be denoted by the symbol S (treated as a type), and, if necessary, the various kinds of reference will be distinguished by subscripts: S_1, S_2, etc.

If I say 'this token' with reference to any expression written here, I am acting similarly to when I form the name of any expression. This is usually done by placing the name in quotation marks, that is, by forming its quotational function. The difference is that in the latter case I am forming the name of an expression, that is, a type for a type, whereas here we are concerned with a token-reflexive, single indication of such an inscription, that is, with forming a token for a token. Let us therefore introduce quasi-quotation marks in the form of reverse parentheses (which we shall call 'arrows'):

(1) $)a($

What we have written here is a token for a token, i.e., the individual sign of the token placed between the arrows. Likewise, the number (1) written to the left of the same line denotes only the token written in that line; if we write an equiform token a second time, we must denote it by another number:

(2) $)a($

since it is a new token for a new token, equiform with the former.

The name of a token, that is, the set of tokens which are equiform with it, we shall form in a manner analogous to the operation of defining. Thus the definition

(3) 'a^2' is equisignificant with '$a \cdot a$'

could have the form:

(4) Every token which is equiform with $)a^2($
is equisignificant with every token which
is equiform with $)a \cdot a($

('equisignificant' being used in the sense of 'synonymous' to indicate the analogy with 'equiform', these being the original Polish terms — Tr.).

The above pattern serves us to form the definition of a name *n* for a specified token:

(5) Every token which is equiform with $)n($
is equisignificant with $))b(($

The inner arrows transform the inscription which they contain into its token, that is, an analogue of its name, while the outer arrows correspond to the quotation marks which contain the definiens in (3), that is to say, they indicate that we refer here not to the inscription within the arrows, but to the whole which consists of that inscription and the inner arrows containing it.

In order to define the name of the token of (1) in the same way, we should have to precede it by a formula analogous to (5) and contain that token within a second pair of arrows. We could also consider the type (1), written to the left of the token designated by that number, to be its name; to do so we would have to interpret the fact that the token being placed there is an equivalent of formula (5). But this would not be consonant with the ordinary usage of numbers of symbolic formulas, since such a number is usually a type of a type and serves to denote every equiform symbolic formula, whenever that formula is written, whereas in this case it would have to be a type of a token, namely of the inscription placed there.

The introduction of a name of a token-reflexive expression enables us to transform it into an equivalent expression which is not token-reflexive. The token-reflexive name:

(6) The sheet on which I have written this token

denotes that sheet, and only that sheet, which bears the above inscription, and were we to repeat it on another sheet, it would then be another inscription denoting another sheet. If, on the other hand, the type (6), written to the left of that inscription, is used as the name of that inscription, then instead of (6) we may write, equivalently,

(7) the sheet on which I have written (6)

on any sheets we choose, any number of times, and the same inscription will always denote one and the same sheet, namely that which bears the original inscription.

Likewise, if the expression 'this token', which denotes an individual inscription or an individual utterance, is replaced by a name s, defined by a definition in the form of (5), then in accordance with (A) it enables us to eliminate token-reflexive names from that inscription or that utterance. This gives us an apparatus which — as will be shown by the examples given below — is sufficient for an analysis of token-reflexive expressions.

Example 1

The sentence 'it is a cold day today' (s), as uttered by someone on a specified day, may be analysed thus:

1) for some x, s refers to x — $(Ex)\, Ssx$,
2) for every y, if Ssx and Ssy, then x is identical with y — $(y)\, CKSsxSsyIxy$,
3) x is a cold day — Mx.

The type s is the name of the inscription being analysed; sentence 1) states that the day in question is the day to which s refers; it is also clear that this is an existential statement; 2) states that this is an individual day; 3) assigns its predicate to it. On combining the three statements into a conjunction we obtain an analogue of s:

$$(Ex)\ KKSsx\ (y)\ CKSsxSsyIxyMx.$$

Example 2

'This man is my friend' (s)

The above sentence has two token-reflexive names: 'I' ('my') and 'this man', and hence both must be eliminated by reference to the name s and the relation S:

1) $(Ex)\ Ssx$
2) $(Ey)\ S_1sy$
3) $(z)\ CKSsxSszIxz$
4) $(z)\ CKS_1syS_1szIyz$
5) $NIxy$ (i.e., 'I' and 'this man' are different persons)
6) Pxy (i.e., the relation of friendship exists between x and y)

The conjunction of the sentences 1)–6) is the equivalent of (s):

$$(Ex)\ (Ey)\ KKKSsxS_1syNIxyK\ (z)\ KCKSsxSszIxzCKS_1syS_1syS_1szIyzPxy$$

Example 3

'This is the Vistula' (s) — in this example the subject is a token-reflexive name, and the predicative term is a proper name.

1) $(Ex)\ Ssx$
2) $(y)\ CKSsxSsyIxy$
3) IxV (x is identical with the Vistula)

$(Ex)\ KKSsx\ (y)\ CKSsxSsyIxyIxV$

Example 4

'This man is called NN' (s)

1) $(Ex)\ Ssx$
2) $(y)\ CKSsxSsyIxy$
3) Hx (if the range of the variable x is not defined in advance as the set of human beings, then we have to state explicitly that x is a human being)
4) $IxNN$ (x is identical with NN)

$(Ex)\ KSsxK\ (y)\ CKSsxIxyKHxIxNN$

Example 5

'William Shakespeare is the author of *Hamlet*' (s) — in this example we have two proper names.

1) $(Ex)\,Ssx$
2) $(z)\,CKSsxSszIxz$
3) $IxWS$ (x is identical with William Shakespeare)
4) $(Ey)\,S_1sy$
5) $(z)\,CKS_1syS_1szIyz$
6) IyH (y is identical with *Hamlet*)
7) Axy (x is the author of *Hamlet*)

$(Ex)\,(Ey)\,KKKSsxS_1sy\,(z)\,KCKSsxSszIxzCKS_1syS_1szIyz$
$KKIxWSIyHAxy$

***

In modern logic, we distinguish two kinds of singular sentences: those of the first kind have a proper name or a token-reflexive name as the subject and are termed atomic sentences, while those of the second kind have a singular term as the subject and are termed description sentences. Description sentences are held to be compound sentences, while atomic sentences are held to be simple elementary sentences (hence their name). This distinction shows that we can predicate about singular objects in two ways: either as in atomic sentences, where we treat those objects as absolute primitively given elements of reality, or as in description sentences, where the object spoken about is related to the singular term by being defined as subsumed under that term, and it is only after singling it out in this way that we predicate singular sentences about that object.

The analysis carried out above has shown that those singular sentences which are termed atomic are not simple sentences, but compound, as in the case of description sentences. In these sentences, too, the singular object about which a given sentence is predicated is not given as something absolute, but is related to a token-reflexive expression. Simple sentences, of which a singular sentence is formed, are those which consist of a propositional function of a term variable with the existential quantifier $(Ex)\,fx$, and hence they are sentences which predicate something about an object which becomes individualized *only* as a result of the fact that a singular sentence is predicated about it. This conclusion has an ontological significance. Singular objects are substances, and hence we are concerned here with a certain manner of interpreting the concept of substance.

In modern philosophy, the controversy over substance has been due to the definition of substance adopted by Descartes, which radically opposes substance to property. Substance is something abstract, which is obtained through elimination of all the properties which are its attributes. Conceived thus, substance is both an absolute entity (that is, not dependent on anything), because all relationships are properties, and also a transcendent entity (that is, lying beyond the reach of cognition), because cognition has only properties as its objects. This concept of substance was subjected to powerful criticism by the exponents of 18th century empiricism, who replaced it by another definition which states that substance is a relatively permanent set of properties and as such can be singled out as a particular unit. But the latter concept of substance also collapsed in the face of criticism, which pointed to the fact that properties are empirically secondary to individual objects, so that it is not properties but things which must be considered elementary entities.

In contemporary metaphysics the problem is formulated otherwise: both viewpoints which interpret the world as a set of substances or a set of properties are rejected. According to this third viewpoint, the world is viewed not as a heap of sand, consisting of separate grains of sand, but as a uniform stream of water which is in constant motion and exhibits all its properties. According to this view, any detail of the universal process indicated by a token-reflexive or proper name is individual substance. In this interpretation, which remains completely within the bounds of empiricism, the concept of substance is related to the act of singling out a given substance within the universal process. The act which singles it out consists in referring it, in a singular sentence, to the subject of that sentence. This can be done in one of the following two ways: either by subsuming substance under a singular term in a description statement, or by indicating it in a token-reflexive expression.

Notes

[1] Cf. H. Reichenbach, op. cit., Sec. 50, pp. 284ff.

[2] The original has grammatical, and not lexical sense, but the author's intention seems to have been to refer to words as lexical items before introducing the terms 'token' and 'type'. (Tr.)

[3] The translation has been simplified, since 'type' and 'token' are now standard terms in English-language works on semiotics, and their use does not require special explanations. (Tr.)

CONNOTATION AND DENOTATION

by

Tadcusz Czeżowski

Much has been written on what 'to mean' means. In taking up this issue within a limited sphere I will discuss it first — in order not to complicate my analysis — only with reference to the meanings of names of things, in particular of well-defined general names, or of terms, such as scientific and technical terms in the broad sense of the word: 'square', 'house', 'tree', etc. All approaches to the problem under consideration reflect three trends: (1) the psychological trend, according to which meaning is the content of thoughts about the designatum of a given name; (2) the behaviourist trend, which assumes that meaning is a disposition to a certain behaviour with respect to that designatum; (3) the semantic trend, which defines meaning in terms of the properties of the designatum. All three trends go beyond the sphere of logic.

The question arises as to how to interpret meaning within logic itself. To look for an answer we have to start from distinctions between definitions. In formulating the definition 'a square is an equilateral rectangle' we define an object which is a square, but at the same time we impart meaning to the term 'square', because the intra-linguistic, definition given above has its perfect analogue in the metalinguistic definition 'a square means an equilateral rectangle'. The intra-linguistic, or real, definition refers to the square, considered as a geometrical figure, and to its properties, i.e., equilaterality and rectangularity; the metalinguistic, or nominal, definition, on the contrary, refers to the term 'square' and to its meaning, the latter being the logical product of the terms 'equilateral' and 'rectangular'. It is sometimes said that the meaning of the term 'square' is the set of the essential properties which are attributes of a square, but that is incorrect and results from a confusion of the two kinds of definitions mentioned above: the set of properties which are attributes of squares under the intra-linguistic definition, which refers to the object called a square, defines that object, and not the meaning of any terms; on the other hand, the meaning ascribed to the term 'square' in the metalinguistic definition is not a set of properties, but the logical product of the terms which occur in the definiens and which are names of the properties of a square.

This analysis suggests that we should interpret the name 'the meaning of a term' as the relational name of the first term of the relation of defining (as between terms), just as 'husband' is the relational name of the first term of the relation of marriage (as between human beings), and 'greater than' is the relational name of the first term of the relation of 'being greater than' as between numbers. The logical product 'equilateral and rectangular' (and not 'equilaterality and rectangularity', as is sometimes said) is the meaning of the term 'square'. Unfortunately, we do not have in our language any name to correspond to the relational name 'meaning' in the same way as 'marriage' corresponds to the relational name 'husband', and 'being greater than', to the relational name 'greater than'.

Having thus singled out one of the many various interpretations of the term 'meaning' we should now select a distinctive name for it. I suggest using the term 'connotation', which also may be interpreted in various ways, but, being an artificial term and hence not burdened with associations stemming from everyday language, can easily be adjusted to our needs. We shall accordingly speak about the meaning of a term as its connotation, and the relation which exists between the meaning of a term and the term itself will be called the relation of connotation.

A definition establishes the relation of equisignificance, or synonymity, between the term being defined (the definiendum) and its definiens; this relation allows us to replace one term by the other, and this leads us to conclude that the two are coextensional. Each term which forms part of a given connotation (the latter being considered as a logical product of these terms) is hierarchically higher than a term with that connotation; thus, the terms 'equilateral' and 'rectangular' are hierarchically higher than the term 'square '.There are many other terms that are hierarchically higher to the latter, such as 'plane quadrangle with two diagonals of the same length and perpendicular to one another', 'plane quadrangle with four axes of symmetry', 'plane quadrangle with angles whose sum equals 360 degrees', 'plane quadrangle which can be inscribed in and circumscribed on a circle', etc. They all form a logical product which is equivalent to the logical product 'equilateral rectangle', and hence, from the extensionalistic standpoint adopted in logic, we include all of them in the connotation of the term 'square' thus settling the sterile controversy over how broadly connotation is to be interpreted. Those terms which are hierarchically higher than the term 'square' can be truly predicated about a square; hence the connotation of the term 'square' covers all those terms which can be truly predicated about a square.

A term which is hierarchically higher than a given term is obtained by generalization, i.e., by logical addition; for instance, 'rectangle' as a term which is hierarchically higher than the term 'square' is a 'square or non-equilateral rectangle'. A term which is hierarchically lower than a given term is a result of determination, i.e., logical multiplication. The preceding example will help us to visualize this. A 'square' means an 'equilateral rectangle'; if for 'rectangle' we substitute 'rectangular quadrangle', then square means 'equilateral and rectangular quadrangle'; both terms in the definiens, 'rectangular quadrangle' and 'equilateral quadrangle', are hierarchically higher than the term 'square', and hence are logical sums with the common term 'square'. This is so because 'rectangular quadrangle' means 'equilateral' or 'non-equilateral rectangular quadrangle' (an 'equilateral' one being a 'square'), and 'equilateral quadrangle' means 'rectangular equilateral quadrangle' (i.e., 'square') or 'non-rectangular equilateral quadrangle' (i.e. 'rhombus'). When we form the product of these two logical sums, then 'square' means 'equilateral or non-equilateral rectangular quadrangle and rectangular or nonrectangular equilateral quadrangle'. By multiplication we obtain the sum of four terms, three of which are mutually contradictory and are accordingly deleted, and the fourth is 'equilateral rectangular quadrangle' and is the desired connotation of the term 'square'. Using the simple symbolism of the algebra of logic let *a* stand for 'square', *b* for 'quadrangle', *c* for 'rectangular', and *d* for 'equilateral'. The connotation of the term is equivalent to it extensionally, hence

$$a = bcd + bcd' \cdot bcd + bc'd,$$

which on multiplication yields

$$a = bcd.$$

In general terms, the simplest schema of connotation is in the form

$$(1) \qquad a = a+b \cdot a+b',$$

which shows that connotation is a product of hierarchically higher terms and is equivalent to the term being defined. For instance, let a stand for 'white ball', and b for 'blue ball', and let us suppose that a box A, which stands for the factor $(a+b)$, contains white and blue balls, and that a box B, which stands for the factor $(a+b')$, contains white and non-blue balls. Now the term 'white ball' has, in equation (1), the following connotation: 'ball of the colour represented in both boxes'.

The meaning of a term, interpreted psychologically or semantically, i.e., as mental content or as a set of properties, determines its extension. The extension of a term is the set of its designata, that is, the set of those objects about which that term can truly be predicated. But, when interpreted in this way, the relation between a term and its extension is extra-logical in nature, since it is non-homogeneous in that it links logical terms with objects of a different kind, which are outside the sphere of logic. (The same observation holds true when we consider the relation existing between a term and its psychological meaning, i.e., our notion concerning a designatum of that term.) That is why instead of the extension of a term a understood as the set of those objects about which the term a can be truly predicated, we take into consideration the set of those terms which can be subjects in those true statements in which a is the subjective complement or, in other words, the set of terms that are hierarchically lower than a. The set of terms defined in this manner is the set of names of subsets of the extension of a. For instance, let a stand for the term 'quadrangle': the set of its designata, or its extension, is the set of individual quadrangular figures; a square is a quadrangle, but so is a rhombus, a trapezoid, etc. The logical sum of all those terms which denote the subsets of the set of quadrangles covers the extension of the term 'quadrangle': a quadrangle is a square, or a rhombus, or a trapezoid, or ..., etc. Let us call this logical sum — denotation, and its terms — the denotata of the term 'quadrangle'. It may be supposed that in classical logic, too, the extension of this term was interpreted as the set consisting of kinds of quadrangles, and not of individual quadrangles, since classical logic worked with genera and species, and not individual objects; but doubtless no distinction would be made between the set of designata and the logical sum of the subsets of that set. Those terms which are hierarchically lower than the term quadrangle can be obtained by multiplying that term logically by other determining terms, such as 'rectangular', 'non-rectangular', 'equilateral', 'non-equilateral', 'parallelogram', 'non-parallelogram', etc. The denotation of a term is the logical sum of such products; in its simplest form, analogous to the formula of connotation, it is written in the form of the equation

(2) $$a = ab+ab',$$

which can be expanded into any number of terms. On comparing (1) and (2) we see that a term can be defined either by its connotation or by its denotation (by its intension or by its extension, as we used to say), since denotation is equivalent to connotation:

(3) $$ab+ab' = (a+b)\cdot(a+b').$$

All three formulas are Boolean expansions of the term into its constituents, these expansions being interpreted as the connotation and the denotation of the term. Understood in this manner, connotation and denotation satisfy the law of inversion of intension and extension first formulated in *Port Royal Logic* and later questioned by Bolzano because the concept of intension was not clearly defined. Let a stand for 'square', b for 'rectangle', c for 'equilateral'; the logical equation

$$a = bc$$

may be read as: 'a square is an equilateral rectangle'. The product bc is the connotation of the term 'square'. If we expand the denotation of a by joining to it, through logical addition, bc', which stands for 'non-equilateral rectangle':

$$\begin{aligned} a+bc' &= bc+bc', \\ bc+bc' &= b, \\ a+bc' &= b. \end{aligned}$$

A square or a non-equilateral rectangle is simply a rectangle; the connotation bc has been decreased by c. If we now expand the connotation of b by multiplying both sides of the equation by the factor c:

$$(a+bc')\,c = bc,$$

$$(a+bc')\,c = ac+bcc' = ac = bcc = bc = a,$$

$$a = bc.$$

The denotation $a+bc'$ has again been reduced to a.

The connotation of term a is the conjunction of certain other terms, b, c, which in turn have connotations of their own. The old question about the primitive terms of definitions recurs: what are the respective connotations of the primitive terms which constitute structural elements of the connotations of all other terms in the language of a given discipline? In the mathematical sciences, primitive terms are defined axiomatically. Deictic definitions define the primitive terms used in the empirical sciences. Hence the connotations of primitive terms must be sought in axiomatic and deictic definitions.

In order to answer the question posed above let us also refer to a distinction between the various syntactic categories of terms. Those terms which are predicated about individuals and which denote their properties, such as 'square' (or 'square-shaped': 'this is a square', 'this is square-shaped'), 'green', 'house', 'object', etc., belong to the syntactic category of the first degree; those which are predicated about properties of individuals (in answer to such questions as: "What is the meaning of to be a 'square'?", etc.) belong to the syntactic category of the second degree. Syntactic categories of higher degrees are defined in a similar manner: the syntactic category of the third degree includes those terms which are predicated about properties of properties of individuals, etc.

We have hitherto confined ourselves to the terminology of ontology and classical logic, where the primitive terms include the categories of object and property, and the attributive relationship of a property to an object. Definitions of these categories are included in the set of axioms of a theory and in the statements deducible from these axioms, e.g., 'no two contradictory properties may be attributes of any object', 'every object has some property as its attribute', etc. Such statements can be transformed into sentences with a subject-and-predicative term structure, if adequate predicative terms are formed (there are few of them in everyday language), for instance: 'no object is contradictory', 'every object is definable in some way', etc. Primitive terms occur in declarative sentences of the form 'a is an object (or a property)' and function as predicative terms. The term 'to be an object' is a name of that property of the object denoted by the subject of the sentence about which we predicate: 'to be an object' means 'to be non-contradictory, to be definable in some way', etc. If we extend the concept of connotation to cover this case, too, then we have to call the connotation of the term 'to be an object' the logical product of the predicative terms mentioned above. The connotation of the term 'to be a property' can be defined analogously. Such a connotation belongs to the syntactic category of the second degree as compared with the syntactic category of such terms as 'square', 'house', etc. In general terms, connotations of the primitive terms of a theory belong to the syntactic category which is one degree higher than the syntactic category of the terms of that theory.

This can be verified by other examples. Terms in the sentential calculus are names of truth functions, among which implication and negation are primitive terms (in Łukasiewicz's system). Connotations of those terms in the sentential calculus which are not primitive are given in equivalences valid in that calculus; for instance, *Kpq* is equivalent to *NCpNq* or *NANpNq*, and *Apq* is equivalent to *CNpq*. Generally speaking, the connotation of any such term is a truth function of other terms. On the other hand, the connotation (as determined by the system of axioms) of any of the primitive terms, is reducible to a conjunction of terms such as 'transitiveness', 'asymmetry' (implication is a transitive, asymmetrical, etc., relation), etc., which belong to the syntactic category which is one degree higher than the category of names of truth functions. In other words, truth functors are names of intersentential relations, and 'to be an implication' and the like are names of properties of those relations. When we predicate that an implication has such and such properties, in order to denote those properties we introduce names whose syntactic category is one degree higher than that of the category of truth functors.

Numbers are properties of sets which are characterized by properties common to all elements of a given set. Thus, 'triplets' is a name which is predicated about a set of three children which share the property of having been born at one birth. Hence a numeral is a name in the syntactic category of the first degree in relation to the name of the set about which a given number is predicated, and in the syntactic category of the second degree in relation to the names of the elements of that set. The term 'natural number', which is primitive in arithmetic, has the connotation 'the property shared by equinumerous sets' (Frege); for instance, 3 is the property shared by all sets such that if w, x, y, z are elements of one of those sets and if w, x, y are different from one another, then z is identical either with w, or with x, or with y. The syntactic category of this connotation is one degree higher than that of the numerals.

Deictic definitions serve, among other things, to define simple qualitative elements of sensory perception, such as colours, and provide primitive terms of empirical knowledge. A term derived from such a definition by abstraction (e.g., the 'colour green') is given connotation by a psychological description which characterizes it as located in the spectrum between yellow and blue, contrasting with red, corresponding to light waves of a specified length, etc. The terms included in that connotation belong — as in the previous cases — to the syntactic category one degree higher than that of the term 'green.'

The foregoing analyses are intended to draw a demarcation line between logic and psychology in that section of their common frontier in which such a demarcation line is perhaps least perceptible, namely in the logical theory of names. This demarcation line is determined by the difference which has been interpreted as the opposition between the psychological meaning of a term (also called the intension of a concept) and its connotation, and also between its extension and its denotation. Connotation and denotation have been defined here exclusively in terms of logical relations, which has also made it possible to establish the relationship between connotation and denotation in classical logic. It has also been shown that the difference between the connotations of terms in a given theory and the connotations of the primitive terms of that theory is a difference of syntactic category.

PROPOSITION AS THE CONNOTATION OF SENTENCE

by

Kazimierz Ajdukiewicz

1. Every meaningful sentence, whether true or false, states something. What is stated by a sentence is called in German '*Sachverhalt*'. A literal translation of that term, e.g., 'state of things' or 'state of affairs', is seldom used in English. In the latter language, what is stated by a sentence is usually called a 'proposition'. In this sense, a proposition is neither a linguistic expression, nor a psychological act of thinking, nor any 'ideal meaning', but something that belongs to the sphere of objects to which a given sentence refers. The relation of stating, which holds between the sentence and the proposition, is therefore a semantic one, but should be distinguished from the semantic relation of denoting as it was understood by Frege. According to Frege, every true sentence denotes one and the same object, namely a mysterious object called 'truth', and every false sentence denotes one and the same object, still more mysterious, called 'falsehood'. But different true sentences may state different propositions, and different false sentences need not state the same state of things. Hence it follows that the semantic relation of stating differs from another semantic relation, namely that of denoting (in Frege's sense), and the proposition stated by the given sentence is neither the truth nor the falsehood.

2. The concept of proposition seems to be important for all study which has sciences as its object. In such a study the notion of theorem is one of the fundamental notions. Theorems of a science are usually identified with the sentences in which they are formulated, and metasciences adopt the notion of sentence as a fundamental concept. Yet such an identification of theorem with sentence leads to undesirable consequences. For instance, it implies that there is not *one* Newtonian theorem of gravitation, but as many theorems as there are languages in which the Newtonian idea can be formulated. A theorem is, therefore, not a sentence, but something that is common to all

those sentences which state the same thing, or, in other words, which are connected with the same proposition. The simplest answer to the question, what are theorems, seems therefore to be: theorems are propositions.

The need for a definition of proposition can also be seen in connection with the puzzling question, what is stated by a false sentence. A true sentence states some fact, but what is stated by a false sentence is not a fact, although it seems to be something. The difference between a meaningless sentence and a false one consists just in that the former states nothing whereas the latter states something — which, however, is not a fact. The definition of proposition should provide an answer to the question, what is stated by a sentence, whether true or false. Such a definition would have to point to a kind of entities which in the case of true sentences may be called 'facts'.

The distinction between such propositions which are facts and such which are not is the same as the distinction between true and false propositions. Since propositions are conceived as objective entities, their truth or falsehood are properties of objective entities and as such belong to the objective world. They may, therefore, be understood to be those mysterious objects which Frege called 'truth' and 'falsehood' and related to sentences as their denotations. A definition of true and false propositions would therefore help to elucidate the fundamental semantic ideas of Frege.

It is to be hoped that the definition of proposition will also be helpful in solving another perplexing metalogical problem, namely that of intensional sentences, but this question will be the subject matter of another paper.

3. All this seems to justify an attempt to define proposition as an objective entity. Carnap in his *Meaning and Necessity* has given his own definition of the term 'proposition', but his definition characterizes proposition by means of syntactic terms alone, without resorting to any semantic concepts. Consequently, what Carnap calls a proposition belongs to the linguistic sphere and not to the sphere of objects, and does not promise a solution of the problem stated above.

In this paper it is our intention to give such a definition of proposition which conceives it as an objective, and not a linguistic or psychological, entity, connected with the sentence by a semantic, and not a syntactic or pragmatic relation, viz., the relation of stating. That is why, with reference to sentences, beside the semantic relation of denoting, another semantic relation, namely that of stating, must be established.

4. To begin with, the traditional semantics of names distinguishes two semantic relations for them. In conformity with this traditional semantics, every name (i.e., such a name which may be used as a predicate in a sentence and consequently may be called a predicable name, as distinguished from a proper name) denotes its extension or denotation, and connotes its intension or connotation. The extension or denotation of a predicable name is the class of all individual objects with reference to which that name may be truthfully used as a predicate. Those objects are called designata of that name. The intension or connotation of a name is a set of properties whose conjunction holds with reference to each of those, and only those, objects which are elements of the extension of that name. In other words, the connotation of a name is a set of properties which univocally determines the extension of that name. The properties of the objects designated by a name belong to the objective, and not to the linguistic sphere, and that is why the relation between a name and its connotation is a semantic and not a syntactic one.

Now, in connection with distinguishing two semantic relations for names, one is tempted to look for such a generalization of those relations that we might speak not only of denotation and connotation of names, but also of denotation and connotation of other expressions. Further, one might try to generalize them so that the denotation of sentences would be, in conformity with Frege, their truth-value, and the connotation of sentences would be an entity which we might call the proposition, in conformity with our unanalysed, intuitive interpretation of that term. Such a solution will be sought in this paper.

5. First of all, the concept of denotation will be broadened so as to cover proper names, sentences, and operators which bind no variables. Thus, we define: the denotation of a proper name is the object named by it, the denotation of a sentence is its truth-value, the denotation of an operator binding no variables is the relation between the denotations of its arguments, and the denotation of the expression formed by that operator together with its arguments. Thus, for instance, the denotation of a truth-operator (e.g., of the implication symbol) is its matrix. The denotation of the sentence-forming operator 'shines' is the

relation between any individual object and the truth value of the sentence obtained from the sentential formula 'x shines' when the proper name of that object is substituted for the variable 'x'. The concept of denotation is applied neither to variables, nor to formulae containing variables, nor to operators binding variables. Other semantic notions, analogous to the notions of designating and denoting, are applicable to those expressions; for instance, variables represent their values but do not designate them, and they range over the set of their values but do not denote it. An analysis of the semantic notions referring to variables and to expressions connected with them would go beyond the limits of this paper. The fact that these notions are disregarded means that what is said in this paper refers only to those languages which include neither variables nor operators binding variables. This undoubtedly reduces the value of the results presented here and imposes upon the author the duty to endeavour to formulate a concept of proposition which would be applicable to languages that make use of variables.

6. It is much more difficult to generalize the concept of connotation, because the notion of the connotation of names itself, which is to be generalized, has no univocal definition in traditional logic. The formulation that the connotation of a name is the set of properties which univocally determines its extension, cannot be considered as a definition of the connotation of a name, since a given class of objects, forming the extension of a name, can be univocally determined by different sets of properties. And the connotation of a name is not just any set of properties which univocally determines its extension, but a set distinguished among those which satisfy that condition.

The connotation of a name is understood so that between the denotation, the connotation and the meaning of a name the following relations take place: the meaning or the sense of a name univocally determines its connotation; the connotation of a name univocally determines its denotation; but the denotation of a name does not univocally determine its connotation, i.e., two names with the same denotation may have different connotations; and the connotation of a name does not univocally determine its meaning, i.e., two names with the same connotation may differ in their meanings because, e.g., they may differ in their emotional tinge which also belongs to the meaning of a name. Hence a name that has a definite meaning has a definite connotation. Therefore it does not suffice to say that the connotation of a name taken in some meaning is the set of properties which univocally determines the extension of that name, because such a formulation does not ascribe to anything definite, anything which is univocally determined.

In order to define properly the connotation of a name let us see how the connotation of a name is formulated in practice. In traditional logic, this practice was confined to names which either are conjunctions of two or more other names, e.g., 'circular and red', or are conventional abbreviations of such conjunctions, e.g., 'square' which may be considered as a conventional abbreviation of the conjunction: 'tetragon, equilateral, rectangular'.

Now, in considering such names which are either explicitly or implicitly conjunctions of other names, the practice was to give as their connotations the sets of properties corresponding to the component names forming a given conjunction. For instance, the connotation of the name 'circular and red' was said to be the set of the property of circularity and the property of redness; the connotation of the name 'square' — if it is considered as a conventional abbreviation of the name 'tetragon, equilateral, rectangular' — was said to be the set of properties of tetragonality, equilaterality and rectangularity.

This practice permits us to formulate a rule of building connotations for those names which are either explicit or implicit conjunctions of other names, but does not allow formulatation of a general rule that would be applicable to all names. For there are names which are composed of other names, but not on the conjunction principle. E.g., the name 'circular or red' is based not on conjunction but on alternative. The connotation of the name 'circular or red' may not be considered to be the set of properties of circularity and redness. Should one think so, one would ascribe to the name 'circular or red' the same connotation as to the name 'circular and red'. Since the two names differ in their extensions, that is, in their denotations, this would be in contradiction with the principle that the connotation of a name univocally determines its extension.

If as the connotation of a name composed of simple names one wants to give something which would univocally determine its extension, then it is not enough to pay attention only to the objective references of the component names alone, but it is necessary to take into consideration the objective reference of that component of the compound name which binds these component names together. The connotation of such names as 'circular and red' or 'circular or red' is influenced not only by the component names, in which these two compound names agree, but also by the words which bind the component names and form with them together the compound name, in which they differ.

Further, there are names which are composed of other names in a non-symmetrical way, i.e., in such a way that if the order of the component names is changed, the extension of the compound name changes too. E.g., the extension of the name 'the brother of John's mother' differs from the extension of the name 'the mother of John's brother'; the extension of the name 'red but not circular' differs from that of the name 'circular but not red'. Hence, if the connotation of a name is to be interpreted as something which univocally determines the extension of that name, it does not suffice to define connotation as the set of certain properties or other objective references of members of the given name, but it is also necessary to take into consideration the syntactic role, or syntactic place, of those members within the compound name in question.

It follows from the above that in looking for the definition of the connotation of expressions it is necessary: 1) To determine the connotation of the expression E in such a way that its component parts should be some objective referents of all component expressions of the expression E, and not only its component names. To take into consideration not only those component expressions of the expression E which are contained in it explicitly, but also those which are contained in it implicitly and which can be seen clearly when all conventional abbreviations contained in the expression E are expanded. 2) To determine the connotation of the expression E in such a way that it should reflect not only the words contained in that expression, but also the syntactic places which those words occupy in the expression E.

This paves the way for so general a definition of connotation as to cover not only names, but any expressions as well. That, however, must be preceded by an explanation of the concept of the syntactic place of a given component expression in a compound expression. This concept seems very important — not only from the point of view of the analysis contained in this paper.

7. In explaining the concept of syntactic place it must be stated

first that every compound expression is articulated, i.e., consists of members which form a certain hierarchy. If only those expressions are taken into consideration which contain no binding operators, it may be said that in each such expression one can distinguish the principal operator and its successive arguments. That operator contained in the expression E is called its principal operator which together with its arguments exhausts the entire expression E. The principal operator of the expression E and its arguments are called *first-order members* of the expression E. First-order members of first-order members of the expression E are called *second-order members* of the expression E. In a general way, first-order members of n-th-order members of the expression E are called $n+1$*-th-order members* of the expression E. Further, the expression E itself will be called its own *member of* 0-*th order*.

As can be seen from the above, every n-th-order member of the expression E, for $n > 0$, either is the principal operator of an $n-1$-th-order member or one of the successive arguments of that operator. That $n-1$-th-order member, if it is not the expression E, is in turn either the principal operator of an $n-2$-th-order member, or one of the successive arguments of that operator, and so on, until the expression E is reached.

Now, the description of the syntactic place occupied by the expression A in the expression E (A being an n-th-order member of E) consists in the information stating both for the expression A and for every member of the expression E, which includes the expression A, whether it is the principal operator of a member of an order lesser by 1, of which it is a component part, or the first, second or k-th argument of that operator. Let, for instance, the expression

$$3 \cdot 4 = 5+(8-1)$$

be analysed. The syntactic place of the expression '5+(8−1)' in the expression E is described by the statement saying that it is the second argument of the principal operator of the expression E. The syntactic place of the expression '8−1' in the expression E is described by the statement saying that it is the second argument of the principal operator in the second argument of the principal operator of the expression E. Still more complicated is the description of the syntactic place occupied by the figure '8' in the expression E. The figure '8' is the first argument of the principal operator in the second argument of the principal operator in the second argument of the principal operator of the expression E.

As can be seen from the above, the indication of the syntactic place by means of a verbal description is very complicated. To simplify it a special symbolism will be introduced. The syntactic place of an expression E as such will be marked with a symbol of a natural number, e.g., the symbol of the number 1. Further, if the syntactic place occupied by the expression A in the expression E is marked ⌜m⌝, then the place of the principal operator of the expression A in the expression E will be marked (m, 0) and the place occupied in the expression E by the n-th argument of that operator will be marked ⌜(m, n)⌝.

Thus, if the syntactic place of the sentence E:

'Plato est philosophus et Aristoteles est philosophus'

is as such marked with the figure "1", then the places occupied in that sentence by the members of that sentence will be marked as follows:

```
(1,1,1)   (1,1,0)       (1,1,2)            (1,2,1)       (1,2,0)    (1,2,2)
 Plato      est      philosophus    et    Aristoteles      est     philosophus.
\_____________________________________/  \_/  \_______________________________________/
               (1,1)                   (1,0)                 (1,2)
```

This symbolism makes it possible to indicate in a simple way even the most remote syntactic places.

For a univocal description of a compound expression it is not enough to enumerate (write out) the simple members of which it consists, but it is also necessary somehow to describe the syntactic places they occupy in that expression. For instance, from the '5', '>', '3' one can build the sentence '5 > 3' and the sentence '3 > 5'. In arithmetical symbolism the order of the elements and sometimes the use of brackets serve to indicate the place occupied by simpler expressions within more complex ones. Such languages in which the syntactic places of members of a compound expression are indicated by their order and by the semantic categories of simple elements are called positional languages. There are, however, languages which indicate syntactic places of words not only by means of their order, but also by their inflexional forms. For instance, in the Latin sentence

'Petrus amat Paulum'

the fact that the word 'Petrus' is the first argument of the operator 'amat', i.e., the subject, and 'Paulus' its second argument, i.e., the object, is indicated by this that 'Petrus' is in the nominative case, and 'Paulus' in the accusative. Should even the order be changed into

'Paulum amat Petrus',

the syntactic role of the elements of that sentence would not undergo any change.

Such languages which resort to inflexion to indicate the syntactic places of words are called inflexional. Yet no natural language indicates the syntactic places of words exclusively by means of inflexional forms, but resorts to word order, punctuation, etc., so that no natural language is purely inflexional.

Now, a purely inflexional language can be built by means of the symbolism indicating the syntactic places of words, as described above. Those symbols, when attached to words, can be treated as their inflexional endings which themselves, regardless of the order of the words in question, indicate their syntactic places. Words with such endings can be scattered in an arbitrary way, and yet on the strength of these endings they will continue to form definite wholes.

Such a purely inflexional language will be used later in an attempt to define true proposition.

8. After this digression on syntactic places, the previously announced definition of the connotation of expressions will be given.

For that purpose a given expression *E* must first be transformed so that all the conventional abbreviations contained in it are expanded and consequently none of its simple members are abbreviations of compound expressions. The syntactic places occupied by simple members in such an expanded expression *E* will be called ultimate syntactic places of the expression *E*. Thus, in the expression *E* there is a one-one correspondence between those ultimate syntactic places and separate words. Such a one-one correspondence between words and syntactic places in the expression *E*, that is the function, determined for those places, which establishes a one-one correspondence between those places and words, is the characteristic function of the expression *E*. But there is a one-one correspondence between the words occupying the ultimate places in the expression *E* and their denotations. If there is agreement to this, it follows that for every expression *E* there exists a function establishing a one-one correspondence between every ultimate syntactic place in the expression *E* and a certain object, namely the object which is the denotation of the word occupying that place.

It is suggested that that very function should be considered as the connotation of the expression E. Consequently, the following definition is adopted:

The connotation of the expression E is the function determined for the ultimate syntactic places of the expression obtained from the expression E by the expansion of all the abbreviations it contains, which establishes a one-one correspondence between those places and the denotations of the words occupying such places in the expanded expression E.

For instance, the connotation of the expression

'circular and red'

is the relation establishing a one-one correspondence:

1) between the syntactic place of the principal operator (1,0) and the denotation of the word 'and',

2) between the syntactic place of the first argument (1,1) and the denotation of the word 'circular',

3) between the syntactic place of the second argument (1,2) and the denotation of the word 'red'.

Every binary relation, and consequently every function of one argument, can be identified with the class of ordered pairs of objects between which that relation holds. If that class is finite, it can be symbolized by means of enumeration of those pairs. Thus the function which is the connotation of the expression

'circular and red'

may be written

$$\langle (1,1) \text{ — circular; } (1,0) \text{ — and; } (1,2) \text{ — red} \rangle.$$

If bound variables and the abstraction symbol were used, the above expression might be replaced by the equivalent one:

$$(\hat{x}\,\hat{y})\,\{[x = (1,1) \wedge y = \text{circular}] \vee$$
$$[x = (1,0) \wedge y = \text{and}] \vee [x = (1,2) \wedge y = \text{red}]\}$$

Such a way of symbolizing the relation, which resorts to an operator binding the variables, is, however, not necessary if the domain of the relation is finite and may be replaced by enumeration.

It can easily be seen that the definition of connotation of expressions, as suggested above, satisfies the conditions previously formulated with regard to such a definition.

First, in the case of the above definition of connotation, connotation of an expression univocally determines its denotation. If the name of the connotation of the expression E is given by enumeration, the name of its denotation is obtained by the juxtaposition of the second elements of every pair contained in that name and by writing below every such element the first member of its pair. For instance, from the symbol of the connotation of the expression "circular and red", which has the form

$$\langle(1,1) — \text{circular};\ (1,0) — \text{and};\ (1,2) — \text{red}\rangle$$

the symbol of its denotation is obtained by writing

$$\underset{(1,1)}{\text{circular}}\ \underset{(1,0)}{\text{and}}\ \underset{(1,2).}{\text{red}}$$

It can also be noticed that although the connotation of an expression, defined in this way, determines its denotation, it is not determined conversely by its denotation.

Secondly, the definition of connotation, as given above, makes the connotation of an expression depend on the objective referents of all its component expressions, and not only of its component names; moreover, not only those components of a given expression are to be taken into consideration which appear in it explicitly, but also those which will appear in it after the abbreviations have been expanded.

Thirdly, the above definition of connotation takes into account not only the objective referents of the words of which a given expression consists, but also their syntactic roles, i.e., their syntactic places.

9. The above definition of connotation is so general that it may be applied not only to names, but to any expressions as well. Consequently, it may also be used with respect to sentences. Now, *the connotation of a sentence will be called the proposition stated in that sentence*. This is the definition of proposition which this paper suggests.

Now, in the light of that definition, what is the proposition stated, e.g., in the sentence

'Socrates likes Alcibiades'?

Transcribed with the help of symbols indicating the syntactic places, that sentence can be written thus:

Socrates likes Alcibiades.
(1,1) (1,0) (1,2)

The connotation of that sentence, i.e., following our definition, the proposition stated by that sentence, is the function establishing a one-one correspondence between the syntactic places of its words and their denotations, that is

$$\langle (1,1) \text{ — Socrates; } (1,0) \text{ — likes; } (1,2) \text{ — Alcibiades} \rangle.$$

Is such an interpretation of the concept of proposition in conformity with our intuitive interpretation of proposition as that which is stated in a given sentence? It seems so. Objects exist in the world regardless of whether and what people think about them. There exists Socrates, there exists Alcibiades, and — in a sense — there exists the relation of liking. These objects are related to one another, or not — for instance, the relation of liking occurs between Socrates and Alcibiades, or does not — regardless of what people think about them. The thought that Socrates likes Alcibiades, does not by itself create the objective fact consisting in that the relation of liking occurs between Socrates and Alcibiades. But by thinking that Socrates likes Alcibiades we think about those objects as being related to one another in that way. And it seems the best way to describe in what the thought about those objects as so related to one another consists, when it is said that a one-one correspondence is established between the relation of liking and the syntactic place of the principal operator, between Socrates and the syntactic place of its first argument, i.e., the subject, and between Alcibiades and the syntactic place of its second argument, i.e., the object. Thus, when one comprehendingly states the sentence 'Socrates likes Alcibiades', one establishes a one-one correspondence between the syntactic places in that sentence and the objects denoted by the members of that sentence occupying such places.

And such a correspondence between the syntactic places and the objects in question is what is being stated by that sentence, and consequently, a proposition.

So much for winning an intuitive approval for the definition suggested in this paper.

Following this definition of proposition, to every sentence, whether true or false, corresponds the proposition stated by that sentence. The puzzling question, what is stated in a false sentence, seems to have been solved in this way.

The correspondence between syntactic places and objects may, or may not, agree with the relative positions of those objects in reality. If the sentence stating a given proposition is true, then relative positions occupied with respect to one another by the objects referred to in the sentence correspond to these syntactic positions which are allotted to them in the proposition stated in that sentence. In that case it seems natural to call the proposition stated in a true sentence a fact.

10. The propositions stated in true sentences may be called true propositions. But here the question arises, whether true proposition might not be defined directly, without reference to the concept of true sentence. It is known that in defining the truth of a sentence one must resort both to object language and to metalanguage, which involves considerable difficulties. Perhaps a definition of the truth of proposition, i.e., of an entity which can be referred to in an object language, might itself be built in an object language too.

For the time being I cannot build a satisfactory definition of the truth of proposition, or even a general schema of such a definition, and I do not know whether this is possible at all. That is why only partial definitions of the truth of proposition will be given here, to serve as a possible starting point for the formulation of a general definition.

It has been shown above how to pass from a symbol of connotation of an expression to a symbol of its denotation. If, e.g., the connotation of the expression

'circular and red'

is written in the form

$$\langle (1,1) — \text{circular};\ (1,0) — \text{and};\ (1,2) — \text{red} \rangle,$$

in a language which makes use of symbols of syntactic places, the name of its denotation is obtained in the form

$$\underset{(1,1)}{\text{circular}} \quad \underset{(1,0)}{\text{and}} \quad \underset{(1,2).}{\text{red}}$$

This will help to understand the following partial definition:

(1) ⟨(1,1) — Socrates; (1,0) — likes; (1,2) — Alcibiades⟩ is true

$$\equiv \underset{(1,1)}{\text{Socrates}} \quad \underset{(1,0)}{\text{likes}} \quad \underset{(1,2)}{\text{Alcibiades}}$$

The expression in pointed brackets in the left part of the above equivalence is the name of a proposition the truth of which is stated; the right part is occupied by the sentence stating that proposition.

The above equivalence resembles partial definitions of the truth of the sentence, e.g.,

(2) 'Socrates likes Alcibiades' is true ≡ Socrates likes Alcibiades.

The difference between (1) and (2) consists in that in (2) reference is made both to words and to the objects named by those words, whereas in (1) reference is made to objects only, but not to words. That is why the partial definition (1) may, *cum grano salis*, be considered as formulated exclusively in an object language. The reservation is due to the fact that although the symbols of syntactic places appear in the right part not as words but as inflexional endings of words, yet in the left part they are taken as words, namely as names of syntactic places. Consequently, the definition (1) may not be considered as formulated exclusively in an object language. This problem, very important for the present analysis, will be discussed later.

Starting from the partial definitions of the truth of propositions, here represented by the formulation (1), one might consider the following formulation as the first approximation of a general definition of the truth of propositions:

(3) $S \in \text{true}$

$$\equiv \sum_{n \in \text{Nat}} \sum_{\substack{x_1, x_2, \ldots, x_n \in \text{object} \\ m_1, m_2, \ldots, m_n \in \text{synt. place}}} \left\{ (S = \langle m_1 x_1, m_2 x_2, \ldots, m_n x_n \rangle) \wedge \begin{pmatrix} x_1, x_2, \ldots x_n \\ m_1, m_2, \ldots, m_n \end{pmatrix} \right\}$$

Many objections may be raised against this formulation. First of all, it does not hold for all S, because for those S which are not propositions the right side of the definition is a syntactic nonsense. For, if S is not a proposition, i.e. a connotation of a sentence, and S is a connotation of the expression $\begin{bmatrix} x_1, x_2, \ldots, x_n \\ m_1, m_2, \ldots, m_n \end{bmatrix}$, then that expression is not a sen tence and consequently may not meaningfully appear as an argument of the symbol of conjunction in the right part of the definition. The only way of avoiding that difficulty seems to change that definition into a rule permitting, if it is already known that S is a proposition, adoption of the equivalence (3).

The second objection refers to the fact that this equivalence is written in a mixed language: everything in it is written in a language which does not use special symbols for marking syntactic positions, except for the last member, which is formulated just in such a language. One might think of re-formulating that definition in such a language in which its last member is written. But to do so one would have to expand the system of syntactic positions so as to make it possible to mark the syntactic places of quantifiers, whereas the above analysis referred to languages which do not include quantifiers.

The third objection refers to the claim of defining the truth of propositions exclusively in an object language. The above definition is not built in this way, because it contains names of syntactic places.

The second and third objections point to the direction in which research should be continued. The concepts of connotation and proposition must be made more general, so as to make them applicable to languages which include operators binding variables. Further, the concepts of connotation and proposition must be freed from the metalanguage concept of syntactic place. I see a possibility of solving both these problems, but my ideas are still too vague to be presented here. Yet there seem to be prospects for a solution.

INTENSIONAL EXPRESSIONS

by

Kazimierz Ajdukiewicz

1. Intensional expressions can both be expressions containing free variables and expressions which do not contain such variables. The expression *E*, which contains no free variables, is an intensional expression if it can be transformed into a non-equivalent expression *E'* by replacing one of its members by an expression which is equivalent to that member.

The expression *E*, which contains one or more free variables, is an intensional expression if two non-equivalent expressions can be obtained from it by substituting for each of these variables two different but equivalent constant expressions. As it is well known, expressions which are not intensional are called extensional.

The sentence 'Newton knew that $8 > 5$' (Z_1) may be given as an example of an intensional expression (for a certain interpretation of its syntactic structure). For if in that sentence we substitute for '8' the equivalent expression 'the atomic number of oxygen', we receive the sentence 'Newton knew that the atomic number of oxygen > 5' (Z_2). The first of these two sentences (Z_1) is true, and the second (Z_2) is false, which means that they are not equivalent, although the second is obtained from the first by substituting for '8' the equivalent expression 'the atomic number of oxygen'.

An example of an intensional expression containing a free variable is the formula 'Newton knew that $x > 5$', because we obtain from it non-equivalent sentences (Z_1) and (Z_2) by substituting for the variable respectively the expressions '8' and 'the atomic number of oxygen', which are equivalent.

The notion of intensional expression has been defined here in a general way, so that it can be referred to expressions belonging to different semantic categories. We may thus speak of intensional sentences, intensional predicates, intensional operators, etc. The most disturbing problem, however, is that of intensional sentences and intensional sentential formulae, and only they will be the subject matter of the main part of this paper.

Intensional sentences are mostly seen in sentences resorting to the so-called indirect speech, such as 'John believed that...', 'John said that...' etc. Further, the sentences including the so-called modal terms, such as 'it is necessary that...', 'it is possible that...', etc., are also considered intensional sentences. For various reasons, chiefly those of time limits, only the sentences making use of indirect speech will be discussed in this paper.

2. Two terms used in the definitions given above require additional explanations, viz. the terms 'equivalent expressions' and 'a member of an expression'.

The expression *A* is equivalent to the expression *B* if either *A* and *B* are proper names of the same object, or *A* and *B* are sentences having the same truth-value, or *A* and *B* are (non-binding) operators with the same number of arguments, which operators, when combined with equivalent arguments, form equivalent expressions.

We assume that all expressions between which the relation of equivalence may hold have a certain objective referent which will be called their denotation, and further, that two equivalent expressions denote the same thing.

In particular, we assume that the denotation of a proper name is the object named by it, the denotation of a sentence is its truth-value, i.e., truth or falsehood. As far as not binding operators are concerned, we assume that the denotation of an operator of n arguments is the relation of $n+1$ arguments, holding between the denotations of its arguments and the denotation of the expression which that operator forms jointly with its arguments.

As for the term 'a member of an expression', we say that the expression *A* is a member of the zero-th order of the expression *B* if *A* is the same as *B*. In other words, every expression is its own member of the zero-th order.

The expression *A* is called the first order member of the expression *B* if *A* is the main operator in the expression *B* or if *A* is an argument of that operator.

In general: the expression *A* is a member of the $n+1$-st order of the expression *B* if *A* is a first order member of an *n*-th order member of the expression *B*.

The expression *A* will be called simply a member of the expression *B* if it is a member of any order of the expression *B*.

The following notation will be used to indicate the syntactic place occupied by a member in a given expression:

The syntactic place occupied by a given expression in itself will be marked with the figure 1. To indicate the syntactic place of other members the following principle will be adopted: if the place occupied by the expression *A* in the expression *E* is symbolized as (*m*), the syntactic place occupied in the expression *E* by the main operator of the expression *A* will be symbolized (*m*, 0), and the place occupied by the *n*-th argument of that operator will be symbolized (*m*, *n*). For instance, in the following expression the syntactic places of its members of different orders will be indicated thus:

$$\underbrace{\underbrace{\underbrace{2}_{(1,1,1)}\quad\underbrace{+}_{(1,1,0)}\quad\underbrace{3}_{(1,1,2)}}_{(1,1)}\quad\underbrace{=}_{(1,0)}\quad\underbrace{\underbrace{6}_{(1,2,1)}\quad\underbrace{-}_{(1,2,0)}\quad\underbrace{1}_{(1,2,2)}}_{(1,2)}}_{(1)}$$

3. It must be emphasized that for an expression to be intensional it is not enough that it can be transformed into a non-equivalent expression by replacing any of its parts by an equivalent expression, but it is necessary that it should change into a non-equivalent expression as a result of replacing such of its part which is its member. E.g., the true sentence

$$5+3\cdot 4 = 17$$

changes into a false one if its part

$$5+3$$

is replaced by the equivalent expression

$$2+6$$

because the sentence

$$2+6\cdot 4 = 17$$

is false. Yet the sentence under discussion is not intensional, because its transformation did not consist in replacing any of its members.

Paying attention to that detail is essential for further analysis, because any expression quoted as an example of intensional expression will be such an example only if that part of it whose replacement has transformed that expression into a non-equivalent one, is its member. Now the syntactic structure of colloquial expressions, from among which examples are usually drawn, is not univocal enough to allow only one way of dividing them into members. If a given expression allows two different ways of dismembering it, such that in one case that expression must be considered intensional, but not in the second, then by giving more precision to the language which determines its syntactic structure we may choose a syntactic structure for which intensionality will not occur, and in this way we may free our language from intensional expressions.

4. There are several reasons why it may be desirable to free our language from intensional expressions, in particular from intensional sentences and sentential formulae.

Some of them will be recalled here, and attention will be drawn to some others, which — as far as I know — have been disregarded so far.

First of all, the presence of intensional formulae in a language prevents us from adopting Leibniz's definition of identity. That definition states

$$a = b \leftrightarrow \underset{F}{\Pi} \{F(a) \leftrightarrow F(b)\}\,.$$

If, however, there is an intensional formula $\ulcorner F_1(x) \urcorner$, then there are two expressions a and b such that $a = b$ and yet $\sim\{F_1(a) \leftrightarrow F_1(b)\}$, which is in contradiction with the above definition.

Further, every extensional sentential formula of one variable establishes a correspondence between the value of that variable and the truth-value — namely the truth for those values of the variable which satisfy that formula, and the falsehood for those which do not. In other words, every extensional sentential formula $\ulcorner(\ldots\, x \,\ldots)\urcorner$ determines the functional relationship

$$(\ldots\, x \,\ldots) = t$$

in which the variable $\ulcorner x \urcorner$ is the indenpedent variable ranging over the entire universe of a corresponding order, and the variable t the dependent variable the values of which may be truth or falsehood. This means that every extensional sentential formula determines a class of those values of the variable x for which that formula becomes true. This fact is reflected in the rule, adopted by the logical systems based on the type theory, which permits stating for any sentential formula

$$(\ldots x \ldots)$$

that

$$\sum_f \{f(x) \equiv (\ldots x \ldots)\}.$$

Now in contrast to the extensional formulae, which always determine functional relationships between the values of the variable and the truth or the falsehood, the intensional formulae determine no such functional relationship.

Let us take, for instance, the intensional formula

$$\ulcorner \text{Newton knew that } x > 5 \urcorner .$$

That formula does not establish any correspondence between a definite value of the variable $\ulcorner x \urcorner$ and a definite truth-value. E.g., in the case of the number '8' it establishes its correspondence both to the truth and to the falsehood according to the name of that number which we choose to use in a given case. Thus, an intensional formula does not determine any class of the values of the variable which satisfy that formula. Consequently, if the formula

$$\ulcorner (\ldots x \ldots) \urcorner$$

is an intensional one, then it is not true that

$$\sum_f \{f(x) \equiv (\ldots x \ldots)\},$$

which may also be written thus

$$\sum_f \{(x \in f) \equiv (\ldots x \ldots)\}$$

and which states the existence of such a class f to which belong all those, and only those, objects which satisfy the formula

$$\ulcorner (\ldots x \ldots) \urcorner .$$

It follows from the above that any logical system which adopts the rule permitting to state for any sentential formula

$$(\dots x \dots)$$

that

$$\sum_f \{f(x) \equiv (\dots x \dots)\}$$

confines the range of sentential formulae to extensional formulae and justifies that by adopting the rule of extensionality.

What has been said above requires, as it seems, a revision of the universally adopted view that the presence of intensional formulae in a language prevents us from adopting Leibniz's definition of identity. That definition, which states that

$$a = b \leftrightarrow \prod_F \{F(a) \leftrightarrow F(b)\},$$

is formulated exclusively in the object language and does not refer to any expressions, in particular to any sentential formulae. The quantifier $\prod_F$ should not be read 'for any predicates F', but on the object language level, 'for any classes F' or 'for any properties F'. Now, in the light of what has been said above, the enriching of the language with intensional formulae does not lead to the enriching of the set of all classes or the set of all properties, because — as we have just seen — no classes correspond to the intensional formulae. This means that if

$$\ulcorner(\dots x \dots)\urcorner$$

is an intensional formula, then it is not true that

$$\sum_f \{(x \in f) \equiv (\dots x \dots)\}.$$

Hence, in conformity with Leibniz's definition, a may be identical with b, i.e., the name $\ulcorner a \urcorner$ may denote the same object as the name $\ulcorner b \urcorner$, in spite of the fact that there exists an intensional sentential formula $\ulcorner(\dots x \dots)\urcorner$ which becomes truth when the name $\ulcorner a \urcorner$ is substituted for the variable $\ulcorner x \urcorner$, and becomes falsehood when the name $\ulcorner b \urcorner$ is substituted for the variable $\ulcorner x \urcorner$.

The third peculiar property of intensional expressions, which makes us adopt a critical attitude toward them, is the following.

It seems obvious that the denotation of a compound expression which along with first order members has members of the second order, and possibly of still higher orders, cannot change as a result of a change in the denotation of a member of a higher order if that change does not involve a change in denotation of members of lower orders. In other words, the denotation of an expression having the form

$$F\{a, \varphi\,[\chi\,(b), c]\}$$

cannot change as a result of a change in the denotation of the member b (here a member of the third order) if such a change does not involve a change in the denotations of the second order member $\chi\,(b)$ and of the first order member $\varphi\,[\chi\,(b), c]$. For within expressions there is no action in distance which could make a member influence some other member, separated from it by members of intermediate orders, without causing any changes in those intermediate members.

The extensional expressions comply with that obvious principle which, however, is violated by certain expressions if they are syntactically interpreted as intensional expressions.

If we, for instance, take the sentence 'Caesar knew that the capital of the Republic lies on the Tiber' and interpret it syntactically as shown by the brackets below: {Caesar} knew that {[the capital (of the Republic)] lies on [the Tiber]} we see that it will be an intensional sentence because from a true one it changes into a false one when its member

'[the capital (of the Republic)]'

is replaced by an equivalent member

'[the capital (of the Popes)]'.

In other words, the sentence quoted above changes its truth-value when its third order member '(of the Republic)' is replaced by the expression '(of the Popes)'. The change in the denotation of the third order member, resulting from such a replacement, brings about a change in the denotation of the entire expression, that is, also of the zero-th order member, although it does not influence the denotation of the intermediate, second order member 'the capital of the Republic', or the intermediate, first order member 'the capital of the Republic lies on the Tiber'.

The fact that the obvious principle, mentioned above, which excludes *action in distance* between members of expressions is violated by certain expressions if their syntactic structure is interpreted so that these expressions become intensional, raises misgivings about such a syntactic interpretation of their structure and induces one to go searching for some other interpretation of their syntactic structure, for which neither will the principle in question be violated, nor will it be possible to consider such expressions as intensional ones.

5. Such are the motives which may induce us to try to eliminate from our language the intensional expressions, and in particular the intensional sentences and sentential formulae. We shall now analyse the problem of how to eliminate them. It has been stated above that the intensional sentences appear either under the form of sentences making use of such linguistic terms as 'said' or such psychological terms as 'thought', 'knew', etc., connected with indirect speech. Further it seems that those sentences which include modal terms also are intensional sentences.

Now we can get rid of intensional sentences and sentential formulae by eliminating from our language those linguistic, psychological and modal terms which give rise to intensionality. That is the way mathematics gets rid of intensional sentences and sentential formulae: it does not introduce such terms into its language.

It would be difficult to suggest such a radical measure with regard to colloquial language which, following the elimination of all those terms, would be deprived of means of expression required by its practical function.

But there is another method of eliminating intensional expressions from the language, which would not demand giving up all such statements that someone thought, or believed, that something is so and so, or statements saying that something is necessary or possible. That method, already hinted at above, consists in interpreting the syntactic structure of sentences in a different way from that which would lead us to consider them as intensional.

That is precisely our task in this paper. It consists in pointing to the possibility of such a syntactic interpretation of sentences considered to be intensional, for which they would cease to be intensional. But our aim is not only to indicate such a possibility, but also to analyse the meaning of those allegedly intensional sentences in order to show that such a syntactic interpretation which deprives them of their intensional character is in conformity with the meaning of those sentences. As mentioned above, the analysis carried out in this paper will be confined to sentences with psychological terms.

6. As an example of such a sentence, which for a certain interpretation of its syntactic structure is an intensional sentence, we take the sentence (Z_1)

'{Caesar} believed that {[Rome] lies on [the Tiber]}'.

That sentence is an intensional one because the word 'Rome' is one of its members, and the replacement of that member by an expression, which is equivalent to it, 'the capital of the Popes', transforms that sentence from a true one into a false one.

Attempts have been made to deprive that sentence of its intensional character by interpreting it as a sentence stating a certain attitude of Caesar towards the sentence 'Rome lies on the Tiber', and not towards what is stated by that sentence. That interpretation implies that Caesar accepted the sentence 'Rome lies on the Tiber'. Such an interpretation makes intensionality disappear, and yet it would be difficult to adopt it, because the interpreted sentence (Z_1) 'Caesar believed that Rome lies on the Tiber' is true, and the sentence "Caesar accepted the sentence 'Rome lies on the Tiber'" is false because Caesar did not know English and did not accept any sentence in that language.

But the following interpretation (Z_1^t) will be free from that objection: "Caesar accepted some sentence which is a translation of the sentence 'Rome lies on the Tiber' from English into some other language".

This interpretation does not substitute any false sentence for the true sentence (Z_1), but is not satisfactory, because for the sentence (Z_1), which is an object-language sentence, it substitutes the sentence (Z_1^t), which is a metalanguage sentence. In the sentence (Z_1), 'Caesar believed that Rome lies on the Tiber' we have clearly to do with Caesar's mental attitude with regard to Rome, to the relation of lying on, and to the Tiber. and not with Caesar's attitude towards some verbal entity. And our intention is to give such an interpretation of the object-language sentence (Z_1) and of other sentences of that type which would itself be an object-language sentence but at the same time would no longer be an intensional sentence. We shall strive for such an interpretation through the analysis of the interpretation (Z_1^t), which refers to the notion of translation.

To begin with, what does it mean that the expression *A* is a translation of the expression *B* from the language L_1 into the language L_2? First of all, let us notice that the language L_1 need not be different from the language L_2, so that it is not only about two expressions belonging to different languages that we may say that one is a translation of the other; we may say that also about two expressions belonging to the same language. To simplify our formulations we shall hereafter disregard relativization to languages, which can always be easily added.

To define the notion of translation we shall not refer to the notion of meaning. In particular, we shall not say that the expression *A* is a translation of the expression *B* if both these expressions have the same meaning, because such a definition would burden the notion of translation with all those obscurities which are connected with the term 'meaning'. The definition of translation will be based on the notions of denotation and of syntactic place.

Before proceeding to define the notion of translation we must realize that we may speak of translations of varying degrees of precision, more or less literal.

We say that the expression *A* is a literal translation of the expression *B* if, and only if, after the expansion of all the abbreviatons which they contain they are transformed into two abbreviation-free expressions A' and B', such that there is a one-one correspondence between all the members of one expression and the members of the other expression, and the members between which such a correspondence is established, 1°, occupy in A' and B', respectively, the same syntactic places, and 2°, are equivalent, i.e., they denote the same thing.

We say that the expression *A* is a translation of the *n*-th degree of the expression *B* if the relation between them is such that they might be considered reciprocal literal translations if we disregarded their members of degrees higher than the *n*-th, i.e., if their members of the *n*-th order were treated as simple members.

More precisely: the expression *A* is an *n*-th degree translation of the expression *B* if, and only if, after the expansion of all the abbreviations which they contain they are transformed into two abbreviation-free expressions A' and B', such that there is a one-one correspondence between all the members of the n-th and lower degrees of one expression and the members of the other expression, and the members between which such a correspondence is established, 1°, occupy in A' and B', respectively, the same syntactic places, and 2°, are reciprocally equivalent, i.e., they denote the same thing.

It can be seen from these definitions that a literal translation of a given expression A is the same as its translation of the highest degree of precision possible for a given expression, i.e., its translation of the degree equal to the highest order shared by the members of the expression A.

This will be illustrated in the following expressions:

$$(A) \qquad 2+(3\cdot 5) = 3+(2\cdot 7)$$

$$(B) \qquad 2+(5+10) = 3+(10+4),$$

it being assumed that the single symbols which appear in A and B are not any conventional abbreviations of any compound expressions. It can easily be verified that these expressions are reciprocal translations of the zero-th order, for each expression is its own member of the zero-th order. Thus by establishing a one-one correspondence between the expression A and the expression B we establish a one-one correspondence between the members of the zero-th order of these two expressions, and these members, 1°, occupy respectively the same syntactic places, namely, the places occupied by each of these expressions themselves, and 2°, are equivalent since both are true sentences.

If we establish a one-one correspondence between the first order members of the expressions concerned:

A	B
$=$	$=$
$2+(3\cdot 5)$	$2+(5+10)$
$3+(2\cdot 7)$	$3+(10+4)$

we establish a one-one correspondence between the zero-th and the first order, such that, 1°, the members between which such a correspondence is established occupy in A and B, respectively, the same syntactic places, and 2°, are equivalent, since they denote the same things.

It can further easily be seen that the expressions *A* and *B* are reciprocal translations of the second degree, because if the above correspondence between members of the zero-th and the first order be supplemented, as below, by one-one correspondence between second order members:

A	*B*
2	2
+	+
$(3 \cdot 5)$	$(5+10)$
3	3
+	+
$(2 \cdot 7)$	$(10+4)$

we obtain a one-one correspondence of members of the zero-th, the first and the second order, in which the members between which such a correspondence is established, 1°, occupy in *A* and in *B*, respectively, the same syntactic places, and 2°, they denote the same things.

But the expressions *A* and *B* are not reciprocal translations of the third degree, because if we establish a one-one correspondence between the third order members of these expressions, occupying the same syntactic places:

A	*B*
3	5
·	+
5	10
2	10
·	+
7	4

we find that the members occupying the same syntactic places are not equivalent.

Since the expression *A* is not a translation of the expression *B* of the third degree, whereas the highest order which the members of these expressions can have is the third order, the expression *A* is not a literal translation of the expression *B*.

Similarity between the definition of literal translation, as given above, and Carnap's definition of intensional isomorphism (*Meaning and Necessity*, first ed., p. 56) can easily be noticed. The difference between these two definitions consists above all in the fact that Carnap requires that if two expressions are to be intensionally isomorphic, then their members occupying respectively the same places must be *L*-equivalent, whereas our definition is satisfied with ordinary extensional equivalence.

To pave the way for answering the question, whether the sentences of the type (Z_1), 'Caesar believed that Rome lies on the Tiber', remain intensional if they are interpreted as (Z_1^t), "Caesar accepted some translation of the sentence 'Rome lies on the Tiber'," the following analysis will be made.

Every sentence is characterized by a correspondence between its syntactic places and the expressions which occupy those places. And since each such expression has its definite denotation, then every sentence is characterized by a correspondence between its syntactic places and certain definite objects. For instance, the sentence 'Rome lies on the Tiber', which has the syntactic place of the entire expression (1), of its principal operator (1,0), of the first argument of that operator (1,1) and of the second argument of that operator (1,2), is characterized by the following correspondence: (1) — Verum, (1,0) — the relation of lying on, (1,1) — Rome, (1,2) — the Tiber.

Let the correspondence, characteristic of a given sentence, between the syntactic places of its members and certain definite objects, namely those objects which are the denotations of the members occupying their respective places, be called the full content of that sentence.

It can easily be noticed that all the sentences which are literal translations of the same sentence have the same full content. For, in conformity with the definition of literal translation, the sentences which are literal translations of the sentence *A* have at the individual syntactic places expressions having the same denotations as the expressions occupying the corresponding syntactic places in *A*. We may say therefore that a literal translation of the sentence *A* is the same as a sentence having the same full content as has the sentence *A*.

Let the correspondence between the syntactic places occupied in the sentence *A* by simple expressions and the objects denoted by those simple expressions be called the basic content of the sentence *A*. Now, to determine univocally the full content of a given sentence, it suffices, except for the cases to be discussed later on, to give the basic content of that sentence. In other words, if the denotations of the simple expressions contained in a sentence are given, this fact univocally determines the denotations of the compound expressions contained in that sentence as its members. For instance, determination of the denotations of the simple expressions which appear in the sentence

$$2+3 = 5$$

— i.e., the expressions '2', '+', '3', '=' and '5' — determines the denotations of the compound expressions which are its members, i.e., its first order member '2+3', and the denotation of its zero-th order member, i.e., the entire sentence itself.

But it is not always so. It is not so if the sentence *Z* is itself an intensional expression or contains an intensional expression as its member. For intensional expressions are characterized by this: that the denotations of their members do not univocally determine the denotation of the expression consisting of those members. Thus the theorem stating that the basic content of a sentence univocally determines its full content is applicable only to such sentences which neither are intensional expressions nor include such expressions as their members.

It has been stated above that a literal translation of the sentence *Z* is the same as a sentence which has the same full content as has the sentence *Z*. Now it has been stated that as far as sentences free from intensional members are concerned, they have the same full content if, and only if, they have the same basic content. Consequently, with respect to the sentences free from intensional members, we may say that a literal translation of the sentence *Z* is the same as a sentence which has the same basic content as has the sentence *Z*.

The sentence 'Rome lies on the Tiber' (which hereafter will be written in the abbreviated form 'R *l* T') does not include intensional members. Therefore a literal translation of the sentence 'R *l* T' = a sentence which has the same basic content as has the sentence 'R *l* T'. But the basic content of the sentence 'R *l* T' is the correspondence between the syntactic places occupied in that sentence by certain words and the objects denoted by those words. Consequently, the basic content of the sentence 'R *l* T' = the set of ordered pairs: (1,1) — Rome, (1,0) — the relation of lying on, (1,2) — the Tiber, or in an abbreviated form

[(1,1) — R, (1,0) — *l*, (1,2) — T].

It follows from the above that instead of

(t) "a literal translation of the sentence 'R *l* T'"

we may say

(c) 'a sentence with the basic content [(1,1) — R, (1,0) — *l*, (1,2) — T]'.

Let it be recalled, for the sake of avoiding any misunderstandings, that the basic content of a sentence is the correspondence between the syntactic places in that sentence and the objects denoted by the simple expressions occupying such syntactic places in that sentence, and not a correspondence between the syntactic places and the words which occupy them. The correspondence between the syntactic places in a sentence and the words which occupy them is sufficient for a univocal description of that sentence. That is why such a correspondence between the syntactic places in a given sentence and the words which occupy them shall be called the basic description of that sentence. The basic description, e.g., of the sentence 'Rome lies on the Tiber' is the following correspondence: [(1,1) — 'Rome', (1,0) — 'lies on', (1,2) — 'the Tiber']. Consequently, the basic description of a sentence should not be confused with the basic content of that sentence. In the former, a correspondence is established between the syntactic places and the words which occupy them, in the latter, between the syntactic places and the objects denoted by the words occupying their respective places.

Let us now draw attention to the following fact which is of essential importance for further analysis. The expression (c) is extensional. In particular, it does not change its extension if its members 'R', '*l*', 'T' are replaced by other but equivalent expressions. Thus, for instance, the class of sentences with the basic content [(1,1) — R, (1,0) — *l*, (1,2) — T] is identical with the class of sentences with the basic content [(1,1) — the capital of the Popes, (1,0) — *l*, (1,2) — T]. The former includes sentences which in the syntactic place (1,1) have some simple expression denoting Rome, and the latter, sentences which in (1,1) have some simple expression denoting the capital of the Popes. But since Rome is the same as the capital of the Popes, this means that every and only such expression which denotes Rome denotes *eo ipso* the capital of the Popes.

7. What has been stated above will now be used to expand the above interpretation of the sentence

(Z_1^t) 'Caesar believed that R *l* T'

as

(Z_1^t) "Caesar accepted a translation of the sentence 'R *l* T'".

Attention has been drawn to the fact that when we speak about translations we may mean translations of different degrees of precision. It seems that the interpretation (Z_1^t) can be correct only if we mean a literal translation. So we reformulate our interpretation as

(Z_1^{tlit}) "Caesar accepted a literal translation of the sentence 'R *l* T'".

But we have stated above that a literal translation of the sentence 'R *l* T' is the same as a sentence with the basic content [(1,1) — R, (1,0) — *l*, (1,2) — T], and this leads to the following version of our interpretation:

(Z_1^c) "Caesar accepted a sentence with the basic content [(1,1) — R, (1,0) — *l*, (1,2) — T]".

There is an essential difference between (Z_1^{tlit}) and (Z_1^c). To realize it let us state first that the expression "'Rome lies on the Tiber'", built of the quotation marks "' '" and the sentence contained therein, was used as an abbreviation of the following expression:

— "the English language sentence in which the place (1,1) is occupied by the word 'Rome', the place (1,0), by the word 'lies on' and the place (1,2), by the word 'the Tiber'." —

Making use of the term 'the basic description of a sentence', introduced above, we may say that the expression

'Rome lies on the Tiber'

was treated as an abbreviation of the expression

— "the English language sentence with the basic description (1,1) — 'Rome', (1,0) — 'lies on', (1,2) — 'the Tiber'." —

Consequently, the interpretation (Z_1^{tlit}), after expanding the abbreviation contained there in the quotation marks, may be written as follows:

(Z_1^{tlit}) "Caesar accepted a translation of the English language sentence with the basic description: (1,1) — 'Rome', (1,0) — 'lies on', (1,2) — 'the Tiber',"

(where the quotation marks enclosing the inscriptions 'Rome', etc., are to be understood as ostensive names of those inscriptions).

The interpretation (Z_1^c) has the form:

(Z_1^c) "Caesar accepted a sentence with the basic content: [(1,1) — Rome, (1,0) — lies on, (1,2) — the Tiber]."

In order to realize clearly the difference between the formulations ($Z_1^{t_{lit}}$) and (Z_1^c) we must briefly explain the quoting expressions which appear in ($Z_1^{t_{lit}}$) and which consist of the quotation marks and the words contained therein.

Quotation marks may play various roles. One of them was explained above. The quotation marks which occur in the formulation ($Z_1^{t_{lit}}$) play a quite different role: here they cannot be treated as an operator which together with the word which it contains as its argument forms the name of the name of the object denoted by that word. In other words, we may not assume that for instance

'Rome' = the name of Rome.

For the extension of the term on the right-hand side of the equation includes such expressions as '*urbs aeterna*', 'the capital of Italy', etc., since both the first and the second expression quoted above is a certain name of Rome. Yet these expressions do not belong to the extension of the left-hand side of the equation since neither of them is the word 'Rome'. The quoting expressions must here be treated like ostensive names which consist of an indicating gesture and the object which that gesture indicates. The quotation marks play here the role of the indicating gesture, and the inscription which they contain is the object which they indicate. And as the ostensive name, which consists of the indicating gesture and the object indicated by it, is one simple word, and not an expression composed of an operator and an argument, so is the quoting expression not an expression with a compound syntax, but one simple word of which the word contained in the quotation marks is a physical part, but not a syntactic member. The inscription inside the quotation marks, which on other occasions may be used as a separate word, is part of the entire quoting expression in the same way as 'lack', which also may be used as a separate word, is part of the word 'lackey'. But as in the word 'lackey' the inscription 'lack' does not occur as a separate word, so the inscription which occurs in a quoting expression does not function as a separate word but only as physical part of one simple word.

If the quoting expressions which occur in the formulation ($Z_1^{t_{lit}}$)

must be understood in this way, then we must conclude that that formulation does not include the word 'Rome' as its member. Should we then like to transform (Z_1^{tut}), by replacing some of its members with other expressions, into:

(Z_2^{tut}) "Caesar believed in a translation of the English language sentence with the basic description: (1,1) — 'the capital of the Popes', (1,0) — 'lies on', (1,2) — 'the Tiber'", —

we might do this only by replacing the quoting expression 'Rome' by the quoting expression 'the capital of the Popes'. But one sees immediately that those quoting expressions are by no means equivalent since one of them is a name of a quite different inscription than is the other. By drawing attention to this fact we ward off the invasion of intensionality, since the word 'Rome' does not occur as a member of a supposedly intensional sentence in the case of the interpretation (Z_1^{tut}).

But we cannot get rid of intensionality in the same way if we adopt the interpretation (Z_1^c), since in that interpretation the word 'Rome' occurs as a member of the sentence and not as part of a quoting expression. And that is the point of difference between the interpretation (Z_1^{tut}) and the interpretation (Z_1^c). In the latter, the words 'Rome', 'lies on' and 'the Tiber' occur as members of the sentence, and that interpretation states thereby a certain relation between Caesar and the objects denoted by those words. Thus the interpretation (Z_1^c) better suits our purpose, which consists in finding an object language interpretation, and not a metalanguage interpretation, of the sentence (Z_1) 'Caesar believed that Rome lies on the Tiber', which is an object language sentence. The interpretation (Z_1^c) is still not quite satisfactory since it retains a certain metalanguage aspect: reference is made to the fact that Caesar accepted a certain sentence. Yet it comes nearer to our objective since it refers to Rome, to the relation of lying on, and to the Tiber, and not to any names of those objects. In the concluding part of our analysis we shall see that this admixture of the metalanguage aspect can easily be eliminated. But before proceeding to that point we shall demonstrate that the interpretation (Z_1^c) is not an intensional sentence.

8. Let us bear in mind that the intensional character of (Z_1) 'Caesar believed that Rome lies on the Tiber' was proved by stating that: (Z_1)

is a true sentence, but if we replace its member 'Rome' by its equivalent expression 'the capital of the Popes' (hereafter abbreviated as 'c (P)'), we obtain the sentence

(Z_2) 'Caesar believed that the capital of the Popes lies on the Tiber', which is false, and that is why the sentence (Z_1) is intensional.

Let us examine whether that interpretation may be repeated when the sentence (Z_1) is replaced by its interpretation (Z_1^c), and the sentence (Z_2), by its interpretation (Z_2^c). Now if in the sentence

(Z_1^c) 'Caesar accepted a sentence with the basic content: (1,1) — R, (1,0) — *l*, (1,2) — T'

the member 'R' is replaced by its equivalent expression 'c (P)', we obtain

($\overline{Z_1^c}$) 'Caesar accepted a sentence with the basic content: (1,1) — c (P), (1,0) — *l*, (1,2) — T'.

Now, as stated above, the class of sentences with the basic content: (1,1) — R, (1,0) — *l*, (1,2) — T, is identical with the class of sentences with the basic content: (1,1) — c (P), (1,0) — *l*, (1,2) — T. Consequently, both the sentence (Z_1^c) and the sentence ($\overline{Z_1^c}$) state about the same class of sentences that Caesar accepted a sentence belonging to that class. Therefore, if (Z_1^c) is true, then ($\overline{Z_1^c}$) is also true. The sentence (Z_1^c) does not change its truth-value, i.e., its denotation, if one of its members is replaced by another, equivalent expression. Consequently, there are no reasons to consider that sentence to be intensional.

Let us further take note of the fact that the sentence ($\overline{Z_1^c}$), obtained from (Z_1^c) which — in conformity with the accepted principles of interpretation — is an explication of the sentence (Z_1), may not be, in conformity with those principles, interpreted as an explication of the sentence (Z_2), because in the sentence

(A) 'the capital of the Popes lies on the Tiber'

the syntactic places occupied by the simple words are: (1,1,0), (1,1,1), (1,0), (1,2). The basic content of that sentence thus requires that a correspondence be established between all those syntactic places and the objects denoted by the words occupying those syntactic places in that sentence. Consequently, the basic content of the sentence (A) is as follows: (1,1,0) — the capital, (1,1,1) — the Popes, (1,0) — lies on, (1,2) — the Tiber. Consequently, in conformity with the principles we have adopted, the sentence (Z_2) is interpreted as

(Z_2^c) 'Caesar accepted a sentence with the basic content: (1,1,0) — c, (1,1,1) — P, (1,0) — *l*, (1,2) — T'.

First of all, (Z_2^c) is false since Caesar never believed in any sentence containing any word which denotes the Popes, because he did not know that notion at all. On the contrary, $(\overline{Z_1^c})$ is true. Secondly, $(\overline{Z_2^c})$ includes as one of its members the expression 'c (P)' ('the capital of the Popes'), whereas (Z_2^c) does not contain any such member at all. This reveals the basic idea of our method of eliminating the intensionality of sentences of the type '*x* believes that...'. This will be still more clearly evident when we proceed to analyse another course of reasoning intended to demonstrate the intensional character of certain sentences.

Let us start from the sentence not analysed so far:

(Z_2) 'Caesar believed that the capital of the Popes lies on the Tiber',

and let us replace in it the expression 'the capital of the Popes' by the equivalent expression 'the capital of the Republic'. We thus obtain the sentence

(Z_3), 'Caesar believed that the capital of the Republic lies on the Tiber'.

Now (Z_2) is false, and the sentence (Z_3), obtained from it by means of substituting one equivalent for another, is true. Thus, in this way (Z_2) changes its truth-value, i.e., its denotation. Consequently, (Z_2) is an intensional sentence.

What about this argumentation, if we apply to the sentences (Z_2) and (Z_3) our principles of interpretation? (Z_2) will then take on the form

(Z_2^c) 'Caesar accepted a sentence with the basic content: (1,1,0) — the capital, (1,1,1) — the Popes, (1,0) — lies on, (1,2) — the Tiber',

and (Z_3^c) will become

(Z_3^c) 'Caesar accepted a sentence with the basic content: (1,1,0) — the capital of, (1,1,1) — the Republic, (1,0) — lies on, (1,2) — the Tiber'.

Now (Z_2^c) does not include as one of its members the expression 'the capital of the Popes'; it only has as its members the simple words 'the capital' and 'the Popes'. Consequently, if we in (Z_2^c) replace the expression 'the capital of the Popes' by the expression 'the capital of the Republic', we do not replace any *member* of (Z_2^c) by its equivalent, but we merely replace a loose string of words in the sentence (Z_2^c), which is not a member of that sentence, by its equivalent. In this way we proceed as if in the sentence

(C) $$5+3\cdot 2 = 11$$

we replaced the expression '5+3' with its equivalent, e.g., '1+7' and thus obtained the sentence

(D) $$1+7\cdot 2 = 11$$

In this way, by means of replacing one equivalent with another, we passed from a true sentence to a false one, but this fact does not in the least prove the intensional character of the sentence in question. For a sentence is intensional if it can change its logical value following the replacement of any of its *members*, and not just of any string of words, by its equivalent.

Thus the basic idea of our method of eliminating intensionality is as follows. The intensional operator, e.g., 'believes that...', is the analyser which breaks the syntactic structure of the subordinate clause into separate simple words, with the consideration of the original syntactic role they played in that subordinate clause. It breaks that structure in the sense that all those expressions which were members of the subordinate clause are not members of the entire compound sentence; it is only the simple words of the subordinate clause, characterized by their syntactic places which they occupied in that clause, that are members of the compound sentence. This can be explained more clearly with the help of our notation of syntactic places.

The usual interpretation of the syntactic structure of the sentence (Z_3), for which the intensional character of that sentence can be proved, is as follows:

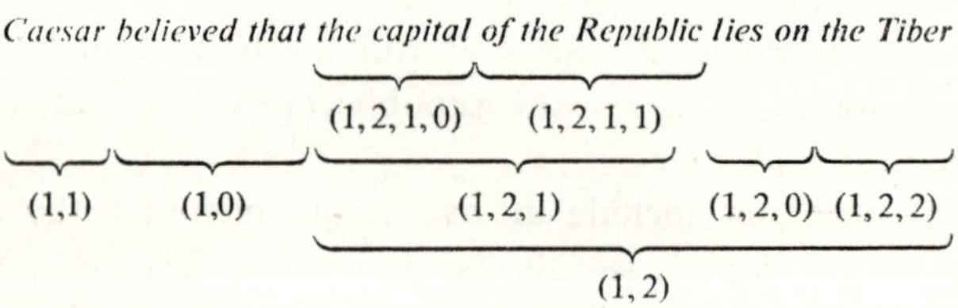

In our interpretation, as in the usual one, the word 'believes that' is the principal operator of the entire sentence (Z_3); similarly, the word 'Caesar' is its first argument according to both interpretations. But whereas in the usual syntactic interpretation of the sentence (Z_3)

the second argument of the principal operator, 'believes that', is the entire subordinate clause 'the capital of the Republic lies on the Tiber', in our interpretation that clause is not the second argument of the operator 'believes that'. The second, third, etc., arguments of that operator are, in our interpretation, the various words constituting the subordinate clause, together with the syntactic places they had when the subordinate clause was still an independent sentence. What are the syntactic places of the various words in the sentence 'the capital of the Republic lies on the Tiber'? They can be indicated by means of our indices for the syntactic places in the following way:

the capital	of the Republic	lies on	the Tiber
(1,1,0)	(1,1,1)	(1,0)	(1,2)

Consequently, the syntactic structure of the compound sentence (Z_3) is in our interpretation as follows:

Caesar	believed that	the capital	of the Republic	lies on	the Tiber
		(1,1,0)	(1,1,1)	(1,0)	(1,2)
(1,1)	(1,0)	(1,2)	(1,3)	(1,4)	(1,5)

In such an interpretation of the syntactic structure of sentences of the type '*x* believes that...' their intensionality, which can be proved in the case of the usual interpretation of their syntactic structure, disappears — as we have shown above. Thus, to get rid of intensional expressions of the type in question, it is enough to pay a very low price, namely that of choosing a different syntactic interpretation of the expressions concerned. We face the alternative: either to treat the syntactic structure of expressions of the type '*x* believes that...' in the usual way, with all the paradoxical consequences of intensionality, or to choose a different interpretation of the syntactic structure of such expressions and to get rid of intensional expressions.

I do not presume to claim that that interpretation of the syntactic structure of the expressions concerned in which intensionality disappears is the correct one. We have here to do with expressions of a natural language in which no syntactic interpretation is the only one 'correct' one. After all, there may be controversy as to whether the sentence

'John likes Peter'

consists of two members: the simple subject 'John' and the compound predicate 'likes Peter', or of three members: the principal operator 'likes' and its two arguments of the same order, 'John' and 'Peter' — and there is no way of settling that dispute in a way that would not admit of any arguments to the contrary. All this refers *a fortiori* to the syntactic structure of such complicated sentences which are usually adduced as examples of intensional sentences. A natural language leaves room for different syntactic interpretations. That is why I abstain from claiming that my interpretation is the correct one, but I confine myself to demonstrating its possibility and the advantages that can be obtained therefrom.

Someone might raise the objection that the problem has not been presented in a sufficiently general manner, since I have shown on two examples only that if we abandon the usual syntactic interpretation of sentences of the type '*x* believes that...', etc., the intensionality which characterizes them in the case of such usual interpretation is eliminated. So let us now try to give a general formulation.

Let the general schema be taken into consideration:

(A_1) '*x* believes that (...)'.

We interpret it syntactically so that the arguments of its principal operator 'believes that' are not '*x*' and the clause (...) which follows that operator, but that these arguments are '*x*' and the simple words of which that clause consists. Thus the sentence (A_1) includes as its members, apart from the expression '*x*', only the simple words which form the subordinate clause (...). The operation of transforming the sentence A_1 into some other sentence A_2 by means of replacing a member of the sentence A_1 by some of its equivalents may consist only in replacing a simple word, but may never consist in replacing an expression which was a member in the subordinate clause but is no longer a member of the entire compound sentence. Consequently, such operations as replacing the expression 'the capital of the Republic' by its equivalent, 'the capital of the Popes' are excluded. The equivalent with which we replace a simple word which is a member of the entire sentence '*x* believed that (...)' may be either a simple word or a complex expression. When we analysed the transition from (Z_1), 'Caesar believed that Rome lies on the Tiber', to (Z_2), 'Caesar believed that the capital of the Popes lies on the Tiber', we discussed the latter case. We have shown that in a different interpretation of the syntactic structure of the sentence (Z_1), expressed by the sentence (Z_1^c), by replacing the word 'Rome' by the expression 'the capital of the Popes', we do not obtain the sentence (Z_2) in its interpretation (Z_2^c), but we obtain the sentence (Z_1^c), which always has the same logical value as the sentence (Z_1^c)

We have not, however, examined the former case when a simple word which is a member of a sentence of the type '*x* believed that (...)' is replaced by another simple word which is its equivalent. This case will now be examined. We take as our example the sentence

(A_1) 'John believed that Dr Jekyll was a gentleman',

and replace in it the simple word 'Dr Jekyll' by another simple word, 'Mr Hyde', which is its equivalent since it denotes the same.

We thus obtain the sentence

(A_2) 'John believed that Mr Hyde was a gentleman'.

In the usual interpretation of the syntactic structure of these sentences (if John did not know that Dr Jekyll was identical with Mr Hyde) the first is true and the second is false. Let us now see whether we eliminate intensionality if we interpret these sentences as was suggested above, i.e., if we interpret (A_1) as

(A_1^c) 'John accepted a sentence with the basic content: (1,1) — Dr Jekyll, (1,0) — was, (1,2) — a gentleman',

and (A_2) as

(A_2^c) 'John accepted a sentence with the basic content: (1,1) — Mr Hyde, (1,0) — was, (1,2) — a gentleman'.

Here, too, it can easily be seen that the class of sentences described as sentences with the basic content: (1,1) — Dr Jekyll, (1,0) — was, (1,2) — a gentleman, is identical with the class of sentences with the basic content: (1,1) — Mr Hyde, (1,0) — was, (1,2) — a gentleman. If it is so, then the sentences (A_1^c) and (A_2^c) state about the same class of sentences that John accepted a sentence belonging to that class, and consequently, if one of them is true, then the second is true as well.

If the sentence (A_1) seems to us to be true, and the sentence (A_2) to be false, this happens probably because we understand them not as sentences stating a certain relation between John and the man whose name was both 'Dr Jekyll' and 'Mr Hyde' and the kind of people called gentlemen, but as sentences stating a certain relation between John and the corresponding words. More strictly, if we think that (A_1) is true and (A_2) is false, it happens so because we do not understand them as the object language sentences (A_1^c) and (A_2^c), but as metasentences.

9. It has been proved, as it seems, in a general way that sentences of the type

(Z_1) '*x* believes that (...)',

interpreted as

(Z^c) '*x* accepts a sentence with the basic content (...)'

cease to be intensional. Yet the interpretation (Z^c) still contains a certain metalanguage residue since it refers to accepting a sentence, whereas the sentence (Z) under consideration seems to be free from any admixture of metalanguage elements. Consequently, we must attempt such a transformation of the interpretation (Z^c) in which no reference would be made to the assertive attitude of a given person toward a certain sentence, i.e., a certain linguistic entity; in such a transformation reference would be made to the assertive attitude of that person toward something taking place in the objective sphere.

Now, we say both 'I accept the fact that the sun shines' and "I accept the sentence 'the sun shines'". In these two cases the words 'I accept' have a different meaning. Let us differentiate between them by writing: 'I accept$_1$ the fact that the sun shines' and "I accept$_2$ the sentence 'the sun shines'". What is the connection between the object of the belief$_2$, i.e., the sentence 'the sun shines', and the object of the belief$_1$, i.e., the fact that the sun shines? Everyday speech has a certain term for naming that connection: in conformity with the colloquial interpretation of words, one may say that the sentence 'the sun shines' states that the sun shines. The term 'states that' has in colloquial speech such a meaning in conformity with which we may say that the sentence *A* states the same as the sentence *B* if, and only if, the sentence *A* is a literal translation of the sentence *B*. But we have seen above that the sentence *A* is a literal translation of the sentence *B* if, and only if, either, generally speaking, the two sentences have the same full content, or, when these sentences include no intensional members, if, and only if, the two sentences have the same basic content. Hence it follows that two sentences state the same thing if, and only if, they have the same full or basic content. This is the result we have obtained by basing ourselves on the theorems which have been dictated by the colloquial meaning of the term 'states that'. We shall continue to be in agreement with the above result if we assume that what is stated by a given sentence is its full or basic content.

Confining ourselves to an analysis of sentences which do not contain any intensional members, we adopt the following definition:

What is stated by the sentence A = the basic content of the sentence A.

Further, we may agree, I think, that between the term 'accepts$_1$' and 'accepts$_2$' there is the following connection which seems to be in conformity with their colloquial interpretation:

(1) X accepts$_1$ what is stated by the sentence $A \equiv X$ accepts$_2$ a literal translation of the sentence A.

But in conformity with the definition just adopted, what is stated by the sentence A is the same as its basic content; and a literal translation of the sentence A is the same as a sentence whose basic content is identical with the basic content of the sentence A. Let, e.g., the basic content of the sentence A have the form $[m_1—x_1, m_2—x_2, \ldots, m_n—x_n]$. In that case

(2) What is stated by the sentence $A = [m_1—x_1, m_2—x_2, \ldots, m_n—x_n]$

(3) A literal translation of the sentence A = a sentence with the basic content $[m_1—x_1, m_2—x_2, \ldots, m_n—x_n]$

By substituting (2) and (3), respectively, for the two sides of the equivalence (1) we obtain:

(4) X accepts$_1$ $[m_1—x_1, m_2—x_2, \ldots, m_n—x_n] \equiv X$ accepts$_2$ a sentence with the basic content $[m_1—x_1, m_2—x_2, \ldots, m_n—x_n]$.

The last equivalence will be applied to the formulation

(Z^c) 'X accepts$_2$ a sentence with the basic content $[m_1—x_1, m_2—x_2, \ldots, m_n—x_n]$' which we did not think satisfactory as an interpretation of the sentence

(Z) 'X accepts$_1$ that $\dfrac{x_1, x_2, \ldots, x_n}{m_1, m_2, \ldots, m_n}$',

because the sentence (Z) is formulated exclusively in the object language, and (Z^c), which refers to accepting a sentence, contains a certain metalanguage element. Now, in conformity with the equivalence (4), (Z^c) will be transformed into

(Z^h) 'X accepts$_1$ that $[m_1—x_1, m_2—x_2, \ldots, m_n—x_n]$'.

This is that interpretation of sentences of the type (Z) which we finally adopt as fully satisfactory. The interpretation (Z^h), like the interpretation (Z^c), is not an intensional sentence, which may be proved by repeating, *mutatis mutandis*, what has been said above to demonstrate that the interpretation (Z^c) is not an intensional sentence.

In this interpretation there is still one disturbing detail, namely that of introduction, in the interpretation we have adopted, of ordered pairs, each consisting of a syntactic place and a word, as members of sentences of the type

'*X* believes that Rome lies on the Tiber',

since in the ordinary interpretation of such a sentence its members are words and expressions, and not ordered pairs consisting of syntactic places and words. Owing to the notation we have adopted for syntactic places, the ordered pairs consisting of those places and of words which occupy them can easily be represented. But what corresponds to such pairs in ordinary languages, which have no special symbolism for syntactic places?

To answer that question let us pay attention to the ambiguity of the term 'word'. For that purpose, let us consider the sentence

$$2 = 2$$

and let us ask the question, of how many words does it consist. According to what we understand by the term 'word', our answers will differ correspondingly. If we answer that this sentence, understood as an inscription localized *hic et nunc*, consists of three words, viz. two *two's* and one symbol of equality, we thereby reveal that by the term 'word' we understand a name the designata of which are definite inscriptions, localized in time and space and separated from other inscriptions by spacing. When interpreting the term 'word' in this way we shall say 'word *in concreto*'. But someone may say that the sentence in question consists of two words only, viz. of *two* which occurs twice, and of the symbol of equality which occurs once. Such an answer will prove that the speaker understands the term 'word' so that its designata are not concrete inscriptions but certain classes of concrete inscriptions, i.e., certain species, which may become concrete entities in the form of the various inscriptions occurring *hic et nunc*. When interpreting the term 'word' in that way we shall say 'word *in specie*'.

This ambiguity of the term 'word' is universally known. But there is still another ambiguity, namely an ambiguity of the term 'word *in specie*'. To demonstrate it let us consider the following Latin sentence

'Petrus amat Petrum'

and ask whether the word *in concreto* 'Petrus' and the word *in concreto* 'Petrum', both of them occurring in that sentence, are concrete occurrences of the same word *in specie* or of two different words *in specie.* The word *in concreto* 'Petrus' differs from the word *in concreto* 'Petrum' by its inflexional ending, which in Latin serves to indicate the syntactic place in a sentence. Consequently, if we answer that the words *in concreto* 'Petrus' and 'Petrum' are concrete occurrences of the same word *in specie*, we state thereby that we understand the term 'word *in specie*' so that its designata (individual word species) may concretely occur in the various syntactic places. If we, however, answer that the word *in concreto* 'Petrus' is a concrete occurrence of a different word *in specie* than is the word *in concreto* 'Petrum', we state thereby that we understand the term 'word *in specie*' so that its designata are such word species which may concretely occur only in certain definite syntactic places. In the former case we shall say 'word *in specie*, not determined as to its syntactic place', in the latter, 'word *in specie*, determined as to its syntactic place'. The individual word species which are designata of the term 'word *in specie*, determined as to its syntactic place' require for its univocal characterization two, so to say, co-ordinates: a definite word *in specie*, not determined as to its syntactic place, and a definite syntactic place. In other words, a given word *in specie*, determined as to its syntactic place, is precisely an ordered pair, consisting of a syntactic place and a word *in specie*, not determined as to its syntactic place.

For instance, if we say that in the sentence '2 = 2' the word '2' occupies the places of the first and of the second argument of that sentence, we understand the term 'word' as 'word *in specie*, not determined as to its syntactic place', because only in the case of such an interpretation of the term 'word' may we say that the wofd '2' occupies in a certain sentence two different syntactic places. In general, when we say that a certain word occupies in a certain sentence a certain syntactic place, we understand the term 'word' as 'word *in specie*, not determined as to its syntactic place', because a word *in specie*, determined as to its syntactic place in a certain sentence *A*, is itself an ordered pari consisting of a certain syntactic place in that sentence and a certain word *in specie*, not determined as to its syntactic place in that sentence, and such a pair is not a member of that sentence and does not occupy any syntactic place in the sentence *A*.

If, however, that sentence *A* becomes a syntactic part of some other sentence *B*, it may happen that the members of the sentence *B*, which occupy in it certain syntactic places, are words *in specie*, determined as to their syntactic places in the sentence *A*. This occurs, for instance, in our opinion, when a given sentence, e.g., the sentence 'Rome lies on the Tiber', becomes part of another sentence, e.g., '*X* believes that Rome lies on the Tiber'. The syntactic structure of the simple sentence (if the place which that sentence occupies itself is marked '2') is

(Z_1) 'Rome lies on the Tiber'
(2,1) (2,0) (2,2)

and the syntactic structure of the compound sentence is

(Z_1^h) '*X* believes that (2,1)-Rome (2,0)-lies on (2,2)-the Tiber'.
(1,1) (1,0) (1,2) (1,3) (1,4)

The above analysis permits us now to answer the question, what corresponds to the ordered pairs, consisting each of a syntactic place and a word which occupies that place, in the ordinary languages which have no special symbolism to mark syntactic place. What corresponds to such a pair is the word *in specie*, not determined as to its syntactic place. For such a word *in specie* is something which for its univocal indication requires both the specification of a syntactic place and a word *in specie*, not determined as to its syntactic place, which word is assigned a syntactic place of its own.

Different languages have different means to indicate the syntactic places of words. In the inflexional languages this function is performed by the ending. E.g., in the Latin sentence

'Roma sita est ad Tiberim'

the fact that the word 'Tiberis' (word *in specie*, not determined as to its syntactic place) plays the role of the second argument of the main operator 'sita est', i.e., the role of object, is indicated by putting that word in the accusative case. That word 'Tiberis', not determined as to its syntactic place, by taking on the ending of the accusative is in that sentence located in the place of the second argument of the main operator. That word 'Tiberis', not determined as to its syntactic place, is a member of that sentence. But in the sentence

'Caesar credidit Romam ad Tiberim sitam esse'

it is not the word 'Tiberis', not determined as to its syntactic place, which is a member of that sentence, but the word 'Tiberim', determined as to its syntactic place and assigned a certain place in that compound sentence.

So much to explain how our interpretation (Z^h), which resorts to symbols of syntactic places, is expressed in those languages which have no such symbolism.

10. Our analysis has been confined to sentences with such psychological terms as "believes that", etc., but I think that the results obtained here are applicable to sentences containing modal terms as well. The intensional character of the sentence

(M_1) 'It is necessary that $\sqrt{81} = 9$'

is usually proved by pointing to the fact that if in the true sentence (M_1) we replace the expression '$\sqrt{81}$' by an equivalent expression 'the number of planets', we obtain a false sentence

(M_2) 'It is necessary that the number of planets = 9'.

Hence it follows that (M_1) is an intensional sentence, but only under the assumption that the expression '$\sqrt{81}$', for which we have substituted its equivalent 'the number of planets', is a member of the sentence (M_1).

Now, this assumption is not satisfied if, like before, we interpret the operator 'it is necessary' as an analyser which breaks up the sentence '$\sqrt{81} = 9$' into separate words, in the sense that not all the members of the sentence which follow the operator 'it is necessary' are the arguments of that operator, but that the arguments are only simple words together with their syntactic places. In such a case the expression '$\sqrt{81}$', which is a member of the simple sentence '$\sqrt{81} = 9$', is not a member of the compound sentence 'It is necessary that $\sqrt{81} = 9$'.

But to take such a stand and to interpret the modal operator as an analyser we would have to start from such an interpretation of the modal sentences which would be in agreement with our intuition, and to demonstrate, as has been done with the sentences containing psychological terms, that in the case of such an interpretation the modal operator becomes an analyser. This, however, will be demonstrated in another paper.

CONCERNING THE SO-CALLED EMPTY NAMES

by

Izydora Dąmbska

The traditional dichotomous division of names into general and singular has been replaced in contemporary logic by a trichotomous division: general, singular, and empty. As Professor Kotarbiński says in *Gnosiology*, "There are singular terms, which denote one and only one object, there are general terms, which denote more than one object, and there are empty terms, which denote no objects at all".[1] We usually find cited as examples of empty names such self-contradictory names as 'square circle' or 'son of a childless mother', or names of mythical deities — fictitious figures that exist only in legends, poems, novels, etc. The singling out of empty names is important in the structure of logical calculus, and especially that part of it which the handbooks of logic term traditional logic. This logic, granted the traditional interpretation of propositions in the square of opposition, proves *in toto* valid only with regard to those propositions which do not include empty names.

Despite this important consideration, the method of singling out the empty names as used in contemporary logic gives rise to certain doubts which, if they prove justified, should induce us to make certain corrections in this section of semantics.

These doubts are as follows:

(1) The basic semantic function of names consists in denoting. By introducing the concept of an empty name as one which does not denote anything, we either arbitrarily change the meaning of the term 'name' or become involved in a contradiction. "(...) to denote a given object in a given language is, as it were, to supply a term for that object in that language (...)." Denoting "is a property of terms", says Professor Kotarbiński,[2] who in his previously quoted statement defined an empty name as one which does not denote anything. He cannot be accused of being inconsistent as he made the reservation that to be a name

of an object is the same as to denote that object, and an empty name is not a name of any object; nor did he quantify the statement that denoting is a property of names: he did not say that it is a property of every name. Nonetheless, for the everyday interpretation of the term 'name', a name which does not denote anything is practically a non-name.

(2) Hence, in order to avoid this contradiction, we have to modify that definition of a name which is in agreement with everyday usage. This has been done by Professor Kotarbiński, who replaces a semantic definition by a syntactic one: "to be a term is to be usable as a predicative word in any sentence '*A* is *B*' with the primary understanding of the copula 'is'."[3] Now an empty name is usable as a predicative word in such statements, and hence is a name, but there is no *true* statement in which an empty name might occur as a predicative word. No one is a werewolf, and nothing is a square circle. It seems to me, however, that statements of the type '*A* is *B*' must be false only if we substitute an empty name for *B*, and a singular non-empty name for *A*, for instance, 'John is king of the French Republic'. If, however, we substitute empty names for both *A* and *B*, or an empty name for *A* and a non-empty one for *B*, we obtain statements which may be true or false. The statement 'Erato is a Muse' is true; 'Erato is one of the Fates' is false; 'Zeus was a deity in Greek mythology' is true; and 'Zeus was a deity in Chinese mythology' is false. The latter case (a statement in which an empty name occurs only as subject) is not an argument against the definition of empty names, but the former is. For to questions about who Erato and Atropos were, I can give the true answer: Erato was a muse, and Atropos was one of the Fates, even though supposedly neither the Muses not the Fates exist. Something is wrong here. Am I to say that Erato is nothing and Atropos is nothing since they do not exist in the same sense in which chairs and horses exist, for allegedly only the last-named entities may be said to 'be this or that' when the copula is understood in its primary sense. But why only these entities? Once we pose this question we become involved in a whirlpool of metaphysics. And here is the third difficulty.

(3) The extension of the term 'empty name' varies according to one's *Weltanschauung*. For a pious ancient Greek the word 'Zeus' was not an empty name, but for an 18th century atheist the word 'God' was. Hegelians and dialectical materialists hold that contradictory objects

do exist, and hence for them the names of such objects are not empty. And what are we to do with the singular names of past and future objects? They, too, do not denote anything since their respective objects either have ceased to exist or have not yet begun to exist. Someone might say that this is not an issue at all, since logicians are not interested in the extensions of empty names; they merely claim that the names of non-existent objects do not denote anything. And you honest people have to worry about deciding — under the guidance of philosophers — — which names are empty. But you must not run the risk of claiming that there are no empty names, for then you will have to accept the negation of that statement: for if there are no empty names, then at least the name 'empty name' is empty since it does not denote any object. This antinomial issue can easily be dealt with: it is sufficient to make the reservation that we are concerned only with names taken in their formal supposition, and that we are only interested in such a use of names. On the other hand, those who claim that empty names do exist must concede that by introducing them they are trafficking in obscure terms. For we are not in a position to decide whether every single name is empty or not. Moreover, this impossibility is not merely a point of fact, arising from a temporary imperfection in our methods of cognition. Indeed, the undecidability of this issue is as fundamental as that of many problems of metaphysics.

(4) Finally, when putting the class of empty names on the same level as the class of general names and that of singular names, people overlook the fact that those so-called empty names can themselves be either general or singular. The name 'an inhabitant of Olympus' is general, while the name 'Zeus' is singular, for the same reason for which the name *vir populi Romani* is general, and the name 'Augustus' is singular. In each pair, the first denotes more than one object, and the second, one object only.

These and similar considerations induce one to revise one's views concerning empty names. In my opinion, every name denotes something. If a word does not denote anything, then it is not a name, and if, while not denoting anything, it has the morphological structure of a name, then it is an apparent name. I do not think that empty names are apparent names only. Like other names, they denote the possible objects of our thoughts. And these include both existing and non-existing objects, whether general or singular, timeless, present, past, or future, fictitious, or even self-contradictory.

It is the task of ontology to investigate the formal structure of these possible objects of thought, and that of metaphysics, to decide which of them exist, and what their existence means.

I am far from claiming that whatever can be thought about thereby exists in reality. In this connection, I can make a distinction between existential and non-existential names while admitting that these are obscure terms so long as I do not know what it means to exist in reality. Every denotatum of a name exists as a possible object of thought: some exist only in this way, others also exist otherwise. We could risk the statement that self-contradictory names are non-existential, while those of conscious subjects are existential. But I know that not everyone would agree even with that. Nevertheless I will not refuse denotata to either existential or non-existential names nor will I call them empty, for I want to be consistent with everyday linguistic usage, which prevents me from saying with conviction that any person who talks about gods, works of art, numbers, etc., is talking about nothing. In *Phaedo*, when Crito asks Socrates how he wants to be buried, Socrates, far from identifying himself with his corpse, which will have to be buried, criticizes Crito for using a formulation which is not only unattractive, but also poisons human souls in a bad way.

The claim that empty names do not denote anything is, as I have tried to demonstrate, just such an unattractive formulation which poisons human souls. Of course, like all other names, empty names denote one or more objects each, and in view of this fact we must, I believe, return to the old semantic classification of names into singular and general.

Does it follow from the above that we shall have to recognize as false certain theorems in traditional logic (e.g., the law of subalternation), whose negation can easily be proved in the theory of an apparent variable, unless we eliminate empty names from the range of variability of S and P?

I believe that there is no such danger; on the contrary, we may rid ourselves of many problems and paradoxes.

(1) When constructing a system of traditional logic, if it turns out that there are no empty names at all, we shall not have to introduce an additional axiom regarding the non-existence of empty names or the

elimination of empty names from logic. But, obviously, in logic we are in fact concerned not with empty names, but with non-existent objects, with empty classes, and with the fact that the laws of this logic do not apply to them. This, however, may be conceded — provided that there are no other objections — even by persons who claim that there are no names which denote nothing. Such persons will merely restrict the range of variability of S and P to existential names.

(2) It can easily be demonstrated that with a proper modification of the definition of singular propositions as it might be the propositions about the relations between the extensions of S and P, respectively, and not about existence, we can preserve the validity of the laws of traditional logic for substitutions for S and P of non-existential names, too.[4]

If this is done, then the paradoxical statements that between two names simultaneously there may exist a relationship of exchangeability and exclusion or of subalternation and exclusion, vanish from the theory of relations between extensions.[5] But, as I have already mentioned, this problem requires further analysis which would, I believe, be independent of the method whereby we are trying to resolve the issue formulated here, i.e., whether it is true that empty names do not denote anything.

Notes

[1] Quoted from the English language version of Kotarbiński's handbook (first published in 1929), *Gnosiology*, Pergamon Press 1966, p. 7.

[2] Loc. cit.

[3] Loc. cit.

[4] This is also done in contemporary logic, where propositions from the square of opposition are interpreted as conditionals.

[5] Cf. K. Ajdukiewicz, *Logiczne podstawy nauczania* (The Logical Foundations of Teaching), Lwów 1934, p. 24.

ISSUES IN THE PHILOSOPHY OF PROPER NAMES

by

Izydora Dąmbska

(Fragments)

1. *Introduction*

Among words commonly classed as names, proper names form a certain natural class which is interesting because of its semantic, psychological, and cognitive role. Proper names are *par excellence* expressions used in everyday language, and — in the field of scholarship — expressions used in the language of historiography and other humanities. This is due to the basic function of proper names, which consists in naming individual objects that have a certain personality or else a quasi-personality or pseudo-personality. As such, proper names are accordingly most closely connected with man and his conscious activity, for they are either names of human individuals ('Peter', 'John', etc.), whether real or fictitious (such as names of literary characters), or names conventionally given by human beings to other objects in order to single out their individual character and to bring out their historical continuity. For instance, we give proper names to domestic or domesticated animals with which we have personal contacts: dogs, cats, hedgehogs, etc. We give such names to old trees which are objects of worship (such as the 'Devaitis' oak in medieval Lithuania), to inanimate objects which are of special value to us and have a history of their own (cf. the names of knights' swords, such as Roland's 'Durandal', or precious stones, such as 'Koh-i-noor', etc.). Proper names are also given to topographical objects, such as towns, rivers, mountains, etc., and in such cases serve to denote certain individual objects connected with man and his past.

We can speak about proper names in a strict and a broader sense of the term. In their strict sense, proper names are used to name *persons*. In the broader sense, they include pseudo-personal names, i.e., names of individual objects that are of special value to human beings and have historical continuity. In this paper we shall be concerned exclusively with proper names in the strict sense of the term, that is, personal names.

Personal proper names can occur in various functions, both proper and improper. A personal proper name is used in its improper function when it serves as a general name or when it is used metaphorically as a substitute for a general name. For instance, a proper name serves as a general name when used in the formulation: 'every Sophie has her name-day on the 15th of May'; whereas in the case of the proverb 'what Johnnie has not learned, John will not know' we are confronted with a metaphorical use of proper names. In the former case, the improper use of the proper name consists in the fact that the formulation in which it occurs is a general statement whose actual meaning is: 'every person whose first name is Sophie has her name-day on the 15th of May', where the word 'Sophie' serves to denote a class of human beings singled out in a somewhat artificial manner, since that class is not based on any set of common traits shared by all its members, but merely on the conventional property of having the same first name. In the latter case we have to do with a rhetorical figure, in which proper names, used metaphorically, are substitutes for the names 'boy', 'young man' (i.e., 'Johnny'), and 'adult' (i.e., 'John'), respectively. Proper names are often used in such substitutive roles to denote certain classes of human beings. Thus, during the World War it was common in various languages to use proper names to denote the soldiers of a given country: for instance, 'Tommy' was used to denote British soldiers; 'Hans,' German soldiers; etc. It may also happen that a proper name serves to single out a class of persons because it symbolizes the specific characteristics stereotypically represented by the person who originally bore that name. Thus, the proper name 'Zoilos' has become a synonym for the class term 'sarcastic person', and 'Cato', a synonym for 'person of firm principles'. Likewise, but in the reverse way, general names can, when used in their improper function, become proper names and in the course of time assume such a character permanently. For instance, in Greek, names of various properties or other abstract entities have come to be used as proper names. Thus, 'Sophia' is not only 'wisdom', but also a feminine name, and the same applies to 'Eirene', 'Euphrozyne', etc. In most cases the original *common* meaning of a name is lost in the consciousness of those who use it, unless a person deliberately and, as it were, symbolically calls his newborn child 'Joy' or 'Hope'.

In their improper, substitutive functions proper names have the sense of general names, and as such they denote certain sets of objects. In their proper function, they name certain individuals. Denoting and naming are two different semantic functions that can be performed by words. It is true that both functions consist in the designation or indication of certain objects, but this indication is not the same in the two cases. In the first place, words name individuals, but denote either classes of objects or individuals interpreted as elements of certain classes, and therefore, in the second place, denoting is *bound* in the sense that a name which is assigned to a class (or, in the case of ambiguity, to certain classes) of objects cannot be assigned to another class of objects without a change in the content of the language to which it belongs.[1] On the contrary, naming is *arbitrary* in the sense that one and the same name may be used to designate ever new individuals without any change in the content of the language. In the third place, naming and denoting differ from one another in both their congitive nature and cognitive value. What is the cognitive nature of naming? What is the meaning of proper names when they occur in their proper function? And what is their semantic, syntactic, and psychological role? These are questions which must still be answered.

2. *The syntactic and semantic roles of proper names*

When analysed out of context proper names have no definite meaning. We might suppose that they are essentially ambiguous, since a name like 'Paul' can denote various persons, and hence different objects, and its meaning varies according to the object which it denotes – and hence it is an ambiguous name. Yet we do not seem justified in talking about ambiguity in cases in which the meaning of a word is indefinite merely because it is indefinite, for then we should have to class as ambiguous all variables, all general names which occur as subjects in non-quantified statements, etc. Therefore, while retaining the term 'ambiguous name' to denote names which have various but definite and enumerable meanings, we have to acquire a better understanding of the aforesaid indefiniteness of meaning proper names used outside a context. This is not the same kind of indefiniteness as the indefiniteness of meaning of a general name in a non-quantified statement. A general name which is out of context has a meaning (or meanings, if it is ambiguous) which is (are) essentially definite. For even if we are dealing with vague names, such as 'old man', 'adolescent', etc., we can decide at least with regard to certain objects whether or not those names are applicable to them. On the other hand, a proper name out of context is essentially indefinite as to its meaning. It is only linguistic or situational context (such as, respectively, a sentence or an indication of a person bearing a particular name) that can impart a definite sense to a proper name. In this respect proper names resemble token-reflexive words, such as the personal pronouns ('I', 'you', etc.), whose meanings are in each case conditioned by the situational context in which a given word is used. There is, however, a difference between a proper name and a token-reflexive word, for example a personal pronoun. The personal pronoun 'I' performs a different function whenever the person who uses it with reference to himself changes, whereas for proper names the range of situational contexts is restricted to persons bearing a given proper name. It appears that a proper name, when considered apart from any verbal or situational context, is a variable with a limited range of values, its range of values being the singular names of persons.

The expression 'Peter is British', like the expression '*X* is divisible by 2', is neither true nor false even though the range of values of '*X*' is larger than that of 'Peter'. One might think that an expression which has as its subject a proper name out of context is not a sentential function, but an ill-formed sentence. If we bind its subject with the universal quantifier, we obtain 'Every Peter is British', which is a false statement, and if we bind it with the existential quantifier, we obtain 'A certain Peter is British', which is true. This imputation does not appear to be justified. An expression which has as its subject a proper name out of context is singular in character and as such may be completed with neither a universal nor an existential quantifier. In sentences of the type 'Every Peter is British' Peter is a general name, and hence it occurs in them in its improper function, which we have mentioned above. To be sure, there are expressions which have as their subject a proper name out of context, and which are nevertheless sentences. For instance, the expression 'Peter is a person' is a sentence if we assume that proper names designate persons only. But the same may be said about the expression '*X* is a number' if we assume that *X* is a variable whose range of values is restricted to numbers. In both cases we are dealing with analytical statements, and the variables which occur in them are apparent variables. Having made the foregoing comments, we can define a proper name as a component of a value which satisfies a categorical singular propositional function in which the place of the subject is filled by a variable with a restricted range of values and equiform with a given proper name. If a word is to be a proper name it must be set in a verbal or situational context that determines its use. It is this context which is the value of the variable (singular name), and this is why we say that it is a component of that value, and not the value itself, which is equiform with the variable. A somewhat more precise formulation of our definition would be:

For every *i*, *w*, *z*, *s*, *f*: *i* is a proper name if and only if *i* is a component *s* of the value *w* of the variable *z* and if there is a singular function *f* (*a*) such that it has *z* as its subject, and *i* is equiform with *z*.

Do not the definitions formulated above conceal a vicious circle? A proper name has been characterized as a component of the value of a given variable, and this variable has been defined by reference to its value. It appears, however, that we do not have a vicious circle here. We have distinguished between two forms of proper names: (a) a proper name used outside a situational or verbal context indicating the person called by that name; and (b) a proper name used in a situational or verbal context indicating the person called by that name. In the first instance a proper name is a variable with a restricted range of values; such a variable may be the subject of a singular propositional function, and its range of values consists of singular names. Those singular names may be descriptions as well as proper names in the sense of (b), i.e., they may be set in a situational or verbal context indicating the person called by that name. For instance, the function 'Richard was a king' is satisfied both by the sentence 'Richard the Lion-hearted was a king' and by the sentence 'the brother of John Lackland was a king'.

In the second sentence, the description which functions as the subject is a value of the variable 'Richard', but it is not a proper name as it is not equiform with that variable. On the other hand, in the first sentence the word 'Richard' is a component of the value of the variable and is equiform with that variable, and hence it is a proper name. Thus a proper name used in context has been defined by reference to the concept of a proper name out of context, but not vice versa. There is accordingly no vicious circle here. But there is another difficulty, at least on the surface: can a proper name in context and a description that indicates the same object occur within the range of values of one and the same variable in view of the fact that there are several semantic differences between a proper name and a definite description? The differences are as follows. (1) Like a general name, a definite description has a specified meaning out of context, whereas a proper name indicates an object only in a situational or verbal context. (2) A definite description denotes an object, whereas a proper name names it. (3) A description can occur as the predicative word in sentences of the type 'A is B', in which the subject is a proper name, whereas a proper name cannot occur as the predicative word in sentences of this type. When it occurs in such a context it is an abbreviation of the expression 'a person whose name is such and such'.[2] The differences mentioned so far do not prevent a proper name and a definite description from occurring within the range of values of one and the same variable if that variable occurs as the subject of a propositional function. If, however, as Russell believes, definite descriptions are names of classes (of one element each), while proper names are names of individuals,[3] then the question arises whether being of different logical types, they can satisfy the same function. Now it seems that a definite description which occurs as a subject is the name of an individual, and that it differs from a proper name in that it singles out that individual from among the elements of the class to which that individual belongs, and thus indirectly it indicates that class, whereas a proper name is free from such a class-indicating intention. When we say 'the best student in the Law School' or 'the present Mayor of Paris' we denote certain individual persons while simultaneously indicating the class of which they are elements: 'the students of the Law School', 'the mayors of Paris'. It is only when it is used as a predicative word that a description becomes a class name and differs from a proper name in its logical type. In its former function, however, it designates individuals in the same way that a proper name does, and hence it can occur within the range of values of the same variable as a proper name.

The above distinction between proper names in context and out of context and the definitions suggested for them make it possible to explain certain semantic properties of proper names, and above all the fact that a proper name *out of context* is indefinite as to meaning.

This indefiniteness of proper names has not always been duly taken into account. This can be seen, for instance, in the reasoning intended to demonstrate the difference between meaning and denoting. If these two functions did not differ from one another, then two different names which designate one and the same object could be used in a given sentence alternatively without any change in the sense of the context. Yet this is not the case. A singular sentence in which two such names occur, one as the subject and the other as the predicative word, is not a tautology. The word 'Ptolemaios' denotes the same person as 'the author of *Mathematike Syntaxis*', and yet the sentence 'Ptolemaios is the author of *Mathematike Syntaxis*' is not a tautology, whereas the sentence 'Ptolemaios is Ptolemaios' is. This shows that 'Ptolemaios' and 'the author of *Mathematike Syntaxis*' designate one and the same object, but differ in their meanings. This reasoning tacitly assumes that we are talking about a proper name which refers to a specified person — in our case to 'Klaudios Ptolemaios'. Without this assumption the expression 'Ptolemaios is Ptolemaios' is not a tautological sentence, but a propositional function which may become either a true or a false statement of one kind or another, according to the substitutions made. Thus, substitutions of this function include: the true tautological statement 'Klaudios Ptolemaios is Klaudios Ptolemaios', the true synthetic statement 'Klaudios Ptolemaios is the author of *Mathematike Syntaxis*', and the false synthetic statement 'Klaudios Ptolemaios is the commander who defeated Antigonos at Ipsos'.

Indefiniteness as to content of proper names out of context accounts for the fact that they do not on the whole play a cognitive role, nor do they lend themselves to use as scientific terms, or occur as definienda in real definitions. If we use them for practical purposes or in science, then we do so mainly as names of themselves, that is, in their material supposition. If a linguist studies the origin of Latin names or a historian of culture analyses the effect of beliefs or tradition upon the choice of personal names in certain ethnic milieus, or if folk superstitions make people believe certain personal names to be lucky and others unlucky, then in each of these cases it is the names themselves and their properties, and not their designata, which constitute the object of interest and study. This is so because proper names out of context, being variable words, have neither definite meanings nor designata.

The matter is different when it comes to proper names set in a situational or verbal context. It is only then that they perform certain semantic and cognitive functions with regard to the objects which they name. But what then is their meaning? Here we encounter new difficulties if we do not want to confine ourselves to the syntactic definition of the type given above (see p. 135), a definition which does not sufficiently emphasize the essential function of proper names, which is naming. We might try to represent this function, while still keeping to a syntactic definition, by saying:

A word *W*, in a language *L*, is a proper name if and only if *W* can be used as the predicative word in the expression 'my name is *W*'.

The expression in quotation marks is in the first person so as to ward off the objection that names of certain kinds of objects can also be used as predicative words in appellative statements. For example, when a person is learning a foreign language he often asks the name of this or that object in the language in question, and then the word he wants to know may occur as a predicative word in the sentence 'the name of that object is ...'; likewise, a child learns a language through being told that 'the name of this animal is dog, and the name of that one is cat', etc.

The function of naming performed by proper names, is — as we have said above (see p. 133) — different from that of denoting. We say that 'he is a student', but not 'he is John' (this is perhaps more obvious in Polish than in English — Tr.); instead of the latter formulation we say 'his name is John' or 'this is John'. In the formulation 'he is John', if and when it is used, the proper name replaces a description, and the sentence means 'he is the person whose name is John'. Likewise, a sentence which has a proper name as its subject can often be replaced, without any change in its meaning or logical value, by a sentence in which the subject is replaced by the phrase 'he whose name is ...'. The appellative function of proper names — we use this term to emphasize that it differs from the function of denoting, which is often confused with naming — is linked with the important role of proper names as words used to call, order, ask, etc., other persons, or, to put it in general terms, to establish a mental contact with other persons. It is true that this role can be played by other words as well, but it is played above all — *ex officio*, as it were — by proper names.

Proper names in context — we might call them concretized — have, unlike proper names out of context, a full meaning, which can be established by formulating a description of a given individual object, or, more precisely, of the person who bears the name in question. For instance, 'Socrates is the philosopher who was executed in Athens in 399 B.C.' According to Russell, "a proper name is meaningless unless there is an object of which it is the name".[4] The same view is held by Czeżowski.[5] We thus face here a new philosophical problem: what is the relation between a proper name and the object which it names, and what is the ontological structure of objects which are named proper names as their attributes.

(...) What is the relation between the proper name of a historical person and that of a hero of a literary work? Both are given proper names. Marys and Dianas populate not only towns and villages, but poems, comedies, and books of history as well. In order to solve this problem correctly we have to understand the semantic function of the proper names of literary characters. In the definition of the proper name given earlier, we took into account the proper names of persons, and by 'person' we meant a cognizant subject, who has a history of his own, acts consciously, and is at least potentially self-conscious. Literary characters do not in fact satisfy these conditions, and yet they too have proper names applied to them. Is there no inconsistency in this? It appears as if proper names used to denote literary characters share the fortunes of all other names of real objects when they occur in literary narratives about imaginary or fictitious places and beings. In poems we hear the beat of the fictitious hoofs of fictitious horses, which does not prevent the name 'horse' from denoting a class of real quadrupeds. Imaginary horses in poems and novels behave like imaginary Dianas, and this dismisses any objection of inconsistency. Russell would think otherwise, as he sees an essential difference between a proper name and the name of a class. As he says, "a proper name is meaningless unless there is an object of which it is a name", whereas the name of a class is not subject to that restriction. For instance, the name 'people who have heads below shoulder bones' has a meaning, even though such people do not exist.[6] On the contrary, proper names of fictitious persons are not names at all. "We can define him (Socrates) as the philosopher who drank hemlock, but such a definition does not warrant that Socrates exists, and if he does not, then 'Socrates' is not a name." A word '*X*' is a proper name if there exists a true statement which is a substitution for the function 'this is *X*'. When the name of a fictitious person is substituted for '*X*', the function 'this is *X*' (for the proper interpretation of the copula 'is') never turns into a true statement.[7] Russell's opinion that a proper name is meaningfully applicable only to real individuals who are perceivable in a sensory manner, and can be indicated in the same way, results from the arbitrary assumption that only those words which designate objects in the world around us have a meaning and that only such objects exist. Both the methodological and metaphysical components of this statement may arouse doubts. The metaphysical statement is neither substantiated nor self-evident, and the methodological one renders the language of science, literature, and everyday life unnecessarily poorer. We can meaningfully name both individual objects which exist or have existed in time, and individual fictitious objects which are dependent on someone's existence and merely intentional. The name 'Zeus' names the supreme god of Greek mythology, even though he has never existed, and the name 'Solon' likewise names a Greek lawmaker, who did in fact exist. The name 'Solon' cannot now be used to formulate a true statement which would be a substitution of the function 'this is *X*', if that statement is to be taken in its ordinary, indicatory, empirical sense which allows only those objects to be indicated which exist at the time of indication, and hence not those which belong to the past. Russell, as if sensing this objection, admits that the role of such an indication can be played by certain statements. For instance, in our example it would be the statement 'the man about whom we read on the *n*-th page of the encyclopedia is Solon'. Since he accepts such an extended concept of indication, Russell must look for some other, non-syntactic method of distinguishing the names of real individuals from the pseudo-names of fictitious persons. For Russell, something which does not exist as a real individual is not an object at all. Hence any distinction between objects which exist independently of anything else and those which do not is pointless. And yet we are continually dealing with the latter category, too: it includes what are called merely intentional objects,[8] a specific type of which consists of fictitious characters invented by creators of works of art, in particular by authors of literary works. It is not difficult to distinguish them from real persons without depriving them of the right to have proper names used in a specifically modified function. It is more difficult to cope with the ontological structure of real historical personages when these are: (a) considered from a historiographic point of view, and (b) treated in a literary manner in poetry and novels, and (c) treated as literary characters which are representations of supposed real persons.

Let us consider by way of example the following series of designata of proper names:

(1) Prince Joseph Poniatowski (= a real person, 1762–1813),

(2) Prince Joseph Poniatowski in Szymon Askenazy's monograph (= a historical character considered from a historiographic point of view),

(3) Prince Joseph Poniatowski in Stefan Żeromski's novel *Popioły* (The Ashes) (= a historical character considered from a literary point of view),

(4) The trumpeter from Jabłonna in a poem by Artur Oppman (= a literary character with a supposed real model),

(5) Rafał Olbromski in Stefan Żeromski's *Popioły* (= a fictitious literary character).

The persons mentioned above under (1) and (5) undoubtedly belong to structurally different ontological categories. The first exists regardless of whether other conscious persons have any representation of him, whereas the fifth exists only as the counterpart of the author's creative conception and its reception by the reader. The common property of (2), (3), and (4) is that each of them corresponds to a certain individual model of the same type as (1) and the differences between them consist in the approach to that model. Now, (2) might be compared to a realistic portrait of a person X in his ordinary dress, (3), to an impressionistic portrait of X in an attire given him arbitrarily by the painter, and (4), to a symbolic painting designated in the catalogue by the name of the model used by the painter, although the model had not been taken into consideration. In all three cases one may forget about the model or have no knowledge of his existence, and be interested merely in the intentional phantoms and their fortunes. In such a case one forgets that (2) is a character considered from an historiographic (documentary) point of view, and one then experiences the presented history as a fable, legend or poetic narrative. Or else in all three cases one knows about and remembers the existence of the model and tries to compare with it the character depicted by the author and imagined by the reader, which is easiest in case (2), and sometimes proves impossible in case (4). One then treats all three cases as historical accounts, even though such a comparison is required only in case (2). In the last-named case, too, such a comparison merely consists in confronting one phantom with another, namely with that which we are accustomed to identify with the model. For an '*historical personage*' is a person who no longer exists in the empirical world as conceived by Russell and Carnap.[9] Hence the model, of which that "historical personage" is a reflection, is a construction based on documents, recollections, etc. Even when we are dealing with a living person, what is given empirically to a few people at most is a certain fragment of a person's spatio-temporal form of existence, whereas the whole, which is used as the model, is also a conceptual construction. If we treat the descriptions contained in cases (2), (3), and (4) as documentary, we may consider each proper name as a word which primarily names a real person, and hence a consciously acting cognizant subject that has independent existence, and which secondarily names a literary character. It is only in case (5), where a real model is lacking, that the proper name plays only a secondary role: the name, not entered in any register of births, represents a quasi-cognizant subject supposed to act consciously, whose existence is conditioned by the conception of the author of the text. But as to the rest, whether we are talking about real or fictitious persons, a proper name in this purely intentional world of theirs performs the function of naming and representing, ensures their continuity of "existence", and underlines the identity of each of them, not only in their various supposed changes of fortune, but also in the various interpretations on the part of the reader. Like real people, fictitious literary characters are concerned with their respective names and partly identify themselves with them; they change their names as a form of penance or in order to avoid persecution, take on new roles as a result of assuming a new name, act incognito, etc. The glamour, magic, mystery, power and beauty of names are known to, and described by, the poetry of all periods and peoples (...).

NOTES

1 By the content of a language I mean the set of constant and variable expressions assigned to the objects of the process and results of cognition which are expressed in that language.

2 The fact that a proper name often functions as a description of this kind is pointed out by B. Russell in Chap. 10 of his *Introduction to Mathematical Philosophy*, London 1919.

3 Cf. B. Russell, *Human Knowledge*, London 1948, pp. 87ff.

4 Loc. cit.

5 *Logika* (Logic), Warsaw 1949, p. 114.

6 B. Russell, *Human Knowledge*, p. 87.

7 Ibid. p. 93.

8 With regard to the analysis of intentional objects, especially non-independent and heteronomous objects of works of art, see Chap. X in Vol. II of *The Controversy over the Existence of the World* by Roman Ingarden, as well as other works by the same author.

9 Carnap's views on the subject of proper names are discussed in those sections of this paper which are not included in the present book. (Ed.)

TRUTH AND THE CONCEPT OF LANGUAGE

by

Izydora Dąmbska

Treating the words 'true' and 'false' as predicates which refer to statements results in accepting them as relative to the language in which such true or false statements are formulated. In Carnap's opinion, the predicate 'true' is a predicate of two arguments, its arguments being a statement '*p*' and a language *L*, in which *p* is true ('*p*', *L*).[1] But this restriction to language is not the only consequence of the semiotic theory of truth. It seems to me that the *meaning* which is given to the terms 'true' and 'false' depends on the general concept of language which is accepted in a given case. The semantic concept of truth which has been worked out by Tarski[2] for those languages in which that concept can be defined, assumes the concept of language as a system of signs which refers to a certain object model. For it is only in such a correspondence theory of language that we can meaningfully use the classical concept of truth, of which the paraphrase for formalized languages is Tarski's semantic concept of truth.

The situation changes if — as with, for instance, Wittgenstein in his *Philosophische Untersuchungen* — language is interpreted operationally as a biological form of human behaviour. In such an interpretation the meanings of words are not determined by *constant* relations of reference between signs and objects, but by the ever changing way in which those words are used in language. And therefore, language is part of the human world, a dynamic system of ever changing "language games" whose rules are formed by the human mind according to its needs and the specific forms by which facts are grasped. The variability and the dynamics of both find expression in the wealth and multiplicity of linguistic operations, so that no theory which strives to define language as a system of signs describing the structure of objects and the relations between them, or as a system of symbols endowed with logical meanings, or as a system of symbols assigned to the data provided by cognition, or as a system of constant forms of human linguistic behaviour, grasps the real nature of language. We can only describe the given language games and thereby bring out the operational and varying nature of the meanings of words in order to facilitate the comprehension of the varied functions of language. Cognitive operations are those which play an important role among linguistic operations. In language interpreted in this way, their success and failure are determinants of truth and falsehood, respectively. In that case, the meanings of the predicates 'true' and 'false' correspond to the pragmatic concept of the operational effectiveness of statements as these occur in cognitive language games of a certain type. The semantic concept of truth, which is the counterpart of the classical correspondence theory of truth, is replaced — in the operational interpretation of language — by a concept of truth that corresponds to pragmatic concepts.

But next to the correspondence theory and the operational theory of language there is another, which might be described as the immanence theory. It is based on a certain methodological requirement which might be formulated as follows: in attempting to define language we should not adopt any transcendent approach either from the standpoint of the model which language is supposed to depict, or from the standpoint of the person who performs linguistic operations in order to express certain meanings or to influence others. In such an interpretation, language is an autonomous system which can be analysed — as has been suggested, e.g., by Ajdukiewicz[3] — by analogy to deductive systems as a set of signs and a set of rules for connecting and transforming signs. The rules include those of sense: axiomatic, deductive, and empirical. In the formalized languages, all the rules of inference can be transformed into certain rules of sense. If the deductive rules of sense assign to each statement of a given structure or form as a premise (or to a class of statements as premises) a statement of some other specified form as the conclusion; and if the axiomatic rules, which in a sense may also be treated as deductive ones, and the empirical rules single out those classes of statements which under specified circumstances must be accepted under the penalty of a modification of language (should such statements be not accepted), then in the light of such a concept of language the truth of statements can be defined as their deducibility, on the strength of certain rules, from other statements which have been singled out by other rules, whether axiomatic or empirical. In this case, the predicates 'true' and 'false' are related to the rules which single out certain classes of basic statements and to the rules of transformation of such statements. This immanent approach in the concept of language thus results in the coherence theory of truth in its syntactic version. The question arises, however, whether this syntactic concept of truth — if we agree to apply this term to the semiotic analogue of the traditional coherence theory of truth — does not assume the semantic point of view to which we should have to turn in selecting the empirical rules. Formally, this is not necessary. The empirical rules of a language L determine which data make us accept or reject certain statements in L if the meanings of words in that language are not to be modified. For we may assume (this idea is also found in Ajdukiewicz) that the empirical rules are *sui generis* axiomatic rules, and we may consider that the choice of a class of statements which we accept as the class of basic statements on the strength of the empirical rules is determined by a linguistic convention. It is only such an analysis of language, taking its pragmatic functions into consideration, i.e., an analysis of language interpreted as a system of signs used in human communication and in acquiring, making objective, and perpetuating the results of cognitive operations, which makes us consider the semantic point of view as logically antecedent.

But the semantic definition — when it comes to the concept of truth — seems to be derivative, too. For if we were to relate the concepts of truth and falsehood to language *only*, and only in the sense that a choice between the classical (semantic), coherence (syntactic), and pragmatic (operational) theories of truth depends on the adoption of this or that concept of language, then we should have to give up solving the very problem of truth. This is because the construction of the concept of language is also made in a language, in which — since the concept of language is not yet established, and it is this concept which alone determines the meanings of the predicates 'true' and 'false' — nothing can be predicated about the truth and the falsehood of statements; it is therefore impossible to decide whether that language is such that its statements are true or false in one of the meanings listed previously. But then we come back to the epistemological issue of the definition of truth. I do not think that this can be evaded by assuming, as A. J.

Ayer has suggested, that the predicate 'true' is a pseudo-predicate because of the statement in a language *L* that *p* is equivalent to the statement that *p* is 'true' in *L*. In such an interpretation, the predicate 'true' would not carry any logical meaning and would merely be an expressive word (as a manifestation of one's getting rid of one's doubts), endowed with merely psychological functions.[4] Without denying that among its numerous meanings the word 'true' may have that expressive sense, too, and that, when used in that sense, it can be eliminated when linguistic formulations are considered in their cognitive sense, we must bear in mind that the analysis of many formulations in which the predicate 'true' is predicated about statements shows that the role of that predicate is not confined to its expressive function alone. What we have in mind here is, as has been convincingly shown by Pap,[5] the use of that predicate in the rules of reasoning, truth tables, etc.

If this is the case and if our previous comments are correct, then we have to agree that all semiotic concepts of truth which restrict the predicates 'true' and 'false' to the language in which a given statement is formulated, and which are dependent on the concept of language which we are using, are derivative with respect to a non-semiotic concept of truth (which is logical in the broad sense of the term), according to which the predicates 'true' and 'false' refer to propositions (in the sense of the logical content of statements), and secondarily to the statements themselves (in the sense of expressions formulated in a given language). Any further analysis of these predicates must presumably refer to certain fundamental ontological categories, above all to the concept of existence. But here we encounter the age-old source of scepticism which always leaves open the question: τί ἐστιν ἀλήθεια?

Notes

[1] R. Carnap, *Introduction to Semantics*, 2nd ed., Cambridge 1948, p. 20; see also A. Pap, *Analytische Erkenntnistheorie*, Vienna 1955, p. 57.

[2] A. Tarski, 'The Concept of Truth in Formalized Languages', in: A. Tarski, *Logic, Semantics, Metamathematics*, Oxford 1956, and 'The Semantic Conception of Truth and the Foundation of Semantics', in: *Philosophy and Phenomenological Research*, 1944.

[3] Cf. his 'Sprache und Sinn', *Erkenntnis*, IV, 1934.

[4] A. J. Ayer, *Language, Truth and Logic*, 2nd ed., London 1948, pp. 88 ff.

[5] A. Pap, op. cit., pp. 61ff.

AMBIGUITY AND THE LANGUAGE OF SCIENCE

by

Seweryna Łuszczewska-Romahnowa

I

The issue of what the precision of the language of science and the imprecision of everyday language consist of is resolved in elementary methodology by a simple formulation which states that everyday language includes many unclear and ambiguous words, whereas the language of science does not include such words (or includes a much smaller number of them).

Contrary to this formulation, I believe that any discipline which makes use of everyday language or of a language based on everyday language cannot free itself from all obscurities and ambiguities. I believe, moreover, that to purify the language of science from all such detectable ambiguities and obscurities would be unnecessary with regard to the goals of science and is in fact alien to scientific practice. The more cautious version of the above formulation — namely the one which states that the language of science includes fewer ambiguities and impurities than are contained in everyday language — is perhaps true, but it is so poor in content and so general that its cognitive value is almost nil.

One of the issues which we have in mind when we describe the language of science by reference to the formulation mentioned above may be expressed as follows: the ambiguities in which everyday language abounds make it inefficient as an instrument of both communication and the pursuit of truth. This is so because ambiguities result in misunderstandings, verbal controversies, and errors in analysis carried out by persons working independently of others. In those spheres of human activity where the aim is the solution of problems in a manner which is both without error and universally convincing, i.e. the sciences, controversies arising from misunderstandings and errors in reasoning due to shifts in the meanings of words during analyses are much less numerous than in other spheres of human activity. How is this state of things achieved in science? The simplest answer, which is, or so it would seem, supported by the observation of the practice (common in science) of defining terms and sticking strictly to a given definition once it has been introduced, is as follows: the language of science is purged of all ambiguities.

Now, as I have said, I believe that science neither can do that nor attempts to do that. The greater efficiency of the language of science as compared with everyday language, when language is treated as an instrument of communication and of the pursuit of truth, must therefore depend on something other than the use of strictly unequivocal terms alone.

The opinion stated above — for the time being unsubstantiated — does not pretend to be original. Attention has frequently been drawn to the fact that the meanings of all words used in everyday language are fluid and far from being well-defined, and that these defects must affect the language of all those disciplines which make use of everyday language, even if only to some extent (cf. L. Chwistek, *The Limits of Science*, ..., pp. 11 f).[1] The inevitable ambiguity of scientific terms has also been cited by K. Popper (*The Open Society and Its Enemies*, London, Vol. II, p. 15 f), who also outlines an answer to questions about the nature of the precision of the language of science. According to Popper, "We are always conscious that our terms are a little vague (since we have learned to use them only in practical applications) and we reach precision not by reducing their penumbra of vagueness, but rather by keeping well within it, by carefully phrasing our sentences in such a way that the possible shades of meaning of our terms do not matter. This is how we avoid quarrelling about words.

The view that the precision of science and of scientific language depends upon the precision of its terms is certainly very plausible, but it is none the less, I believe, a mere prejudice."

What has been said above gives rise to a number of doubts and queries. The first objection is that the claims formulated above have not been substantiated: how do we know that in science researchers do not strive to eliminate all detectable ambiguities? Why is it impossible to purge the language of science of all ambiguities? And, assuming that scientific terms are in fact ambiguous, how is it possible to communicate efficiently by using such terms, and what is the meaning of Popper's claim that in science we cautiously formulate our statements in a manner which somehow counterbalances the ambiguity of the terms we use?

The aim of this paper is to outline answers to those questions. Our point of departure in the analyses which will lead to a formulation of these answers, will be to state and analyze a very simple theoretical argument chosen at random. Our examination of this argument with regard to the ambiguities inherent in it will result in the formulation of certain concepts and statements which will facilitate our approach to the set of problems in which we are interested.

II

The argument which we shall choose may be expressed as follows: "The vocabulary of the French language includes the names of shades of colours which have no names in the Polish language." "For instance, the words '*beige*' and '*mauve*' are names of shades of colours, and occur in the vocabulary of the French language; they denote shades of colours which have no names in the Polish language."

"This can be proved by the fact that a certain French-Polish dictionary (which is both comprehensive and reliable and a copy of which we are using here) includes these words as words occurring in the French language which name, in that language, certain shades of colour; at the same time this dictionary does not include any Polish words which might be translations of these French words."

The above example includes numerous ambiguities. Let us consider some of them.

(1) When we refer to the French (Polish) language, we sometimes mean the language of the educated classes, whereas on other occasions we include within this term the various French (Polish) dialects and/or argots. This ambiguity of the phrases 'the French language', 'the Polish language' leads to a corresponding ambiguity in such formulations as 'word of the French language', 'the vocabulary of the French language', etc.

(2) The formulation 'word of the French language', 'word of the

Polish language', 'the word *beige*' (and similar words), 'the vocabulary of the French language', etc., are marked by certain oscillations of meanings which are due to the fact that we may both speak and write French (Polish).

A. Sometimes by words of the French or the Polish language we mean only spoken words, that is, certain types of acoustic products, and in that case we interpret French or Polish inscriptions as strings of signs which merely register the corresponding phonetic "forms" in a manner similar to that in which musical notes register tunes, melodies, etc.

B. Sometimes the word 'word' is interpreted in such a way that alongside acoustic word types we distinguish graphic word types. In such an interpretation of the word 'word', in view of the strict correspondence between speech and writing, the various statements about spoken words have exact analogues in statements about written words. Hence, instead of making such pairs of statements we can interpret certain statements about words as being, as it were, programmatically ambiguous and stating jointly the states of things that correspond to both meanings. For such an interpretation of the word 'word', a dictionary of a given language will be treated as a work which lists simultaneously the totality of spoken words and that of written words.

C. But, in such a context, we may encounter a third interpretation of the word 'word' and of many related terms, whereby the same French, etc. word can be both spoken and written. According to this interpretation there is in French only one word '*beige*', which we use both when we say '*beige*' and when we write '*beige*'. With respect to every word interpreted in this way we can discuss how it is spoken and how it is written, or how it ought to be spoken and written.

(3) Another ambiguity of the formulation 'word of the French (Polish) language' and other related expressions arises from the distinction between so-called correct and incorrect pronunciations of words of a given language. For we may ask whether a person who — as is commonly said — mispronounces words when speaking French (Polish), does, or does not, utter French (Polish) words. Indeed, we sometimes use the formulations mentioned above in a way which yields a negative answer, and sometimes in a way which yields an affirmative one.

With this we shall conclude our list of the ambiguities inherent in the example discussed above, even though this list is far from complete. We now have, however, enough data to shed some light on the issues with which we are concerned here.

III

We shall now comment briefly on the content and structure of the example which we have discussed above, and on the purpose which it is intended to serve.

Our example consists of three statements, but it is undoubtedly something more than a mere juxtaposition of them. These statements are linked with one another in a certain way, and each of them plays a certain role in the whole set. The first statement is the thesis of the argument (here after to be denoted by *T*), the second (hereafter to be denoted by *P*) is the premise from which *T* is intended to result, and the third (to be denoted by *PP*) is meant to substantiate *P*, or rather to make it more probable, and is thus, as it were, the premise of the premise.

The whole — and at this point we come to the purpose of the argument — is to convince people about the truth of the statement *T*. The question arises, however, as to who these people are supposed to be. The answer might appear simple: they are those people whom the speaker is addressing. But if such an argument is part of theoretical considerations in a scholarly paper, etc., then the author is not addressing any specific persons but rather tries to be, as is commonly said, universally convincing. Does this mean that he expects to convince everyone who happens to read his paper and the argument contained therein? Certainly not.

Usually, when we argue to demonstrate the truth of a statement, we assume a general acceptance of certain other statements, which may be formulated explicitly but are often merely implied, and we do not expect anything more of our arguments than they should be convincing to those who accept our assumptions and are convinced that they are true. This also applies to the case discussed above, if we treat it as an argument proposed seriously. For instance, when resorting to such arguments we assume that dictionaries compiled by competent experts can be relied upon, and hence that if a word is listed in such a dictionary as a word of a certain language, then that word *is* in all probability a word of that language; we likewise assume that if a French word listed in a French-Polish dictionary has no one-word Polish equivalent listed there, then a corresponding Polish word is in all probability not to be found in the vocabulary of the Polish language.

Let Z denote the totality of those assumptions which, if we can guess from the content of the argument discussed here, could in all probability be made in connection with it. The purpose of the argument can now be described thus: to demonstrate the truth of the statement T in a manner convincing to all who accept the assumptions Z.

We know from the preceding section that the argument under consideration includes many ambiguous formulations. Let us now see how, if at all, the ambiguities to be found in that section prevent that argument from serving its purpose.

Our argument may be hampered by the ambiguity described under (1). If the French-Polish dictionary to which our argument refers lists only the vocabulary of literary French and literary Polish, then the argument will be convincing to a person who, when referring to the French or the Polish language, means the corresponding literary languages, but not to a person who considers dialects to be parts of these two languages. It may be that the vocabulary of some Polish dialects includes names of those shades of colours which correspond to the French words '*beige*' and '*mauve*', even though such names are not to be found in literary Polish.

Point (1) in Section II above thus indicates such differences in the interpretation of certain formulations included in the argument, which result in differences in the assessment of the usefulness of the argument under consideration in substantiating the claim made.

The ambiguities listed under (2) and (3) do not share this property. I think that if the argument is purged of the ambiguity listed under (1) and if this is done in a manner that proves advantageous to the argument (i.e., if attention is paid to whether the dictionary mentioned in the argument lists only the literary vocabularies of French and Polish, or the dialects of those languages, too), then the argument as a whole will be universally convincing, in the sense mentioned above, in spite of the ambiguity of many words used in its formulation, in particular, in spite of the ambiguities indicated under (2) and (3) in Section II.

Anyone who understands the formulation 'word of the French (Polish) language' in the sense described in Section II under (2)–(A) and correspondingly interprets the expressions 'the vocabulary of the French (Polish) language', the word '*beige*', etc., will consider the whole argument as referring to *spoken* French and to *spoken* Polish. In particular, the French-Polish dictionary referred to in that argument will be for him a register of the vocabulary of spoken French and spoken Polish. On the other hand, anyone who interprets the formulations in question in the manner described under (2)–(B) will treat the first two statements in the argument as referring jointly to spoken and written words, so that the French-Polish dictionary mentioned in the third statement will be for him a register of both acoustic and graphic word forms.

Both, however, will readily agree that *T* follows (unfailingly) from *P*. Both will also agree — assuming that they accept *Z* — that *P* follows from *PP* with a large degree of probability. Finally, both will be ready to accept the truth of *PP* upon examination of a copy of the dictionary referred to in the argument. Both will therefore be convinced, by the argument as a whole, with regard to the truth of *T*.

What has been said here concerning (2)–(A) and (2)–(B) may be said, *mutatis mutandis*, concerning (2)–(B) and (2)–(C) and also concerning (3). In connection with (3) it is strikingly clear that one's assessment of the usefulness of the argument discussed here for the demonstration of the claim advanced by it is not affected by whether one thinks that a person who mispronounces French (Polish) words does not in fact pronounce them or whether he is ready to accept, in some cases, such mispronounced words as French (Polish) words.

An examination of cases like the one presented above leads us to the conclusion that an argumentation which substantiates a statement may include ambiguous words and nevertheless be convincing to all who accept its assumptions.

The same idea can be reformulated if we refer to the concept of a "controversy over words", which Popper cites in his opinion quoted in Section I. A controversy over words is a difference of opinion which is due to the fact that the parties to that controversy differently interpret a given word which they both use. Now in view of what has been said in this section we may say that the ambiguity inherent in a context need not be a potential source of a controversy over words, a controversy which would refer to that context (which obviously does not deny the fact that all ambiguity is a potential source of some controversy over words).

This is possible because such properties of statements or sequences of statements as *a priori* truth, empiricity, the relation of consequence between statements, the relation of exclusion between statements, etc. are attributes of certain statements or sequences of statements in everyday language regardless of the oscillations, admissible in a given language, of meanings of terms that occur in these statements.

For instance, the statement *T* follows from the statement *P*, used in the argument above, regardless of the wide spectrum of meanings which may be represented, in everyday language, by the formulations 'word of the French language', 'the vocabulary of the French language', and other related expressions that occur in those statements. Regardless of all those ambiguities, the transition from *P* to *T* in everyday language is incontestable. Likewise, the statement 'if *P* then *T*' is *a priori* true in everyday language despite the ambiguities of the words that occur in it. Following Ajdukiewicz's well-known analysis of the concept of *a priori* truth, I believe that the rejection of this statement would be a violation of the rules for using, in everyday Polish, the words which occur in the statement. We cannot reject this statement without simultaneously ceasing to speak Polish.

One of the questions posed in Section I was: how is it possible to communicate efficiently by means of ambiguous words? I think that we can now answer this question. When formulating the answer I shall merely repeat one of my previous statements in a slightly different form: an argument which includes ambiguous words may admit, in the language in which it is formulated, only one decision as to the truth of the claims made in it and as to the value of the substantiations it advances.

In the light of what has been said in this section we can now understand Popper's opinion that precision in science is attained "by a cautious formulation of our statements so that the possible shades of meanings of the terms we use should not play any role." For this suggests in what this caution consists: more likely than not, in the use of ambiguous words only in such contexts in which "the possible shades of meanings of the terms we use" do not result in different decisions as to the truth of the statements made and as to the value of the substantiations they advance.

IV

An examination of the case discussed above with regard to the ambiguities it contains suggests further reflections in connection with the problem of "ambiguity and the language of science."

The case discussed here reveals, on the one hand, the wealth and ramifications of the ambiguity of words used in everyday language, while on the other hand it illustrates the claim that an argumentation which substantiates a statement may be burdened with numerous ambiguities and nevertheless be universally convincing.

The first point shows that it would be extremely difficult, to say the least, to purge the language we use of all ambiguities. One might suppose this to be an impossible task. A systematic analysis of the correctness of this supposition is a vast problem which I shall not take up here. We need not take it up since it is clear from the second point that it is unnecessary — considering the purpose which the language of science is to serve — to purge the language of theoretical discourse of all its ambiguities. This second point we shall now discuss in greater detail.

In science, we use words in order to make statements and to convince others about the truth of statements. For this purpose we also resort to speech outside the sphere of science. We may quote any example we choose: a conversation in which one person informs another about something or attempts to convince him about something; a speech at a public meeting; an article in a newspaper; etc. There is, however, a very important difference between the attitude adopted by a scientist when, in the pursuit of science, he asserts something and convinces others about something, and that adopted by various people in everyday life. The difference becomes obvious when we ask whom we are addressing when making statements and proving something in science, and whom we are addressing in similar situations in everyday life.

In everyday life we usually have a limited number of persons in mind: we want to inform a certain person about something, or we want to convince a certain audience, or we are writing for specific readers. Once a given argumentation succeeds in convincing those for whom it is intended, its role comes to an end. The fact that it fails to convince other persons who have come across it by chance, is quite beside the point. In such cases we express ourselves in a manner which in our opinion is adjusted to what those whom we address know, to what they believe in, to what they would like to believe in, etc

In science it is otherwise. When making public a scientific argument we are essentially addressing everyone: *we want to speak in a manner that is universally convincing.*

This claim (which was formulated and partly explained in the preceding section) may give rise to various doubts. For instance, one may cite the fact that it often happens that when writing a paper a scientist is interested in convincing a group of persons whose decisions may determine his future scholarly career, etc. We may free ourselves of the necessity of explaining this point if we replace the foregoing statement with the following, closely related statement: in science (regardless of the wishes of each given scientist) the *imperative* is to speak in a manner which is universally convincing (as far as that is possible).

A more important issue is that connected with the difficulties which must be overcome when one engages in the pursuit of a given discipline. Scientific argumentations are not, on the whole, easily accessible and hence, one might claim that not only are they not universally convincing, but also that the number of persons whom they may convince is usually extremely small.

This objection makes us formulate our claim in a more explicit manner.

The inaccessibility of scientific works, as far as the average reader is concerned, is due to three factors. First, nearly every scientific argument is based on results obtained and published previously, and hence it presupposes the knowledge and acceptance of these earlier results. Second, scientific and scholarly publications include terms which are either totally alien to everyday language or, if they do occur in everyday language, they are used there in different meanings. Moreover, in scientific publications such terms are usually left unexplained, since it is assumed that the reader knows what they mean. Third, the very course of reasoning in scientific works is often most difficult to follow, and sometimes it is accessible only to those who have been trained in carrying out and comprehending analyses of a given kind.

For all these reasons scientific argumentations are usually accessible only to a small group of specialists and, in a sense, are written for specialists only. But attention must be paid to the fact that points (2) and (3) in our list of the causes of the inaccessibility of scientific publications refer to factors which make it difficult or downright impossible even to *understand* such publications, and hence to *study* them, let alone to become convinced about the truth of what is claimed in them.

All this leads us to formulate our statement as follows. In the pursuit of science, the correct attitude is to attempt to make one's argumentation convincing to all who understand it (and hence know the language in which it is formulated and are in a position to follow its reasoning) and know and accept those statements or theorems which a given argumentation assumes.

Clearly, this positive conclusion yields the following negative one. In science, we have to avoid making statements which can be interpreted in different ways so that, for one interpretation, they may be accepted as true and for another, rejected as false; or for one interpretation, they may be taken to be true *a priori*, and for another, as requiring an empirical substantiation, etc. We also have to avoid producing sequences of statements (intended to serve as substantiations of stated claims) which can be taken to be convincing or unconvincing according to how certain of their elements are interpreted. In other words, in science we have to strive to make our argumentations free from such ambiguities which might, in connection with the problems analysed in them, become sources of controversies over words (in the sense explained in the preceding section).

This, I think, completes our description of the attitude to be adopted in science with respect to the ambiguity of words. I therefore believe that to strive to eliminate all ambiguities is alien to science.

In science, we use words taken directly or indirectly from everyday language, both in posing problems and in solving them. As for solutions, we want them to be uncontroversially convincing, and nothing more. We are concerned with the ambiguity of the words we use only in so far as they hamper the uncontroversial solution of problems.

This occurs very often, as we know, and hence we frequently have to modify the received meanings of the words we use, discuss the possible ways of understanding certain terms, etc.

One of the major tasks of the science of science is to analyse the various disciplines in a comparative manner with regard to the following problems: how have the ambiguities of words encumbered the development of, and advances in, the various disciplines? (It appears that some disciplines have — and have had — fewer troubles in this area than have others.) How do the various disciplines make, or attempt to make, their reasonings and argumentations free from ambiguities which are detrimental to the postulated uncontroversial nature of reasonings and argumentations in science?

NOTE

[1] The English language version of Chwistek's book (unfortunately unavailable to the translator) is reputed not to be an exact rendering of the Polish original, yet it would seem reasonable to refer the English-speaking reader to that version rather than to the original (Tr.).

SIGNIFICATIO 'PER SE' AND 'PER ALIUD' IN ANSELM

by

Maria Ossowska

Volume 158 of Migne's *Patrologie* contains, on pp. 562–82, a brief treatise by Anselm of Canterbury, *Dialogus de grammatico*, of which the title is well known but the content much less so.

This treatise has been devalued by Prantl who, with his usual ardour and unstinting in his use of highly emotional formulations, denies it all value whatever. He looks upon it as a school exercise which attempts to overcome fictitious difficulties artificially accumulated by Anselm himself and touching on certain problems derived from Boetius.

The present writer does not feel competent enough to pass judgements on the historical role of this treatise, its possible sources, and its originality, and is, for the time being, concerned with one particular detail to be found in it, a detail to be considered independently of any historical considerations. This detail refers to the semantic functions of names which are substantival in nature, and of those which are adjectival in nature.

Anselm's brief treatise is written in the form of a dialogue between a disciple and his master. The disciple is a "straight man". He opposes when he can provide his master with an easy opportunity to display his skill in his reply. He agrees when the master has exhausted his supply of arguments and has practically nothing more to say. The opening chapters seem to some extent to substantiate Prantl's severe criticism. The discussion is sterile and unnecessarily contorted and does not inspire the reader to make an effort at interpretation. Beginning with Chapter 12, however, the issues become more interesting. The issue which is to be the focal point of the dialogue is what is the designatum of the word '*grammaticus*'. Is it an object which has a certain property, and hence a substance, or is it that property itself, and hence a *quale*? The conclusion is that the word '*grammaticus*' *per se* signifies a property, and *per aliud*, an object which has that property, an object which is signified *per se* by such a name as '*homo*'.

A lengthier analysis is devoted to the distinction between *significatio per se* and *significatio per aliud*. The words '*homo*' and '*grammaticus*' signify their designata in different ways: the former, substantially; the latter, attributively. '*Homo*' signifies man as a certain whole: *ut unum significat ea ex quibus constat totus homo*; '*grammaticus*' signifies substance through its property (Chap. 12). The disciple, not satisfied with this explanation, presses his master for further comments (*peto ut aperte mihi has duas significationes distinguas*), to which the master replies (in Chap. 14) by quoting an example which in the present writer's opinion is worthy of a moment's reflection.[1]

An issue here is whether the words 'horse' and 'white' signify a white horse in the same, or in a different manner.

"If in a homestead there is a white horse", the master says, "and you do not know that, and if someone tells you only that there is something white in that homestead, will you know that there is a horse in that homestead?

"No", the disciple agrees, "for until I learn of what that whiteness tis an attribute I will not be able to establish the essence of any thing; I will merely know the colour of that thing."

This is the first part of the proof, and is intended to demonstrate that the word 'white' does not signify a white horse in the same manner as does the word 'horse'.

"Let a white horse and a black ox be placed next to each other", the master continues, "and let someone tell me with reference to the horse 'Hit it!', without letting me know to which object he is referring. I will not then know what he means. But if I ask which one he is talking about, and he tells me, 'The white one', I will know to which one he is referring".

This is the second part of the proof, and is intended to demonstrate that while the word 'white' does not signify a horse *per se* it does nevertheless signify it in some way. The disciple now feels satisfied and summarizes the issue thus: *Nomen vero albi equi substantiam significat non per se, sed per aliud, id est, per hoc quod scio equum esse album.* The text goes on to say that we understand the word 'white' to refer to the horse since we know that whiteness is an attribute of that horse, and we know this through the intermediary of something other than the word 'white'.

This is roughly the text on which the present discussion is based. Let us now try to follow the line of reasoning contained therein.

The example quoted above tells us that among all names applicable to a given designatum there are some which indicate it with greater precision than do others. If I want to inform someone that someone else has a white horse, I will inform the first person better if I tell him that the second person has a horse than if I tell him that he has something white. The point here is not that the word 'white' seems to have a greater extension than the word 'horse' has, and that accordingly the former indicates the designatum with less precision than does the latter. The point is rather that the word 'white' signifies objects which have been included in a single class on the strength of a single common property and are otherwise quite heterogeneous. Something white may also refer to a sheet of paper, a field of snow, a jug of milk — in other words things which are generically different. If I name something by means of the word 'horse', I name it as an entirety and describe it more fully. This appears to be Anselm's line of reasoning.

Let us now turn to the conclusion, as formulated by the disciple. In point of fact, this is not so much a conclusion as a new issue entirely. For what could Anselm have meant by making a distinction between names of which the designata are known to us "by the name itself", and names of which the designata become known to us owing to our additional knowledge of those designata? And this is probably how we are meant to understand *per hoc quod scio equum esse album*. Anselm's intention is far from easy to grasp. May I deduce that I may call an object a 'horse' (in a given language) on the basis of certain properties of that object, just as I may call it 'white' on the basis of one of its properties.

Among various plausible interpretations one seems to suggest itself with the greatest force. This reflects certain intuitions which may often be observed in the history of semantics. According to these intuitions, the name 'horse' is somehow more the name of a certain quadruped than is the name 'white' (when applied to the same quadruped). The former is a name as it were in a narrower and more proper sense. When using the name 'horse' (at least when it does not occur in a sentence or as the subject of a sentence) we use it substantially rather than attributively.[2] In such a usage we do not sense the properties on the basis of which that name has been assigned to a particular designatum, and we are inclined to believe that it is, as it were, the proper name of that designatum and, like a proper name, assigned to that designatum regardless of the properties of the latter. The average person with no training in semantics, when asked by a foreigner what that piece of furniture which he is using for writing is called in his language, will consider one answer only to be possible. Only one word will name for him that piece of furniture. It would not occur to him that he could answer by using any of the words which can be predicated about that piece of furniture in a true statement. For him the word 'desk' is the "proper" name of that piece of furniture, which he feels he knows, not on the strength of any properties of that object, but in the same way as he knows that a particular city is called Warsaw. The following therefore represents the essence of Anselm's ideas: the substantial nature of a name which indicates (or: points to) its designatum without referring to any knowledge of the properties of that designatum.

When discussing the aforementioned passage in Anselm's text Prantl pities the reader who has to read such nonsense. The present writer has nevertheless taken the liberty of pausing to consider Anselm's analysis in the belief that not everything which is to be found there is a misunderstanding, and that what is a misunderstanding is a very characteristic misunderstanding and therefore worthy of a brief mention.

Notes

[1] 'Signifies' is used here to render, on an etymological basis, Anselm's word '*significat*', even though on other occasions he uses the word '*appellat*', which seems to come closer to M. Ossowska's notion of denotation. In view of translation problems this footnote both sums up M. Ossowska's explanation, and comments on the translation itself (Tr.).

[2] In her footnote here M. Ossowska refers to a work by S. Szober, a Polish grammarian active between the World Wars (d. 1938). Apparently what is meant here is not the difference between the attributive and predicative use of adjectives, but the fact that nouns are supposed to indicate "substance", whereas adjectives are supposed to indicate "properties" (Tr.).

AN ANALYSIS OF THE CONCEPT OF SIGN

by

Stanisław Ossowski

(Fragments)

> What's in a name? that which we call a rose,
> By any other name would smell as sweet;
> So Romeo would, were he not Romeo call'd,
> Retain that dear perfection which he owes,
> Without that title.
>
> W. Shakespeare, *Romeo and Juliet*, Act II

I. SEMANTIC ENTITIES

1. The present paper is concerned with the concept of sign in a particular interpretation. (...) In order to avoid misunderstanding we shall call signs, in this particular interpretation, 'semantic entities' (...). By 'semantic entities' I mean material objects which have a semantic function: denoting, representing, or meaning. Each of these functions may be an attribute of an object only with respect to someone's intention. If an object is a semantic entity, it is always that for someone (...).

Semantic entities have a peculiar property: the moment we want to predicate something about an object which is a semantic entity we immediately deprive it of its semantic function and adopt another attitude toward that object, i.e., we treat it as an object which is not a semantic entity. Speaking metaphorically we may say that a semantic entity is transparent to such operations as predicating, indicating, and naming, in a manner similar to that in which glass is transparent to sun-rays: through the intermediary of such an object, predication may refer to another object, but we cannot predicate anything about it unless it ceases to perform the function which turns it into a semantic entity.

For example: "'Warsaw' is a disyllabic word".

It would be extremely difficult to formulate a strict definition of semantic entities. We could, I believe, provisionally adopt the following definition, which is, unfortunately, excessively complex:

A material object *P* is a semantic entity for a person *K* if *K* adopts the attitude *S* toward *P*. The attitude *S* (the semantic attitude) is in turn characterized by three conditions: (1) If *K* predicates something about any object *x*, then *K* does not adopt the attitude *S* toward *x*. (2) If *K* predicates anything about any object *x*, then *K* adopts the attitude *S* toward another object *p*. (3) If *K* adopts the attitude *S* toward *p*, then he does so because of a certain relation which he establishes between *p* and some other object[1]. (...)

2. Among the semantic entities we shall single out those whose semantic function consists in their being assigned to certain other objects, namely their respective designata. Two categories of entities come into play here: symbols and icons.

A symbol is a semantic entity which is assigned to its designatum *exclusively* because of someone's intention. The physical characteristics of a symbol are totally irrelevant since no objective relation is a necessary condition of such an act of assigning. The relation that holds between a symbol and its designatum is called *denoting*. This is clearly an asymmetric relation, since otherwise as long as a symbol performed its function nothing could be predicated either about that symbol or its designatum. (...)

A symbol is a replaceable object: the relation that exists between a symbol and its designatum is based solely on an arbitrary convention (...). When we switch, for instance, from Polish to German or French, the word '*człowiek*' is replaced by '*Mensch*' or '*homme*' respectively. The replaceability mentioned above is not to be confused with another issue: *under a given convention* every element of a class of equiform or isophonic objects is a symbol of a given object. (...) Husserl says that a *word* is conceived *in specie.*[2]

3. Icons form the second category of semantic entities. Like a symbol, an *icon* is a material object and semantically assigned to its designatum, i.e., the object it represents. But the relation that exists between an icon and the object it represents—namely the relation of *representing*—has an objective foundation which does not exist in the case of denoting. The relation of representing may be interpreted as the *product of two*

relations, a symmetric and an asymmetric one. The asymmetric relation, analogous to that of denoting, is present as a result of the intention of a conscious being, but the members of that relation are selected as a result of being linked by an objective symmetric relation, which is always one of similarity (...), such as holds between a map and the territory of which it is an image, or beating time with a stick against the edge of a table to imitate a tune (...).

Hence the following consequences: (a) Unlike a symbol, an icon cannot be replaced by *any* other object one chooses. (...) For instance, a photograph of Loch Ness represents that area otherwise than does a map of Loch Ness. (b) If we know an icon, we can know something about its designatum. But a causal relationship is not a necessary condition of the relation of representing, just as it is not a necessary condition of the relation of denoting. In order to explain this point we have to make a distinction between the relation of representing and the relation which exists between a copy and its model.

The relation between a copy and its model is also a product of two relations: the symmetric relation of similarity and a certain asymmetric relation. But the latter is in no way analogous to that of denoting. It is a genetic relation: a copy is an object *made* after a model (...).

On the contrary, the relation of representing is not dependent on the origin of its members. An icon is usually a copy, but this need not always be so (...). And conversely, a copy often becomes an icon, but not always: the sphere of industry abounds in copies which we do not treat as icons. (...) Like a symbol, an icon is "logically transparent". (...)

4. The third category of semantic entities is formed by those objects which we call signs not because of their designata, but because they "mean something" or "have a meaning". Their semantic function does not consist in their being assigned with respect to other objects. We shall discuss this category in another section of this paper.

All semantic entities in the third category are speech signs. Symbols can, but need not, be speech signs.

It does not seem proper to say that, e.g., the word 'man' is a symbol, while the word 'almost' is not. We would like to extend the term 'symbol' to cover the third category of semantic entities, too. Nevertheless I am retaining the extension of the term 'symbol' as determined by the definition above, because I do not have any more suitable term for the first member of the relation of denoting. The term 'name' suggests itself in this connection, but it is used in language systems only, whereas 'symbol' is a broader concept. (...)

If, on the one hand, symbols and icons stand in opposition to the third category of semantic entities, then, on the other hand, symbols and that same third category stand in opposition to icons as semantic entities whose semantic function does not depend on their external form. These two groups of semantic entities are *conventional entities.*

A system of conventional semantic entities which may, according to the constant conventional rules of a given system, be combined into new, complex semantic entities that have independent functions of their own and are not covered by primary conventions, will be termed *language.*

II. Types of Linguistic Entities

5. In the languages of civilized societies we are concerned with double systems of signs: a system of phonetic signs has a system of graphic signs assigned to it in such a way that every phonetic sign has its counterpart in one and only one graphic sign with the same semantic function. If this function consists in denoting, then such a linguistic entity has assigned to it, on the one hand, its designatum, and on the other, a linguistic entity from the other system, an entity which is a symbol of the same designatum. This can be represented by an isosceles triangle whose equal sides stand for the relation of denoting in a given language, and the third side, for the relation of equivalence between the corresponding signs of both systems.

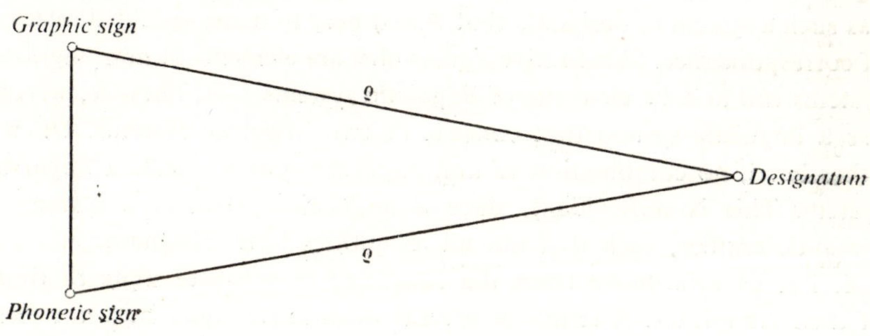

Because of a common designatum, or, more generally, because of an identity of semantic function in a given language, a graphic sign is the semantic equivalent of a phonetic sign. If one of two such corresponding signs loses its normal semantic function in a given language, then the symmetric relation of equivalence may turn into an asymmetric relation of denoting, and the sign which has lost its normal semantic function may become a symbol of its (former — Tr.) analogue in the other system of signs. It is not then a sign in a given language, but it is a sign *with respect* to that language, because of the identity of the semantic functions that are attributes of both objects in that language.

Thus a system of graphic objects may be interpreted in two ways: (a) as a system of speech signs, (b) as a system of symbols to which a system of phonetic objects is assigned so that every graphic object is a symbol of its phonetic counterpart. These two interpretations do not yield systems of the same type. The system of graphic objects which is assigned to the system of phonetic objects is a system of semantic entities, such that each entity has a designatum. On the contrary, the graphic system of speech signs is a system of semantic entities, such that some entities have designata while others have not.

Symbols which are in a system that consists exclusively of semantic entities which have designata will be termed *substitutive signs*. To these we oppose *linguistic entities*, as elements of those systems in which not every semantic entity has a designatum.

Substitutive signs include, for instance, letters as signs of speech sounds. (...) Since perfect correspondence exists between a set of substitutive signs and the set of their designata, operations on substitutive signs can in certain respects replace operations on their designata. Speech signs do not replace their designata because no system of speech signs has such a system of designata that would bear to it any specified relation of correspondence. Substitutive signs which are elements of non-linguistic systems can also be elements of linguistic systems (...); likewise, we can use a linguistic system that consists of two linguistic systems. On the other hand, no combination of non-linguistic systems yields a linguistic system. This is self-evident, since a linguistic system is a system of semantic entities, such that not all its entities have designata.

6. Let us now move from the functions of separate signs to those of their complexes. A string of several substitutive signs does not yield

a new semantic entity that would have a new function, and yet one determined in a certain way by the functions of its components. A complex of substitutive signs is usually merely their sum. Such a complex has as its designatum the analogous complex of the designata of its components. For instance, a string of several letters is a symbol of an analogous string of several (the same number of — Tr.) sounds, each of which is the designatum of a given letter (on the assumption that every letter always stands for the same sound, and also that we disregard the changes which various sounds undergo as a result of being preceded and/or followed by other sounds). If a system of substitutive signs is not, as a semantic entity, the sum of its components, then it is a new sign, but one independent of the functions of its components. These components are physical, but not semantic, elements of the new entity. For instance, in the signal system used in cavalry, raising a sabre once means 'Stop!', and yet raising a sabre three times does not mean 'Stop! Stop! Stop!', but 'Gallop!'

The situation is different when we come to a complex of speech signs. If such a complex is a semantic entity, then it is such because each of its components performs a semantic function. The semantic function of the complex is determined by the semantic functions of its elements.

Semantic entities which are formed of simpler semantic entities that in some way determine the functions of a given complex entity will be termed *second-degree semantic entities.* We contrast them with semantic entities which are not formed by any simpler semantic entities determining the functions of a given whole entity.

When applying these concepts to the definition of the linguistic system described in Sec. I, we say that *a linguistic system is a system of those semantic entities which can combine to form second-degree semantic entities.* Each such system includes not only entities which have designata, but also those which have no designata. And conversely, every system of semantic entities, such that some of those entities have designata while others have none, is a system of semantic entities that can combine to form second-degree entities. Hence, the definition of a linguistic system as *a system of semantic entities, some of which have designata, while others have none,* would be equivalent to the definition now under consideration.

7. (...) Separate words are not said to be meaningful or meaningless

An expression is meaningful if it can be part of a true or a false expression, and if at the same time, all its semantic components perform their semantic functions as are normal in a given language. (...)

Meaningful expressions can be classed into three groups:

(1) *Sentences* — expressions which are independent as to meaning.

(2) *Nominal expressions* — expressions which are independent as to the function of denoting.

(3) *Dependent expressions* — which have a meaning only because they can be parts of sentences. (...)

Any simple or complex linguistic entity which has the function of denoting (...) will be termed a nominal expression. Any linguistic entity which can be part of a sentence, but whose semantic function cannot be explained if that entity is not placed in a context (...) will be termed a dependent expression. (...)

III. Theories of Meaning and Denoting

9. (...) It has been said that nominal expressions are semantic entities because of their function of denoting. What is then the difference between the various nominal expressions that have one and the same designatum? (...) As well as the function of denoting they must (...) have some other semantic function which differentiates them. Moreover, there are examples which seem to indicate that this other function suffices for an expression to be a nominal one, and that the function of denoting is not indispensable. There are expressions which we hold to be nominal and which we know how to use in spite of the fact that the objects which should be their respective designata do not exist; for instance, 'a triangle the sum of whose inner angles is greater than 180 degrees', 'the maritime province of Bohemia'. (...)

This other function, which together with that of denoting is supposed to be an attribute of nominal expressions, is usually termed *meaning*. (...)

IV. Nominal Expressions

13. It seems to me that if we want to clarify the issue of the double semantic function of nominal expressions, namely that of denoting and

meaning, we have to revert to a classification of semantic entities into entities of the first and second degree.

Unlike first-degree semantic entities, second-degree semantic entities have a semantic function dependent on their structure. In this they resemble icons. That is why *understanding* a second-degree entity and *understanding* a first-degree entity are two different things.

To understand a first-degree semantic entity is merely to know the convention on the strength of which that entity has been endowed with a semantic function. When it comes to second-degree entities no such convention has been adopted. Thus, understanding a complex expression covers two stages: first, we must know the functions of its components, and second, we must know how to deduce a new second-degree function from the former functions in accordance with the laws of a given language. This new function, which is acquired by a complex of simple linguistic entities through their very association, is *the meaning of complex expressions.* This new function determines whether a given string of words is a nominal expression, a sentence, or merely a dependent part of a sentence. The conception of meaning makes the corresponding term equivalent with the term 'sense'. In Section II we pointed to the fact that meaningful expressions are always expressions of the second degree. And conversely, a word is, or is not, a sign, but we shall never say that it is a meaningless expression. People speak meaninglessly when they wrongly form complex expressions. Husserl gives two examples of meaningless expressions: '*Abracadabra*' and '*Grün ist oder*'. (...)

The preceding is not the only concept of meaning which we encounter in semantic analyses. It applies exclusively to complex expressions. At the same time there are simple linguistic entities which also have some *meaning*. We shall disregard simple names for the time being, since one may doubt whether in their case a reference to meaning is necessary.

This leaves the dependent expressions. These undoubtedly mean something, even if they are not second-degree entities. When applied to such simple expressions the term 'meaning' is not equivalent with the term 'sense'. Yet their meaning is in some way dependent on second-degree entities: dependent expressions, such as propositions and adverbs, are semantic entities only because they can be parts of independent second-degree entities. The meaning of a dependent expression is always interpreted by reference to that independent complex entity of which a given dependent expression can be part. (...) By analogy to the distinction made between independent and dependent expressions we may speak about *independent meaning* and *dependent meaning*. Both concepts are to a certain extent dependent on the combination of simple semantic entities into complex ones. Hence only linguistic entities can have a meaning. (...)

14. If we consider the class of semantic entities which have independent meanings, and the class of entities which perform the function of denoting, then complex nominal expressions will lie at the intersection of the two classes. They perform both functions, but the function of denoting occurs as something secondary.

The meaning of an expression, which is a function of the semantic functions of the components of that expression, makes us look at the whole expression as the first member of the relation of denoting even before we deal with the designatum as the second member of that relation.

The relation of assigning which exists between the sign and its designatum is in this case the converse of the relation which exists, e.g., in the case of proper names: in the latter case we assign a name to an object, while here we are looking for the designatum of a given expression. *A complex nominal expression must include at least one simple name. The designatum of a complex nominal expression is every element of the class which is the product of the classes of objects denoted by the names which are parts of that expression* (*e.g.*, hero king) *or every element of the class of objects which are in the class covered by a single name forming part of that expression, and which bear a specified relation to the object* (*or objects*) *denoted by another name that forms part of that expression.* (...) An expression may be a nominal expression even if none of its designata exists. (...)

15. Our analysis has started from the distinction between simple and complex nominal expressions. The former (...) I shall term *names.* Since every nominal expression, and only nominal expressions, can be the subject of a sentence, I may define a name as *a simple linguistic entity which can be the subject of a sentence.* (...)

A name differs from those entities which I have termed substitutive signs only by being an element of a linguistic system.

When interpreting a name analogically to a substitutive sign we disregard its case (declensional) forms. (...) A noun in the nominative case, as a name of an object, is a first-degree entity, whereas all other case forms may be treated as dependent second-degree semantic entities. (...) If we were to assume that a noun used in the various oblique cases does not cease to be a name, we should have to drop the assertion that the function of a name – unlike the function of a complex nominal expression — does not depend on the structure of a given entity.

16. In treating *independent meaning* as the special function of second-degree semantic entities we have assigned it to complex nominal expressions. We should accordingly infer that names, being first-degree entities, cannot have meanings. (...)

We can denote objects either regardless of the class to which they belong, or precisely because of their being in a given class. In the former case, denoting an object consists in directly assigning a nominal expression to it. Such an expression is always a name since, as we have said, a complex nominal expression denotes its designata as elements of a given class. And conversely, if we want to denote certain objects by a common name as elements of one and the same class, we have to use a complex nominal expression. We cannot assign a common name to certain objects as elements of one and the same class without first having denoted these elements by a complex expression. But once such a class is formed and its element denoted by a complex nominal expression, nothing prevents us from assigning to that expression a substitutive sign which in a given language will be a semantic equivalent of that expression. As a substitutive sign that entity will be a first-degree entity, but as a semantic equivalent of a complex expression it will take over all the functions of that expression, i.e., both denoting and meaning: furthermore, its designatum in a given language will not be a given complex expression, but the designatum of that expression. As a semantic equivalent of a nominal expression that substitutive sign will also be a nominal expression, and being a first-degree entity it will be a simple nominal expression and hence a name. For instance, the name 'square' may be treated as a semantic equivalent of the complex expression: 'a closed plane quadrangle with equal angles and equal sides'. (...) Names, being semantic equivalents of complex nominal expressions, form part of the same linguistic system to which those complex expressions belong.

As a result of the foregoing pages we shall distinguish two categories of names: (a) names which are directly assigned to designata; (b) names interpreted as semantic equivalents of complex nominal expressions, i.e., names assigned to their designata on the strength of a definition. The former have no meaning, but the latter do. *To say what a name means is the same as to replace it by a complex expression.*

It is above all the proper names which are directly assigned to their designata. But not exclusively the former: (...) we sometimes use general names as if they were names directly assigned to objects. I can therefore place one and the same name either in the first or in the second category, according to whether I consider only the relation which the name bears to its designatum, or whether I insert a complex nominal expression between the two members of the relation. There are, it is true, names such as 'winner', 'dancer', 'keeper', which seem to belong to the second category in all cases and always seem to be endowed with meaning, regardless of the circumstances in which they are used. Let us note that the names have the nature of second-degree semantic entities and are abbreviations rather than substitutive signs of complex nominal expressions: in 'keeper' the suffix '-er' indicates a person who has a certain attribute; hence the word 'keeper' may be treated as an abbreviation of the expression 'the man who keeps'.[3] (...) The semantic function of such words depends on their structure, contrary to such simple semantic entities as 'fast', 'last', 'cast', 'past', 'mast', whose functions cannot be inferred from their external form. Of course, we are not in the least concerned here with the origin (etymology) of these words, but only with whether in everyday usage we feel them to be complex or not.

Certain logical differences exist between names which are directly assigned to objects and those which are assigned to them by definitions.

(a) *Names which are in the first category cannot be predicative terms in sentences* because they do not mean anything. (...) It sometimes happens that a proper name functions as a predicative term, but in such cases it is not treated as directly assigned to an object, but as a semantic equivalent of a complex expression, e.g., 'That man is a Metternich', 'Not every capital city is a Paris', etc.[4] (...)

(b) Names which are in the second category may be empty (...) e.g., 'unicorn'. Names in the first category *cannot be empty*. This is apparent by definition, since they are names directly assigned to objects. We can reach the same conclusion by a different line of reasoning: Names in the first category do not have meanings; the only semantic function of such a name is the relation of denoting which it bears to its designatum. If this designatum did not exist, such a name would have no semantic function at all, it would not be a semantic entity, and it would not be a name. (...)

I think that from the logical point of view we may not permanently place entities of a given language in specified categories: the category in which a given entity is placed is in each case determined by the way in which a given person interprets a given expression under given circumstances. I believe that the static approach, which consists in applying constant linguistic criteria in establishing the semantic functions of expressions accounts for many misunderstandings in papers on the problems of signs.

V. Summary of Results

17. (...) Our principal concept is that of semantic entity. (...) It does not cover, among other things, all those symptoms and indications, whether natural (like a 'rash' which is a symptom of scarlet fever) or conventional (a soldier's uniform), which are signs exclusively because of their indicatory function.[5] If such an indicatory sign is a semantic entity, then it is such only because of its other function, namely that which makes it "logically transparent" (see Sections 1 and 4 above). On the other hand, (...) the concept of semantic entity does cover icons. (...)

We have singled out two (...) categories of semantic entities.

(a) Semantic entities assigned to designata.

(b) Linguistic entities.

These two categories, which are formed according to different principles and which partly overlap, fill the whole extensiou of the concept of semantic entity.

The first category is divided into two groups: (a) *icons*, where the relation which an icon bears to its designatum is based on the objective properties of both; and (b) *symbols*, where the connection of a given symbol with its designatum is purely conventional. The class of symbols partly overlaps that of linguistic entitis; the product of these two

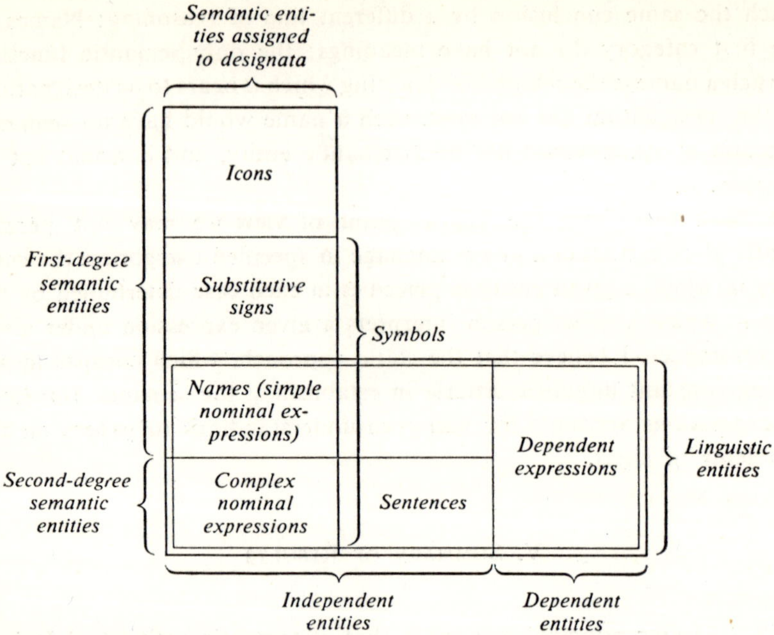

classes forms the class of *nominal expressions*. Symbols which are not linguistic entities have been termed *substitutive signs*.

The second category, that of linguistic entities, divides (...) into: *first-degree semantic entities* and *second-degree semantic entities*. (...) This classification applies to all semantic entities, but is restricted in practice to linguistic entities, since all other semantic entities are first-degree entities. This is true because the ability to combine into second-degree entities is characteristic of linguistic entities. (...) The classification of semantic entities into first- and second-degree divides nominal expressions into *complex nominal expressions* and *names*. Finally, names are either *directly assigned to designata* or *assigned* to them *by definitions*. In the latter case they are interpreted as semantic equivalents of complex nominal expressions. The classification into first-degree and second-degree entities intersects the classification into *independent linguistic entities* and *dependent linguistic entities*. The former include nominal expressions (in consideration of their function of denoting) and *sentences* (in consideration of their property of being true or false).

Dependent expressions are semantic entities only in consideration of their ability to be components of true or false expressions.

Semantic entities may be analysed from two viewpoints: (1) that of their designata, or, more generally, the objects to which they refer because of their semantic functions; (2) that of the mental states of a conscious being, owing to which a given object is a semantic entity. An utterance does not merely denote or mean something; it also *expresses* certain experiences of the person who makes that utterance. It is thus an indication of these experiences. Since the object of which it is an indication is a mental object, we say that it is an *expressive indication*. Semantic entities differ from other expressive indications in that the mental states about which they inform bear a certain relation to the designata of those utterances or to the designata of at least one element of each utterance (a representation of the designatum or a judgement about the designatum). (...)

In this paper I have endeavoured to confine myself to the sphere of logic: I have interpreted signs as elements of classes of equiform or isophonic objects, and I have analysed them exclusively as to their semantic functions. I have deliberately disregarded the *function of expressing*, which makes signs a subject of speech psychology, and also the *function of communicating*, which incorporates the concept of sign into the foundations of sociology.

Notes

[1] That other object need not necessarily be a designatum of a given semantic entity. It may also be a designatum of a semantic entity accompanying the entity in question.

[2] *Logische Untersuchungen*, Vol. II, p. 43.

[3] The linguistic examples in this paragraph have been modified to comply with the English system of word formation while illustrating the author's points. (Tr.)

[4] In English, the use of the indefinite article marks a grammatical difference which is absent in Polish, where there are no articles. (Tr.)

[5] I interpret an 'indication' more or less as an analogue of Husserl's '*Anzeige*'. (Cf. *Logische Untersuchungen*, Vol. II, Chap. 8, p. 25). An indication is an object whose appearance informs us about the existence or properties of another object. (The more or less standard English term is Peirce's 'indexical sign', which has not been used here, so as not to convey the erroneous impression that Ossowski was directly influenced by Peirce. — Tr.)

THE CONTROVERSY OVER THE LIMITS OF THE APPLICABILITY OF LOGICAL METHODS

by

Janina Kotarbińska

1. The controversy which we have in mind is a family quarrel. The parties are representatives of analytic philosophy, a field which embraces all those who regard the analysis of language as the principal task, or at least one of the principal tasks, of philosophy. According to some, the subject of analysis consists of the terminology and statements of philosophy itself; for others, it consists of the terminology and statements which occur in specialized disciplines; while still others see it as including both philosophy and specialized disciplines. This difference in standpoints is not, however, essential for the controversy which is to be discussed here. The essential difference is that which is revealed in the opinions on the nature and tasks of the said analysis. The clash is between two conceptions, the reconstructionist and the descriptionist, or, in other terminology, the formalistic and the linguistic. It may be said that both take as their standard the style of analysis carried out by Russell, Carnap, and to some extent by Woodger, but for the representatives of the former approach that standard is positive, whereas for those of the latter, it is negative.

As for the reconstructionist approach, its characteristic feature is its treatment of linguistic analysis as above all a method of improving language mainly in respect to its logical values. Its adherents therefore endeavour not only to explain, but also to give precision to, the meanings of the terms they analyse, and they do so by using the definitional method on a large scale; they endeavour, by appropriate interpretations, to reduce sentences to statements constructed according to sentential schemata of formal logic; they endeavour to reconstruct reasonings so that the relation of consequence between premisses and conclusions may be seen to hold in one way or another; finally, when analysing theories, they endeavour to construct purely formal systems to map the logical structure of every theory under investigation.

The descriptionist approach, advanced by Wittgenstein in the last period of his activity, as well as by the Oxford school, originated from criticism of the above concept of linguistic analysis. Its adherents blame the reconstructionists mainly for trying to apply logical methods, such as definition, the logical reconstruction of reasonings, construction of formalized systems, etc., in a sphere in which they are not applicable and where endeavours to use such methods are not only useless but sometimes downright harmful. To the reconstructionist programme they oppose a rival one: the aim of analysis is not to improve language as to its logic, but to investigate the ways in which it actually functions; it does not consist in a logical reconstruction of language, but in the most precise possible description of its actual properties.

This is, roughly, the basic controversy. As can be seen, the issue of the limits of the applicability of logical methods comes to the fore. But how are these limits to be drawn? It would be difficult to answer this question unequivocally. We may, however, risk the assertion that according to the descriptionists these limits are meant to divide the region of artificial languages, which have rigorously fixed rules of construction, from that of natural languages, for which, as we know, such rules are not explicitly formulated. Thus, formal logic and mathematics would be found on one side of the boundary, and all those disciplines which make use of natural language — at any rate both philosophy and all the empirical sciences — would be found on the other side. The controversy under consideration is, in principle, about this boundary.

As we follow the literature of the subject we may conclude that the controversy has been finding more and more participants, and that its repercussions are growing wider and wider. It seems that, if only for this reason, it is worthwhile discussing the issues at stake in greater detail and examining the arguments used by both parties. This is exactly what the present writer has in view.

Where shall we begin? It seems that the best starting point is the anti-reconstructionist approach. As the attacking party the anti-reconstructionists are more active and enterprising in the discussion and they have managed to give the course they choose. Their arguments focus mainly on the following three issues: the analysis of concepts; the structure of reasonings; formalized theories. Most attention is paid to the first issue, the point of departure being a criticism of the definitional method.

2. Several objections may be raised against definitions.

(a) The usefulness of definitions is said to be refuted above all by the fact that a definition establishes uniform criteria of the applicability of a given term being defined; such criteria are the same for all cases in which that term is used, whereas in natural language one and the same term is generally used according to different criteria, which are somehow interconnected, but by no means equivalent to one another. Consider, for instance, the term 'game'. As Wittgenstein says,[1] we speak of a game of chess, the Olympic games, a game of football; as games we include patience, noughts and crosses, etc. Do these have any criteria in common? Is not the element of amusement, or rivalry, or the win for one party and the loss for the other such a criterion? But chess is not an amusement, while there is no rivalry in patience, nor any winner or loser. There are no characteristics that could be attributes of all those forms of behaviour which are called games and which could at the same time distinguish any game from any non-game. The term 'game' has many different criteria of applicability, and not just one. The role it plays cannot be grasped by any definition.

This argument is to be found in the works of descriptionists quite often.[2] It may easily be noted that it applies to equivalence definitions only, and even were it correct in its essentials, it would refute at most only some forms of equivalence definitions, such as the classical definition *per genus proximum et differentiam specificam*, and would not oblige us to abandon other variations of equivalence definitions, such as definitions of the type,

$$\bigwedge x\,(Qx \equiv P_1\,x \vee P_2\,x \vee \ldots \vee P_n\,x),$$

which gives a number of alternative criteria for the defined term Q.

(b) The second objection is broader in scope. It emphasizes the discrepancy between the notorious vagueness of the terms which occur in natural language and the criteria of applicability as imposed by definitions. By their very nature vague terms do not have precisely fixed ranges of applicability. Yet definitions, by making the use of defined terms, depend on criteria which are both sufficient and necessary conditions for the use of such terms, assign to those terms extensions which are fixed unambiguously, and this both distorts their role and deforms the language to which they belong.

This objection is certainly applicable to all complete, i.e., equivalence definitions, and hence also to the alternative definitions mentioned above.[3] It does not in the least discredit incomplete definitions (also called partial or conditional definitions), which are characterized precisely by the fact that they only partially define that extension of a given term being defined. They usually consist of pairs of statements of the type

(1) $\bigwedge x\,(Px \rightarrow Qx)$,

(2) $\bigwedge x\,(Rx \rightarrow \sim Qx)$,

where Q is the term being defined, and P and R are mutually exclusive, but not complementary. Such a definition gives some conditions which are sufficient and some which are necessary for Q, but does not give any that are both sufficient and necessary. By way of example we can give the following *ad hoc* definition of leukaemia:

(1′) $\bigwedge x$ (x has more than 15000 white cells per cubic millimetre of blood $\rightarrow x$ has leukaemia),

(2′) $\bigwedge x$ (x has less than 8000 white cells per cubic millimetre of blood $\rightarrow x$ does not have leukaemia).

For certain quantities of white cells in the patient's blood (more than 15000) the above definition enables us to diagnose leukaemia; for certain others (less than 8000), it enables us to diagnose its absence; but for the remaining cases it provides no criteria, whether positive or negative, and leaves the matter undecided.

(c) It might seem that because of these properties incomplete definitions specially lend themselves to the introduction and analysis of vague terms. There is even a tendency to identify vague terms with terms which are only partially definable. This, however, would not suit the descriptionists. It is difficult to find in their writings explicit comments on partial definitions — it seems that they simply do not take this type of definition into account — but they indirectly settle the question in the negative, for they deny to the terms used in natural language not only complete criteria, but even such partial criteria as warrant the use of these terms in a completely sufficient or necessary way. They refer to the incessant fluidity of natural language, which, they claim, is due to two factors.

First, it is virtually a rule that a term which was originally used in a single sense, e.g., to denote objects with the properties *abc*, and in consideration of those properties, in the course of time comes to be used in consideration of other criteria as well, e.g., in consideration of the properties *abd*, *df*, *feg*, etc., if in the cases encountered so far the sets of properties *abc* and *defg* have always accompanied one another. Such a process is essentially never concluded. That is why no criterion in use and none of its components warrant the necessary use of such a term. What links the various cases of its use is not such properties as might be shared by all cases, nor is it similarities in the same respects in all cases, but similarities interpreted more loosely, such as usually occur among members of the same family: a person A may resemble a person B in some respect W_1, B may resemble C in some respect W_2, C may resemble D in some respect W_3, but there may be no clear similarity between A and D. The term 'game' is a typical example of such a situation. Such terms as 'number', 'language', 'sentence', 'acid', and names of the various biological species may also be cited as examples. What is important is either the variety of their semantic functions at a certain time or the variability of such functions in the evolution of a given language.[4]

Second, terms which occur in natural language often change their meanings according to the *circumstances* and *context*, whether situational or verbal, in which they are used. Criteria which determine their applicability under certain circumstances do not do so under other circumstances, and are thus not sufficient conditions of their use. To give a very simple example, '1.4.57' when written by an Englishman usually means *the first of April, 1957*, whereas when written by an American, it usually means *the fourth of January, 1957*.

(d) Note, too, that in connection with the foregoing observation, definitions are also blamed — this being the fourth and last item on our list of objections — for assigning meanings to the defined terms only with reference to the shape of those terms, without taking contexts into account, and hence disregarding the fact that in ordinary speech the semantic function of expressions also depends, in each case, on the context and the pragmatic circumstances, and not merely on their shape and their syntactic properties. In the opinion of the descriptionists, the very question to which a definition is meant to be a proper answer, is wrongly formulated. They are usually formulated very generally. *What is knowledge?*, *What is beauty?* or *What does the word 'knowledge' mean?*, *What does the word 'beauty' mean?*, even though such a definition requires specificity or reference to a given verbal or situational context. No correct answer is possible to a question posed in such a general manner.

3. We already know the conclusion of all these critical comments. The latter tend to demonstrate that in natural language the definitional method of analysis is not suitable. This conclusion is purely negative in nature. The question immediately arises, as to what are the positive requirements of the descriptionists, and what do they suggest as a replacement for definitions. There is no uniform approach in this respect. The only common point is the initial assumption: it is essential to use a method which will not distort the actual functioning of the expressions under consideration, which will not artificially restrict the scope of their use, and will not eliminate the ambiguities that are found in them. But the various authors differ in their attempts to carry this programme into effect.

The most common opinion is that language analysis requires the use of the empirical method, which consists in noting the various cases of the use of the expressions under investigation and in examining their semantic functions in different situational contexts. This method does not yield definitions, but rather what is termed *rules of use* (this concept is of particular importance in the descriptionist approach) or, in other words, *rules of actual use*, or *rules of correct use*. What are these rules of use? They are not always described in the same way, but two points are usually stressed most.

First of all, emphasis is laid on their descriptive nature. They are supposed to be purely descriptive statements, empirical generalizations which report on the actual ways in which given expressions are used. If we list examples of rules of use as found in the works of descriptionists on various occasions we can see that in some cases this is really so. It is so, for instance, when the following statements occur as rules of use: "People who speak Swahili use the word 'hiranu' in 89 per cent of cases to denote brothers", "Expressions of the type 'I know that *p*' are used in English, in most cases, only if one is convinced that *p*", "In everyday language statements of the type '*p* and *q*' are sometimes used so that they do not imply *q* and *p*" (where sequence in time is referred to), etc. Yet one finds it difficult to resist the impression that in a far greater number of cases we are dealing — as far as the final result of a given analysis is concerned — with statements which practically do not differ from definitions as to their methodological status — statements which are not descriptions of received linguistic usage, but are formulations of terminological decisions that are programmatically adjusted to such usage, as in the case of analytic (reporting) definitions, which, as we know, are not empirical statements. If these observations are correct, then rules of use conceived in this manner are nothing other than definitional postulates of a special kind.[5]

It must be noted that when we are dealing with the controversy: definitions *versus* rules of use, the methodological nature of both, and their empirical or non-empirical status are, despite appearances, of no great importance. This is shown by the fact that the arguments adduced here do not refer to the methodological status of such statements, but are concerned merely with their relation to everyday language, to their agreement or disagreement with accepted linguistic usage. We shall, therefore, now leave aside the intricate, but unimportant, issue of the empirical nature of rules of use.

On the other hand, we should now devote some attention to the contextual nature of the rules of use, which is the second of the two characteristics usually mentioned by descriptionists as particularly significant for those rules. The term 'contextual', as used by descriptionists, must always be restricted to a given event, while it covers the totality of the circumstances which accompany the use of a given expression and which co-determine its semantic function. The repertory of such circumstances is very large, and may include, for instance, the time and/or place of use, the user, his intentions, his emotional state, the tacit assumptions which he makes, the verbal context of the expression in question, etc. Now, the view under consideration maintains that, unlike definitions, which as a rule are non-contextual in nature, the rules of use link semantic functions not only with the form of expressions, but also with the contextual circumstances of their use.

How this takes place has not been indicated rigorously. In practice, the rules of use are formulated in various ways. It seems, however, that if we are not to avoid reconstructions, which are so uncompromisingly opposed by the descriptionists, we could subsume at least the majority of those rules under a certain schema, e.g., of the type:

$\bigwedge x$ (if x is an expression of the shape K and if x is used under the conditions C, then x has specified semantic functions) (e.g., means the same as the expression W, refers to the person O, denotes objects of the type P, etc.).

Or:

$\bigwedge x$ (if x is an expression of the shape K and is used under the conditions C, then it is used correctly only if it performs specified semantic functions).[6]

Examples of applications follow:

(a) $\bigwedge x, y, z$ (if x is a sentence of the form 'This is red', and has been uttered by y, and if y, when uttering that sentence, pointed to the object z, then that sentence means the same as 'z is red');

(b) $\bigwedge x, y, t$ (if x is an expression of the shape 'today' and if y used that expression on the day t, then x means the same as the day t);

(c) $\bigwedge x, y, t$ (if x is an expression of the shape 'I know that oviparous mammals exist', and if x is uttered by y at the moment t, then x is used correctly only if x expresses y's conviction at the moment t that oviparous mammals exist);

(d) $\bigwedge x$ (if x is an expression of the shape 'a saw' and if x is used in a sentence of the type 'Every A is B' at a place A, then x denotes a joiner's tool).

In this interpretation, these would be metalinguistic statements which characterize the semantic functions not of all expressions of a given shape, but only of those which in addition satisfy certain specified conditions. On the assumption that we are dealing with statements of a definitional kind, we should have to see in them a variation of partial definitions. They would differ from other partial definitions by being formulated in metalanguage and by the fact that the criteria of applicability which they describe are restricted, not so that they make it possible to decide only with regard to certain objects whether these do, or do not, come under the term being defined, but so that they apply only to certain tokens of the term being defined, namely those which are used in a specified situational context. It would seem that this variation of partial definitions has not been singled out thus far, and yet it would appear to be useful in defining token-reflexive expressions. It must also be added that in the rules of use which refer to expressions whose semantic function does not depend on the circumstances of use (according to the descriptionists, such cases are extremely rare in natural language) the contextual condition may be omitted, and that, on the other hand, if the semantic function of an expression does depend on the circumstances of use, but it cannot clearly be formulated how it depends, then the rules of use take on the form of more general statements, such as "In some circumstances of use, the expression of a specified form is used correctly only if it performs specified semantic functions", etc.

The conception of the rules of use analysed above is represented mainly by members of the Oxford School. Another conception, connected mainly with Wittgenstein and his followers, is also popular. According to the latter approach, explaining the semantic functions of expressions requires reference to the particular cases of their use. Wittgenstein says that if we want to explain to a person what the word 'blue' means we have to show him an object of that colour and say 'This and the like are blue' or 'This and the like are called blue'. Likewise, in order to explain to a person what a 'game' is we have to describe selected examples of the various kinds of games as precisely as possible and to add 'These and the like are called games'. In order to explain what a proposition (in the technical, logical sense of the term) is, we have to give various examples of propositions and to add the comment: 'These and the like are called propositions', etc.[7]

This is not an auxiliary method. Its task is not to provide data for the construction of more definite criteria that would no longer refer to individual cases. Examples are to be taken at their face value. They function as standards according to which we can understand the practice of using the expressions being explained, so that we are in a position to decide whether these standards are, or are not, applicable to specified individual cases. By giving properly chosen standard examples we show samples, as it were, of actual ways of using the expressions in question under various contextual conditions, and we thereby instruct others how to use these expressions under similar conditions in the same way in which they have been used so far. An advantage of the exemplification method is that it does not result in a deformation of the semantic functions of the expressions being analysed, and, in particular, that it does not artificially sharpen the limits of their application and does not eliminate the many shades of meanings which are inseparable from the vocabulary used in everyday speech.

4. So much for the descriptionists. As can be seen from the above, their main endeavour is to preserve agreement with everyday speech and with fixed linguistic usage. Both the requirement that the rules of use be contextual in nature, and the requirement that they be used so as to provide examples are of secondary importance: they indicate means for ensuring or making possible the achievement of that primary goal. That is why two problems arise: are the means which the descriptionists recommend really effective? and does the goal which they want to attain in this way deserve to be treated as primary?

For the time being, let us discuss the first issue (the second, which is more general, will be discussed only in the concluding part of this paper). The requirement that the rules of use be contextual in nature does not provoke any major objections from the point of view under consideration. The fact that expressions used in natural language are token-reflexive and ambiguous is too well known for us to doubt that equiform expressions occurring in that language are mostly used in various meanings which depend on the circumstances of use, such circumstances being extra-linguistic or intra-linguistic in character (the latter case refers to verbal contexts); and hence the rules of use, if they are to comply with accepted usage, must take that fact into account in some way.

Doubts arise, however, as to whether the exemplification method, such as recommended by the descriptionists, does not result, even with the best possible choice of standard examples, in a deformation of the accepted semantic functions of the expressions being analysed. It may be conceded that it does not usually distort these functions in the direction of greater precision. But does it not distort them in the reverse direction? Is it not true that it contributes not only to preserving existing polysemies and ambiguities, but also to the emergence of new polysemies and to the expansion of accepted spheres of ambiguity, i.e., cases where it is not possible to decide whether or not they are within the spheres of applicability of the expressions being explained? Each issue has two aspects, that of the person who explains and that of the person to whom the explanation is addressed. The statement of the person explaining, hence the statement "This and the like are called 'N'", does not deviate essentially from the accepted usage of the expression represented by 'N', provided that it refers to properly chosen standard examples. This, however, is an incomplete statement. As has been pointed out on another occasion,[8] it merely informs that 'N' refers to specified indicated objects, e.g., the objects P_1, P_2, ..., P_n, and to all objects which resemble P_1 or P_2 or ... or P_n, but does not say anything about the manner and degree of resemblance which is meant to exist between these objects, and thus does not sufficiently instruct us in how to decide to which cases the expression 'N' is applicable. Thus the person explaining informs about the criteria of the applicability of 'N' only partially, and the rest must be formulated by the addressee on the basis of this partial information which, as can easily be seen, admits not of one possible solution only, but of a number of such solutions.

Now it is obvious that under such conditions the guesses made by various persons usually differ from one another, and that because of such differences 'N' comes to be given shades of meaning which it did not have before. It is also obvious that the lack of common and clearly fixed standard examples must contribute to the emergence of new polysemies. The relation of resemblance, to which the exemplification method refers, is not transitive (this can easily be perceived if we consider that such resemblance is established by purely subjective criteria); the set of objects which in some specified respect resemble a standard P_n need not be co-extensive with the set of objects which in the same respect resemble a standard P_z, even if P_z in that respect resembles P_n.[9] For all these reasons the exemplification method results in certain changes in the accepted usage of the expressions involved, and hence, or so it would seem, it is subject to the same criticism as the definitional method, which the descriptionists attack. Do they really fail to realize the consequences of explaining meanings of words and expressions by the exemplification method alone? This seems unlikely. It would appear rather that in opposing deformations of natural language the descriptionists merely oppose any disambiguation of its expressions, and are quite tolerant when it comes to changes in the opposite direction — probably because they hold changes in that opposite direction to be in agreement with the trends which prevail in the evolution of natural language. Should this interpretation turn out to be correct, the paradoxicality of their standpoint would be striking.

5. We shall now move on to discussions which pertain to analyses of reasonings. As the descriptionists see it, the various standpoints are more or less as follows. The reconstructionists start from the assumption that in practice our reasonings are usually enthymematic and simplified, and hence they see their task in making the forms of reasonings more precise, so that — in some cases at least — they may be subsumed under the laws of formal logic. Accordingly, they identify the analysis of reasonings with their logical reconstruction, which consists, on the one hand, in completing the set of initial premisses by joining to it the missing links in the form of those premisses which are assumed tacitly, and on the other, in reducing all the components of reasonings, premisses and conclusions — by making use of the appropriate definitions — to statements constructed according to the sentential schemata valid in formal logic (for instance, in reducing statements of the type '*A* exists' to statements of the type '$\bigvee x$ (*x* is *A*)', etc.). The descriptionists of course oppose such an analysis as a matter of principle. They oppose reconstruction as a method which distorts natural language; first of all, they reject the possibility of applying logical laws and schemata to natural language. The primary problem, which is the focus of the controversy, is whether the laws of formal logic are in practice applicable to reasonings formulated in ordinary language; in other words, whether reasonings carried out outside the sphere of logic and mathematics can be taken as particular cases of logical schemata of reasoning.

When it comes to the descriptionists, their main arguments intended to substantiate their answer in the negative may be reconstructed more or less as follows.

Using the laws of formal logic in the process of reasoning as a rule requires substitutions to be made for the variables which occur in those laws. Now the rule of substitution, like the other rules of logic, is purely formal in nature: it refers exclusively to the external properties of expressions, without any reference to their meanings. This legitimates the substitution for a variable which occurs several times in a formula of any expression of the same form, provided that they are within the range of variability of that variable. It is possible to do so without risking errors only under the assumption that the rules of the language involved are rigorously fixed (this applies in particular to the rules which govern the classification of the expressions of that language into syntactic categories); that the categories mentioned above correspond strictly to the categories of the variables that occur in the logical formulas; and that all expressions of that language are used quite unambiguously, regardless of their verbal or situational context. Natural language satisfies none of these conditions. That is why statements obtained by substitutions for well-formed logical formulas are not always true, and why reasonings carried out in accordance with such formulas do not always lead from truth to truth. For instance, the statements 'If today is Saturday, then tomorrow is Sunday' and 'Today is Saturday' do not yield the statement 'Tomorrow is Sunday' if, say, the first two statements were made on Saturday, March 7, 1964, and the third, on Sunday, March 8, 1964. The first token of the statement 'Tomorrow is Sunday' is, in such a case, true, whereas its second token is false. Likewise the statements 'Jack is 17 years old' and 'It is not true that Jack is 17 years old' may both be true if in each case the word Jack is used as a proper name of a different person.[10]

The operation of substitution is the source of another difficulty, too. Substitutions are made for the variables, while the constants remain unchanged. Hence the logical constants which occur in those formulas for which substitutions are made (or the verbal analogues of these constants) must also occur in the substitutes of those formulas (in their unchanged meanings, of course). Now if these substitutes are sentences formulated in everyday language, this condition cannot be met for the simple reason that everyday language has no terms which would have the same meanings as the logical constants have. First of all, it has no sentential connectives in the sense of truth functors. The connectives, or conjunctions, which are used in everyday language to render the constants of sentential calculus are accordingly not the correct translations of the former. For instance, the implication symbol is usually read as 'if ... then ...', but in the ordinary interpretation of this conjunction the truth of the conditional sentence does not depend on the logical values of its components alone, but on their semantic functions as well. In such an interpretation, not all the sentences of the types 'if *q*, then, if *p*, then *q*'; 'if not *p*, then, if *p*, then *q*'; 'if not *p*, then, if *p*, then not *q*', etc. will be accepted as true, even though the formulas $q \rightarrow (p \rightarrow q)$; $\sim p \rightarrow (p \rightarrow q)$; $\sim p \rightarrow (p \rightarrow \sim q)$ are tautologies of sentential calculus. Let us give another example: the conjunction 'and' is the verbal analogue of the symbol of conjunction. But, as the descriptionists claim, that conjunction is often used in natural language to indicate sequence in time between events: not only does the truth of a sentence of the type '*p* and *q*' not result in the truth of the corresponding sentence '*q* and *p*', but, on the contrary, it results in the falsehood of the latter. For instance, from the sentence 'John underwent an operation and recuperated' being true it follows that the sentence 'John recuperated and underwent an operation' is false.

The situation looks no better when it comes to the functors which occur in functional calculus. The copula 'is' is used there in its non-temporal sense, whereas in everyday language it is often used as an abbreviation for 'is now'. Conjunction between terms is interpreted, in functional calculus, as an equivalent of the sentential conjunction, whereas in everyday language such equivalence exists only rarely. The sentence 'Tom and William came' is in no way an equivalent of the sentence 'Tom came and William came', because the former suggests *at the same time*, whereas the latter indicates a sequence in time. The same applies, analogically, to other functors used in formal logic: they determine other conditions of the truth of the statements constructed by means of them than do the conjunctions current in everyday language. Moreover, logical functors have strictly and unambiguously fixed meaning, whereas the conjunctions which occur in natural language are used loosely and in an ambiguous manner.[11]

For all these reasons the laws of formal logic can be used only in languages which, as Ryle says, are governed by military discipline, that is, only in artificial languages constructed especially for this purpose, but not in natural languages, which refuse to follow the rigorous requirements of logic. The latter need an informal logic, adjusted to the much freer and much more varied "civilian" manners of natural languages. This would differ from formal logic in not being constructed as a deductive system and in being concerned with inferential relationships between sentences which depend not on the shape of the sentences alone, but on their meanings as well. If we are to judge from additional explanations, the rules of such a logic would be simply the rules of use, which have become familiar to us from the foregoing sections. According to the view now under consideration, such rules, by informing about the semantic functions of certain words, would thereby inform about the conditions of truth of certain sentences in which these words or expressions occur; in other words, they would inform about inferential relationships between those sentences and certain other sentences. Hence the double role of the rules of use: as rules which explain meanings of expressions, on the one hand, and as rules of reasoning, on the other.[12]

What are we to think about the value of the arguments summarized above? We shall discuss this question by taking these arguments one by one.

In considering the first argument, it is obvious that it undermines the reconstructionist standpoint only if one assumes that the applicability of the laws of logic to everyday language is understood rigorously, making it depend, as a necessary condition, on the possibility of carrying out the formal operation of substitution in the sphere of everyday language with a guarantee of complete reliability. It is obvious that the reconstructionists, when defending the applicability of the laws of formal logic to reasonings carried outside the sphere of logic and mathematics, have a different interpretation in view, namely one which in no way assumes the formal nature of the transformations under consideration. Their interpretation is that the validity of substitutions depends not only on the syntactic properties of the expressions involved, but on their semantic properties as well: a sentence 'Z' is a correct substitution of a logical law P only on condition that in those places where in P there are variables which are equiform with one another, sentence 'Z' has expressions which are not only equiform but synonymous as well.[13]

If this is so, then both parties agree that for a purely formal interpretation of the operation of substitution, the sphere of applicability of the laws of formal logic is naturally restricted to formal languages. The difference of opinion is revealed when it comes to the practical conclusions which they draw from this fact. Descriptionists tend to replace formal logic by an informal one, that is, to deformalize the laws and schemata of logic. Reconstructionists confine themselves to deformalizing the rules of inference, which results in a deformalization of the concept of the applicability of the laws of formal logic to everyday language. It would appear that abandoning the formal nature of the rules of inference is at any rate less embarrassing than abandoning the use, in ordinary thinking, of the criteria of the correctness of reasonings, as supplied by formal logic, the more so as the outline of that informal logic and of the role it would have to perform is still too vague even in its most essential points.

As for the second argument, which refers, as we know, to the discrepancy between the meanings of logical constants and their analogues in everyday language, we must say at the outset that this problem is far from easy to solve. It has been analysed many times. It appears in sharpest outline in connection with what is termed the implication paradox, which has attracted the attention of logicians since ancient times. Discussion focusses on the connective 'if ... then ...' whose paradoxical nature is, however, treated as merely a glaring example of more general problems. Within reconstructionist opinion as a whole we can single out, roughly speaking, three main approaches.

(a) The first approach is marked by a linking of the discrepancy between the interpretation of sentential connectives in formal logic and the current use of corresponding conjunctions in natural language, with the conditions of the proper or correct use of sentences joined by such conjunctions, and not, as is usually the case, with the conditions of the truth of statements constructed through the use of corresponding sentential connectives. The distinction between the truth of a statement and its correct use plays an essential role here. The truth of a statement depends on what that statement asserts and the correctness of its use, on what it expresses. To put it more precisely: a statement '*Z*' is true if and only if the state of things which it asserts exists; a statement '*Z*' is used correctly if and only if the person who utters it is in the mental state which that statement expresses in agreement with linguistic usage. Now, according to the view under consideration, the conditions of the truth of statements are the same regardless of whether the conjunctions which occur in them are interpreted in their logical meanings or in their ordinary ones. It is the conditions of correct use which are different: a statement which is used correctly for a logical interpretation of the conjunctions which occur in it may be used incorrectly if those conjunctions are understood in an ordinary sense, since in such a case the expressive functions of that statement may be different. Such differences, however, do not prevent the laws of formal logic from being used outside its own sphere.[14]

(b) In the case of the second approach the attack is conducted differently. The assumption that for an ordinary interpretation of conjunctions the conditions of the truth of statements differ from those which result from a logical interpretation is not questioned, and the endeavour is to demonstrate that the discrepancy is not wide enough to make formal logic useless in ordinary reasonings. Attention is drawn to the essential fact that even though statements which are true for a logical interpretation of the conjunctions do not always retain their truth for an ordinary interpretation of the conjunctions involved, yet statements which are true for an ordinary interpretation do not cease to be true when the conjunctions which occur in them are interpreted in a logical sense. This accounts for the fact that — as is claimed — the logical schemata of reasonings include many whose reliability does not depend on whether the conjunctions which occur in them are understood in a logical or in an ordinary sense. They include, for instance, schemata of sentential calculus which in their premisses have sentential calculus functors of the types:

$$\frac{\begin{matrix} p \to q \\ p \end{matrix}}{q} \qquad \frac{\begin{matrix} p \to q \\ \sim q \end{matrix}}{\sim p} \qquad \frac{\begin{matrix} p \to q \\ q \to r \\ p \end{matrix}}{r}$$

These schemata remain reliable if the symbol $\to$, interpreted in its logical sense, is replaced by the conjunction 'if ... then ...', understood in its ordinary sense, because the statement of the type $p \to q$ follows from the statements of the type 'if p, then q' (though not conversely). We then obtain derived schemata which we use in ordinary mental operations and whose applicability does not provoke objections. In a somewhat extended sense, this applicability also covers logical schemata corresponding to these derived ones, and hence the laws of formal logic on which those schemata are based.

In this interpretation of the question (which seems to be the most prudent of all) we do not in practice avail ourselves of all the forms of inference provided by formal logic: we use only those which comply with the usage of everyday speech, and hence satisfy certain extra-formal conditions. This is not an exceptional situation. We proceed in the same way when constructing definitions and classifications: we take into consideration not only the formal conditions of correctness, but also extra-formal conditions which ensure the usefulness of a given definition or classification for a specified purpose. In both cases the conditions of formal correctness are necessary, but not sufficient.[15]

(c) The third approach is represented by those who are in agreement with the descriptionists on those points on which both the first and the second approach differ from the descriptionist interpretation, but who prefer to adjust formal logic to the needs of everyday language rather than to abandon its application in the sphere of everyday language. These tendencies yield non-classical systems of sentential calculus, which alongside the symbol of material implication introduce that of strict implication as a semantic analogue of the conjunction 'if ... then ...' in the current interpretation of the latter. According to specialists, the system S5, the most comprehensive of all those constructed by Lewis, suffices to formalize all forms of deductive inference indispensable in practice as well as all necessary logical theorems, and at the same time includes sentential connectives whose meanings come very close to the meanings of certain conjunctions used in everyday speech.[16]

None of these three approaches is free from certain objections, and none of them completely solves the problem in question. However, each of them seems better than the solution suggested by the descriptionists, if only because the idea of an informal logic is still so very vague that it would be difficult to decide whether it indicates the correct direction of research or, rather, leads us astray.

6. We shall make only brief comments on the controversy between the descriptionists and the reconstructionists on the question of the analysis of theories formulated in natural language (above all, philosophical theories). In the discussion of this issue we encounter familiar motifs.

The descriptionists claim that the reconstructionists in this case, too, see the main task of analysis in making the theories under investigation as precise as possible. At the same time they are said to maintain that the possibilities of attaining that goal by resorting to the means provided by natural language are extremely limited, and so they reduce such analyses to methods of constructing formalized systems that have interpretations in the theories in question and reflect their logical structures with at least some approximation. Such systems are treated (by their authors — Tr.) as logical reconstructions of the various theories covered by investigations; they are meant to bring out the logical relationships that exist within given theories, and thereby to make the concepts used in those theories and the reasonings carried out in them more precise. As may easily be surmised, in the eyes of descriptionists such analyses combine the defects of the logical reconstruction of concepts with those of the logical reconstruction of reasonings. Their main objection is that the analyses carried out according to the reconstructionist method deform the conceptual apparatus of the theories under analysis, violate natural language, and yield results which are at variance with accepted linguistic facts. The rival idea advanced by the descriptionists identifies analysis of theories with analysis of the semantic functions of the terms used in the various theories: as we know, these analyses boil down to a formulation of the rules for using these terms in accordance with the uses of everyday language[17] (here again the concept of rules of use comes to the fore, accompanied by the requirement that these rules should agree with the ordinary interpretation of the terms in question).

A discussion of these issues will probably not contribute anything new to what has been said above, and therefore it would not be worthwhile to devote any more time to them.

7. We now pass to the fundamental issue, namely the requirement that the results of analysis should agree with everyday linguistic usage. The importance of this requirement in the descriptionist approach has already been stressed (see above, *passim*). Let us note above all that this requirement (for the sake of brevity, let us call it the requirement of agreement), thus formulated, may reflect widely divergent tendencies, according to whether the agreement to which it refers is interpreted more or less broadly, or more or less rigorously. We should therefore have to determine which is the case we are now discussing. In order to do this it is useful for the sake of comparison to have specific examples of the most typical positions adopted by the various parties to the controversies over the aformentioned requirement of agreement.

We may take Mill's standpoint, which is representative of the moderates, as our point of departure. Mill's statements deserve being quoted *verbatim.*

Reflecting on the conditions which the language of philosophy ought to satisfy (by the way, this is the same initial issue as in the case of the descriptionists), he writes: "In order that we may possess a language perfectly suitable for the investigation and expression of general truths, (...) the first requisite is, that every general name should have a meaning, steadily fixed and precisely determined." (...) "It would, however, be a complete misunderstanding of the proper office of a logician in dealing with terms already in use, if we were to think that because a name has not at present an ascertained connotation, it is competent to any one to give it such a connotation of his own choice. The meaning of a term actually in use is not an arbitrary quantity to be fixed, but an unknown quantity to be sought." (...) "In the first place, it is obviously desirable to avail ourselves, as far as possible, of the associations already connected with the name; not enjoining the employment of it in a manner which conflicts with all previous habits (...). A philosopher would have little chance of having his example followed, if he were to give such a meaning to his terms as should require us to call the North American Indians a civilized people, or the higher classes in France or England savages; (...) Were there no other reason, the extreme difficulty of effecting so complete a revolution in speech would be more than a sufficient one." (...) "The fixed and precise connotation which it receives, should not be in deviation from, but in agreement (as far as it goes) with, the vague and fluctuating connotation which the term already had."[18]

It can now be seen why we have taken Mill as our exponent of the moderate trend. While he strongly emphasizes the need to adjust the suggested use of expressions to their everyday use, he also sets forth a "prime condition" that imposes clear restrictions upon the requirement of agreement. In this connection he does not oppose every deviation from accepted meanings: he does so only with respect to arbitrary, whimsical deviations that are not dictated by considerations of precision. The evil which he wants to curb is the conventionalist approach to language, such as we find later in the post-Millian period, for instance in McKay and in Petrażycki.[19] McKay refers to the purely conventional nature of language signs and defends the position that any signs may be given any meanings. Petrażycki goes even further. He claims that regardless of whether we are dealing with terms which are only just being introduced or with terms which are already in use, a given term is to be treated "as a new and independent linguistic unit, as a conventional sign of the scientific concept we form and of that concept only; and if any earlier terminological habits are linked with it, we must put them aside and not involve them in the issue."

Where then is the descriptionist standpoint to be placed? Now it is obvious that in Mill's controversy with conventionalism the descriptionists would side with Mill. We know, however, that the descriptionists are primarily anti-reconstructionists, and when it comes to that controversy the situation would be quite different, and not only because Mill's attacks are solely aimed at conventionalism, and conventionalism need not, and does not, go hand in hand with reconstructionism. A much more important reason is that on the most essential points Mill's programme of language analysis does not, to all intents and purposes differ from the reconstructionist programme, especially in the form to be found in Carnap's works. To convince oneself of this, it is sufficient to realize that Carnap, like Mill, saw the main task of analysis (he uses the term 'explanation') in making the terms being explained more precise; he even describes explanation as the replacement of a less precise term ('explanandum') by a more precise one ('explanatum'). Like Mill, Carnap wishes to be in agreement with accepted terminology, and he makes the reservation that the 'explanatum' should resemble the 'explanandum', so that it can be used in most cases in which the pre-scientific 'explanandum' has been used previously.[20] The combination of these two requirements — that of linguistic precision and that of the observance of linguistic usage within the limits outlined by the need for precision — seems characteristic of the reconstructionist approach. To make the issue clearer let us add that descriptionists oppose the first of these requirements from the point of view of the second: for the sake of agreement with everyday speech (agreement interpreted, of course, in a much more rigorous manner than has been done by Mill and Carnap), they oppose all measures intended to make accepted meanings more precise, since they see in this striving for precision the principal danger of distortion of linguistic usage. Thus, in the descriptionist approach, the requirement of agreement with everyday language is given an extremely radical formulation. It therefore results in a search for a method of analysis which, unlike the reconstructionist method, will not deform the specific nature of natural language, and which will make it possible to preserve the whole vagueness of the semantic functions of expressions that occur in natural language, the vagueness of extensions of terms, the variability and non-uniformity of the criteria of use of words and expressions. When descriptionists require that the rules of use be in agreement with everyday speech, they have these conditions in view first of all. This is certainly the most essential point of their programme — the point which they stress most.

The basic tendencies of descriptionists have probably become sufficiently clear. They have moved far away from the Socratic tradition, from the Cartesian principle of making ideas clear, and from the position shaped by classical school logic — a position now common among rationalistically-minded thinkers. What can have led the descriptionists to advance such a revolutionary view? Why all that abhorrence of precision? Why the rush to cultivate even those properties of natural language which in theoretical investigations are usually considered to be its disadvantages and defects, and not its merits? Why, to put it briefly, that striving to preserve all the vagueness characteristic of that language? What might be the arguments in favour of such an attitude? Let us take a look at the arguments adduced by the descriptionists themselves.

First of all, they draw attention to the fact that a programme of making natural language precise is basically unworkable. Even if we were to succeed in making some terms precise by resorting to special operations, those terms would remain precise only as long as they were not used again: once they went back into circulation they would be bound to undergo the same processes which they had undergone before being made precise, and which would inevitably result in incessant fluctuations of meanings and in an incessant shifting of extensions in various directions.[21] If this is so, they conclude, the whole business is not worth the effort. It is better to admit at the outset that we are unable to go beyond the boundaries of precision demarcated by the ordinary functioning of natural language, and not to set ourselves tasks that are doomed to failure.[22]

The second argument is that natural language in its evolution adapts itself as well as possible to the purposes it has to serve and does not require any corrections. Language does not evolve by chance. The trend of its evolution is determined by the needs of those who use it, and this accounts for a kind of natural selection: useful concepts remain in use, those which have not passed the test go out of use. As a result of this process we have precise concepts where we need them, and imprecise ones where precision is not required. As Warnock says, it is extremely unlikely that language should contain much more or much less than its tasks require. It has passed the severest test of effectiveness, that of constant use. It is therefore most unlikely that any terminology modified by us (what is meant here is undoubtedly a terminology made more precise) would be much better than that which has emerged as a result of the natural evolution of language. These reflections bring the descriptionists to the conclusion that it is useless to strive for making accepted terminology more precise. This conclusion does not apply only to current terminology. When discussing these issues Scriven emphasizes that all this also refers, and even especially, to terms that are of essential importance in the empirical sciences.[23]

The third argument goes even further. It is intended to demonstrate that giving more precision to terms which are already in use is not only not advantageous in any way, but even has untoward consequences. Several reasons are cited in this connection. (a) Making terms more precise requires the elimination of all shades of meaning of a given term except for a single specified one. This takes place, therefore, at the cost of certain functions of a given term, and hence a debasement of its use value. This happens in the case of conjunctions, for example, if their semantic roles are restricted to those assigned to them by formal logic.[24] (b) Polysemy is necessary if language is to perform all its tasks. Vocabulary is limited by its very nature. If language nevertheless provides an immense wealth of devices which make it possible to render even the subtlest distinctions, if it performs so many and such varied functions, this is only because its component expressions are polysemic and indefinite and function, each of them, not in any single role alone, but in many different roles that change according to the linguistic and extra-linguistic context. Making language more precise would make it poorer by limiting the range of the purposes it can serve. The price to be paid would therefore be very high: language would become less useful as a means of expressing and communicating ideas, and as a means of communication in general.[25] (c) It is only imprecise terms which are suitable for an adequate mapping of the real world, with regard to all the intricacy and indefiniteness of phenomena which take place in that real world, with all their fluidity of forms and indistinctness of boundaries. Terms deprived of this feature can be used at most in discussions about an ideal world, about the extra-temporal and extra-spatial phantasms of Platonism. If we were to want to use such terms, we should find ourselves — as Wittgenstein so picturesquely puts it — in the position of a person who wants to move on an ideally smooth ice surface, which causes no friction, and in a sense therefore in ideal conditions, and who just because of these ideal conditions cannot take even a single step. We have to go back, exhorts Wittgenstein, to a rough surface which provides friction. That rough surface is, of course, everyday language, which is said to owe its usefulness precisely to its being far from the ideal of precision.

Imparting precision to terms which play an essential role in the philosophical disciplines is said to result in additional difficulties. As Wittgenstein says, philosophical problems are born of reflections on matters which we know from everyday life and from ordinary thinking, and that is why they are couched in ordinary terms. By making these terms more precise we distort the meanings that they have had so far, and we thereby modify the problems formulated by means of those terms. As a result, we seek solutions to problems different to those which served as the starting point of our reflections, and we arrive at answers which are not proper answers to the questions originally posed. This is, he adds, the source of the specific difficulties so acutely felt in the sphere of philosophy: numerous unnecessary complications, paradoxes, misunderstandings, apparent problems, endless and sterile discussions. The only effective cure is to restore to the terms involved the current meanings in which they are ordinarily used. Philosophy may in no respect deviate from actual linguistic usage. What it needs is not superterms, but terms of the same kind as 'table', 'lamp', 'door', etc. Accordingly, solving philosophical problems, or eliminating them if they are pseudo-problems, requires that the expressions used be made less, and not more, precise. Whoever believes that ordinary terminology does not suffice because it lacks precision is just mistaken.[26]

These are more or less the arguments which descriptionists adduce to support their primary demands. It is not our intention to analyse these arguments critically in every detail. Besides, that does not seem necessary. Even a cursory analysis of the arguments shows that they are not convincing. We shall therefore confine ourselves to the essential issues.

Without a doubt, the most important role is played by arguments which refer to utilitarian evaluations, whether positive or negative. As we have seen, these arguments question the usefulness of endeavours intended to make accepted terminology more precise; they try to demonstrate that terms which have been made precise are useless or even detrimental. Stress is laid on the fact that the evolution of language establishes the use of serviceable concepts and drives out of circulation those which are useless. These evaluations are made with reference to the purposes which language is meant to serve. It is not, however, stated explicitly what these purposes are, in spite of the fact that they are numerous and differ from one another essentially (suffice it to mention the information functions of language, its expressive, impressive, etc., functions, as distinguished by the descriptionists themselves). Is it necessary to prove that what is useful with respect to some of these purposes may not prove useful with respect to others? Do we especially have to convince others that the conditions which determine the usefulness of concepts from a theoretical point of view (e.g., the conditions which refer to the degree of precision of concepts) do not in the least coincide with those which determine the usefulness of concepts in everyday practice? Moreover, it is probably obvious that the changes which language undergoes during its evolution at the very most adapt it to the needs of everyday practice, and it is at the very most from the point of view of these needs that we may say that the use value of concepts is debased as the precision of those concepts increases. All this is too well known to need to be discussed here in any detail.[27]

The second (and last) issue is that of the concept of precision itself. No great amount of perspicacity is required to observe that the arguments analysed here use this concept in at least two different ways.

It would be difficult to understand the argumentation in proof of the detrimental consequences of precision (see the third argument and its corollaries above) without assuming that what is meant is an absolute precision that excludes all polysemy and all ambiguity. But then these arguments are also used against endeavours at precision in a much looser sense, where it is subject to gradation; and yet we may talk about making expressions precise not only when we impart to them the aforementioned absolute precision, but also when we reduce only to some extent the indefiniteness of their extensions or partially reduce the variety of their meanings. This is another illegitimate step in argumentation, which additionally weakens the force of the arguments used by descriptionists.

Let us take this opportunity to note that this lack of uniformity in interpreting precision seems to have played a significant role in shaping the fundamental ideas of descriptionism. Let us revert for a while to the origin of this trend. The established opinion — which we adopted at the beginning — is that descriptionism is a protest against reconstructionist tendencies. On a closer examination, however, we are inclined to believe that this protest was originally directed only against the most radical variation of reconstructionism, namely the variety represented by Russell and Wittgenstein in the epoch of *Tractatus Logico-Philosophicus*, which deserves to be termed constructionism rather than reconstructionism. It was an appeal to abandon, in philosophy, the everyday language and to replace it by an *ideal language*, constructed according to the pattern of formalized languages. Such a language would have strictly fixed rules of syntax, each word in its vocabulary would be defined precisely and unambiguously, and its structure would adequately map the structure of facts.[28] Now, descriptionism emerged from the criticism of that programme, from an attitude that was hostile to the principle of precision (let us add: perfect precision), which was meant to help to put that programme into effect. It was probably due to the ambiguity of the concept of precision that the conflict, originally waged on narrower ground (we might say, by way of paradox, "the conflict of Wittgenstein *versus* Wittgenstein", that is, the Wittgenstein of the later period *versus* the Wittgenstein of the earlier period), moved imperceptibly on to broader ground to become a campaign against every endeavour at precision, even though the descriptionists could have remedied the evil they had in mind without resorting to such drastic measures.

Be this as it may, it is clear that when we judge the issue on its merits we may say that the descriptionists are right only when they oppose the abuses made by their opponents. They are in the wrong, however, when they themselves are guilty of abuses. If both parties, the descriptionists and the reconstructionists, were to prove willing to rid themselves of extreme positions and (excessive — Tr.) simplifications, they would probably meet somewhere on the territory of reconstructionism à la Mill and Carnap, a reconstructionism which strives for a compromise in the conflict between the need for precision and the necessity of preserving accepted meanings in consideration of the possibility of communication. The descriptionists would then have to submit to a thorough revision of their criticism of the reconstructionists' concept of analysis, in particular their criticism of the definitional method, a criticism which was almost entirely the result of a firm rejection of the requirement of precision in favour of the requirement of agreement with everyday language.

The programme outlined by the reconstructionists (let us add: the moderate ones) seems, therefore, to be the most reasonable. It must be borne in mind, however, that that programme, too, does not preclude the possibility of deviations and abuses. It is by its very nature only an outline, and as such it admits of different implementations. Nor is a correct implementation always chosen. This applies in particular to the analysis of theory. It has been correctly pointed out that reconstructionist analyses occasionally deviate so much from the theories they investigate that it is sometimes even difficult to identify the theory in question.[29] Having recourse to formalizations (and to mathematical language) in cases where such a procedure can at most ensure a semblance of precision is correctly criticized as counterproductive; it is also emphasized, no less correctly, that the semblance of precision is the greatest sin against precision.[30] The descriptionists, however, have worse sins on their conscience, if only because their sins are due to the descriptionist programme itself, and not merely to its faulty implementation. It would be difficult not to be full of reservations toward an attitude which repudiates concern for the logical culture of language, which calls for the abandonment of all endeavours to make concepts and reasonings more precise, and which opposes the campaign against vagueness, thus associating itself in that respect with irrationalistic trends. It would also be difficult to have any doubts as to which of the two trends, reconstructionism and descriptionism, come closer both to what science strives for and to the requirements of common sense.

Notes

[1] L. Wittgenstein, *Philosophical Investigations*, 2nd ed., Oxford 1958, pp. 31ff.

[2] Cf. M. Black, 'Definitions and Presuppositions', in: *Problems of Analysis*, New York 1954; and M. Scriven 'Definitions, Explanations and Theories', in *Minnesota Studies in Philosophy of Science*, Vol. II, Minneapolis 1958.

[3] An alternative definition is not accepted by the descriptionists for other reasons as well. Some reject it because, contrary to the development trends in everyday language, it turns "open" terms into "closed" ones by restricting the alternative of admissible criteria to those which are known from the accepted usage of the term to be defined, which eliminates other possibilities of fluctuation of meanings, fluctuations which are in fact inevitable (cf. L. Wittgenstein, op. cit., pp. 32ff). Others reject it because the various shades of meaning of expressions in ordinary language come so close to one another and are so closely intertwined as to form a nebula as it were, a conglomerate of meanings, a unity that cannot be broken down into elements. Hence the formulation of an alternative definition adjusted to the accepted usage of such words is an impossible task (cf. F. Waismann, 'Language Strata', in: A. Flew (ed.), *Logic and Language*, London 1963, pp. 12–13).

[4] L. Wittgenstein, op. cit., pp. 31ff, 46, *et passim*; M. Scriven, op. cit., pp. 105ff; see also M. Black, op. cit., pp. 24ff.

[5] Cf. J. Kotarbińska, 'Definicja' (On Definition), *Studia Logica*, II, 1955. The examples mentioned above can be found in papers by A. Flew, 'Philosophy and Language', and J. O. Urmson, 'Some Questions Concerning Validity', in: A. Flew (ed.), *Essays in Conceptual Analysis*, London 1956.

[6] The schemata suggested here are naturally not claimed to be completely precise. The point is rather to make them differ as little as possible from the examples which are to serve as generalizations. For instance, there is no restriction here to any specified language — a restriction which must in one way or another be introduced.

[7] L. Wittgenstein, op. cit., pp. 35ff, 52, 13ff, *et passim*. His brief formulation: "One cannot guess how a word functions. One has to look at its use and learn from that" (p. 109) is particularly significant.

[8] J. Kotarbińska, 'On Ostensive Definitions', Philosophy and Science **27** [I], 1960, or in *Twenty-Five Years of Logical Methodology in Poland* (ed. by M. Przełęcki and R. Wójcicki), Warsaw 1977.

[9] In each of these cases we mean a similarity which is sufficient to be taken as a criterion of applicability of the term N (see the paper mentioned in footnote[8]).

[10] Cf. P. F. Strawson, *Introduction to Logical Theory*, London 1952, Chaps. 1–3.

[11] P. F. Strawson. op. cit., pp. 78–90 *et passim*. See also G. Ryle, *Dilemmas*, Cambridge 1954, Chap. VIII.

[12] Cf. P. F. Strawson, op. cit., Chap. 8 *et passim*; G. Ryle loc. cit.

[13] This is clearly formulated by Quine (cf. *Methods of Logic*, New York 1950, pp. xi, xii, 43), whom the descriptionists class as a typical representative of their opponents. This is also the position taken by almost all authors of handbooks of school logic, which is proved by the fact that they warn against the error of equivocation and that of *quaternio terminorum*.

[14] This idea has been developed by K. Ajdukiewicz in his paper 'Okres warunkowy a implikacja materialna' (Conditional Sentence and Material Implication), *Studia Logica*, IV, 1956. Quine's position is similar. It is worth noting that the Oxford School also attaches considerable importance to distinguishing between the conditions of the truth of statements and the conditions of their correct use (cf. P.F. Strawson, op. cit., pp. 18, 82ff, 175–9, *et passim*).

[15] Cf. Z. Czerwiński, 'O paradoksie implikacji' (On the Paradox of Implication), *Studia Logica*, VII, 1958. See also H. Reichenbach, *Elements of Symbolic Logic*, New York 1947, Sec. 7 *et passim*. This relationship between the logical interpretations of sentential connectives and the current interpretation of conjunctions is also borne out by P. F. Strawson (cf. op. cit., p. 86).

[16] L. Borkowski, 'Uwagi o okresie warunkowym oraz implikacji materialnej i ścisłej' (Comments on Conditional Sentences and on Material and Strict Implication), in: *The Book in Honour of Kazimierz Ajdukiewicz*, Warsaw 1964.

[17] Cf. P. F. Strawson, *Construction and Analysis*, pp. 102–4, and 'Analyse, science et métaphysique', in: *La philosophie analytique*, Paris 1962, p. 110. See also J. O. Urmson, *L'histoire de l'analyse*, pp. 15–6.

[18] See J. S. Mill, *System of Logic, Ratiocinative and Inductive*, Vol. 2, Chapter IV.

[19] See McKay, *The Logic of Language*, published in the United States in the 1940's and L. Petrażycki, *Wstęp do teorii prawa i moralności* (Introduction to the Theory of Law and Morals – first published in Russian in St. Petersburg before World War I), Warsaw 1930, pp. 109–10.

[20] R. Carnap, *Logical Foundations of Probability*, 2nd ed., London 1957, p. 7.

[21] The mechanism of those processes has already been mentioned in Sec. 3 above.

[22] Cf. M. Scriven, op. cit., p. 109. It is difficult to refrain from commenting here that one would be equally justified in recommending one to give up weeding, since new weeds will come up, or to stop watering one's flowers, since the soil will dry up again. Note, too, that as far as the language of science is concerned the programme for making it precise – gradually, of course, and in part – has not only proved feasible, but has already yielded excellent results.

[23] See G. Warnock, *English Philosophy since 1900*, p. 150; J. L. Austin, *Presidential Address to the Aristotelian Society, Proceedings 1956–7*, p. 11; J.O. Urmson, 'L'histoire de l'analyse', in: *Philosophie analytique*, Paris 1962, p. 16; M. Scriven, op. cit., pp. 106–7; L. Wittgenstein, op. cit., pp. 33, 42ff. (The first two items are quoted from E. Gellner, *Words and Things*, London 1959, p. 54.)

[24] Cf. M. Scriven, op. cit., pp. 110–1; L. Wittgenstein, op. cit., pp. 46–7 *et. passim*; P. F. Strawson, 'Analyse, science, métaphysique', ed. cit., pp. 106, 112

[25] Cf. L. Wittgenstein, op. cit., p. 46; M. Black, op. cit., p. 28.

[26] Cf. L. Wittgenstein, op. cit., pp. 44, 46, 48–9, 51 *et passim*; M. Scriven, op. cit., p. 167; G. J. Warnock ,'Metaphysics in Logic', in: A. Flew (ed.), *Essays in Conceptual Analysis*, London 1956, pp. 92–3; P. F. Strawson, 'Construction and Analysis', in: *The Revolution in Philosophy*, London 1957, p. 103. See also footnote [28] below.

[27] Cf. L. Petrażycki, op. cit., pp. 96ff; J. Kotarbińska, 'Definicja' (On Definition), *Studia Logica*, II, 1955: M. Przełęcki, 'Prawa a definicje' (Laws Versus Definitions), in: J. Pelc, M. Przełęcki, K. Szaniawski, *Prawa nauki* (The Laws of Science), Warsaw 1957; T. Pawłowski, 'Z logiki pojęć przyrodoznawstwa' (Issues in the Logic of Concepts used in Natural Science), *Studia Filozoficzne*, No. 1, 1957.

[28] The origin of descriptionism is easier to understand if we realize the role played by the fear of all kinds of hypostases and the hostility toward all that savours of idealistic metaphysics in the Platonic spirit. The 'superterms', which, as Wittgenstein claims, would be the result of all operations striving at precision, would have their meanings so sharply defined that no concrete object could meet the conditions of their use; they would also include certain abstract terms introduced to make analyses more precise. Examples: 'ideal gas', 'perfectly isolated system', 'proposition', 'linguistic expression' (understood as an abstract entity), 'denotation', 'meaning', etc. The same motif can be found in Ryle, who is far from pleased by the idealistic theories of language formulated by Meinong and Husserl, and by some ideas of contemporary logicians (cf. G. Ryle, 'The Theory of Meaning', in: C. A. Mace (ed.), *British Philosophy in Mid-Century*, London 1957, pp. 249–51, 255–7). Note, too, that one of the reasons for which Wittgenstein and Ryle turned away from definitions is that definitions are meant to define meanings, and meaning is an ideal, abstract, object whose existence one cannot legitimately assume. The misunderstanding is obvious.

[29] Instructive reflections can be found in J. Giedymin's discussion of Carnap's analysis of the structure of empirical sciences (cf. J. Giedymin, 'O teoretycznym sensie tzw. terminów i zdań obserwacyjnych' (The Theoretical Sense of Observation Terms and Observation Statements), *Studia Filozoficzne*, No. 3(38), 1964).

[30] S. Ossowski, *O osobliwościach nauk społecznych* (The Peculiarities of the Social Sciences), Warsaw 1962, p. 254. On pp. 251–6 are to be found examples of such behaviour, drawn from the sphere of the social sciences.

PUZZLES OF EXISTENCE

by

Janina Kotarbińska

1. The controversy over universals has been revived, the parties being the same that clashed in Antiquity and the Middle Ages: realism, conceptualism, nominalism. The difference is that the universals referred to at present are neither "forms" nor "essences of things", nor are they Platonic ideas of concrete objects, but abstract entities of a special kind: sets of individuals, sets of sets of individuals, sets of sets of sets of individuals, etc., interpreted as objects which are essentially non-perceivable, extra-temporal and extra-spatial. Just as centuries ago, the realistic standpoint is that universals exist; the conceptualistic standpoint is that they exist *in mente*, but not *extra mentem*; and the nominalistic standpoint, in opposition to the other two trends, is that the existence of universals is firmly rejected. Conceptualism has few adherents at present, but the other two movements are very strong.

Their fortunes are, however, variable. Until not so long ago we might have been inclined to think that the nominalistic trend had finally won the upper hand, and that realism, which was almost universally declared to be metaphysical and unscientific, would never recover the ground it had lost. Yet even a cursory knowledge of the current literature of the subject shows that the situation is otherwise. Recently, it is nominalism which has been under fire. It is criticized for being at variance, as far as its basic principles are concerned, with the fundamental assumptions of contemporary mathematics, which is being pursued in a realistic spirit, on the basis of set theory and its conceptual apparatus. As we know, the important concepts of all mathematical disciplines have been defined by reference to the concept of set, and the theorems accepted in those disciplines include existential theorems which assume the existence of sets of specified kinds. The situation is similar in other disciplines, particularly in logical semantics. Matters have thus come to a head. It is obvious that mathematics cannot be the loser in this unequal conflict. It is only nominalism which can be the loser, as it certainly will be unless it succeeds in demonstrating that its inconformity with mathematics is merely apparent or can be eliminated in a manner that will not require far-reaching compromise. Therefore, it is not surprising that in these circumstances nominalists concentrate their efforts on finding a way out of the present crisis.

2. The most radical approach, initiated by Leśniewski in connection with his analysis of Russell's antinomy, goes to the very roots of the evil and calls for a fundamental revision of the foundations of mathematics. His criticism is levelled mainly at the set-theoretical (distributive) concept of set. In Leśniewski's opinion, this concept is as a rule introduced by means of highly unintuitive assumptions which, moreover, impose such conditions on the sets that it would be difficult even to guess what the objects to which the axioms of set theory would apply would be like. As a result, set theory — and the whole of mathematics, if the latter is reduced to set theory — becomes a sphere of speculations concerned with mythical objects invented by logicians, and does not in the least contribute to a scientific interpretation of the real world. If we want to avoid these consequences, there is only one way out: we have to do away with the set-theoretical concept of set — a product of minds demoralized by speculative constructions dissociated from the real world — and to revert to the concept of set in its ordinary, and not denaturalized, meaning, according to which it would be correct to say that Black Forest is the set of those trees which grow in that area, and that Ursa Major is the set of those stars of which it consists. In other words, the classical set theory, i.e., the theory of sets interpreted in the distributive sense, should be replaced by mereology, i.e., the theory of sets interpreted in the collective sense.[1]

The differences between these two interpretations are well known. A set in the distributive sense is supposed to be an abstract object; a set in the collective sense is supposed to be a concrete object whose elements are its component parts, whether proper or improper. According to the first interpretation we also have the following relationships: (a_1) an object x is an element of a set of M's if and only if x is an M; (b_1) a set of M's is identical with a set of N's if and only if, for every x, if x is an element of the set of M's, then x is an element of the set of N's, and conversely; (c_1) for every object x, which is identical either with an individual or with a set, there is a set which is not identical with x and of which x is its only element (hence every object is a starting point for an infinite hierarchy of sets of one element each, sets which differ both from that object and from one another); (d_1) there is an empty set which has no elements. For the second interpretation we have the following relationships: (a_2) if x is an M, then x is an element of the set of M's, but not conversely; (b_2) a set of M's can be identical with a set of N's, even if it is not true that every M is an N and conversely; (c_2) every set of one element is identical with its only element (there are, accordingly, only as many sets of one element as there are individuals; sets of one element thus do not increase the number of existing individuals); (d_2) there are no empty sets; if there is a set of M's, then there is at least one object which is an M.

It immediately follows that the collective interpretation of the concept of set differs essentially from the distributive one. The differences between those two interpretations result, however, not only in intended consequences (the adjustment of the concept of set to the requirements of nominalism), but in unintended and (what is worse) clearly undesirable ones as well. It has become evident that the theory of sets interpreted in the collective sense is much poorer than the theory of sets interpreted in the distributive sense, and does not suffice as a foundation of the arithmetic of natural numbers. It is, therefore, too poor to take over all the functions of set theory as the fundamental mathematical discipline. Accordingly, it has become obvious that the obstacles with which the nominalists have to cope have not been removed by the measures described above.

3. Unlike Leśniewski's idea, the remaining proposals advanced by the nominalists tend to "de-Platonize" mathematics while retaining set theory as the fundamental discipline. The basic method they use is that of semantic analysis. In this connection special attention ought to be paid to the singling out, among utterances in the form of sentences, of three kinds: (1) utterances which are meaningful for the literal interpretation of the expressions which occur in them; (2) utterances which are meaningless for a literal interpretation, but meaningful for a metaphorical, substitutive, interpretation; (3) utterances which are meaningless for a literal interpretation and have no substitutive interpretation, and hence are simply meaningless. Utterances of the first kind are those which

satisfy certain specified criteria of meaningfulness, criteria which, let us add, are chosen so that the extensions of sentences coincide with the extensions of those sentences which comply with the nominalistic requirements. Thus, in a literal interpretation, only those sentences are meaningful which are formulated in the language of individuals. Membership of the second category is determined by the secondary criteria of meaningfulness, namely the reducibility of such sentences, by semantic analysis, to sentences which satisfy the said primary criteria. Finally, utterances of the third kind do not satisfy any criteria of meaningfulness, whether primary or secondary, and are thus not in the stock of expressions of a given language. It can easily be seen that mathematical theorems, like all theorems in which the term 'set' or any terms which are its derivatives occur, are treated as having been formulated in a substitutive language. The problem is to find a method which will make it possible to translate any such theorem into a literal language, and hence into a language in which we can speak solely about individuals.

This is the main outline of the approach now under consideration. As we can see, the problem of the criteria of meaningfulness comes to the fore here. We shall examine the issue in greater detail, taking the reistic (concretistic) doctrine developed and worked out in detail by T. Kotarbiński, as our starting point and basis of analysis. We shall note, too, that an analysis of that doctrine, which is usually believed to be the most representative variation of contemporary nominalism, will be the principal object of investigations in this paper.

In discussing reism, we must first of all make a distinction between its ontological and semantic aspects. Both aspects are, after all, closely connected. As an ontological doctrine reism claims that only things exist, or, in other words, that every object is a thing. It is assumed that every thing is a spatio-temporal object which is in principle knowable by perception, and that the verb 'exist' is taken in its fundamental sense. "In that fundamental sense: *A* exists, is the same as: a certain object is *A*, which (...) can either be shortened into: something is *A*, or expanded pedantically: for some *x*, *x* is *A*. Now the reist deems that only things and persons exist, since it is true only of things and persons that certain objects are things or persons. Should a person further ask about the definition of the term 'object', we should have to refer to the meaning of the copula 'is' in singular empirical statements (such as 'this is green', with an indication of a leaf; or 'the Earth is spherical'; or 'I am gay'; or 'Peter is a carpenter') and say that only that, and all that, is an object about which we may meaningfully formulate a singular sentence (of the type '*A* is *B*') with the copula so understood."[2] Ontological reism is a case of ontological nominalism which adopts the following more general thesis: only individuals exist, in other words: every object is an individual. The reists also assert that every individual is a thing, which the nominalistic approach does not claim.

Semantic reism has undergone certain transformations during its development. Originally, its principal thesis was an inductive generalization which stated that, in ordinary linguistic usage, sentences which include nouns that are not names of things are interpreted as substitutive abbreviations of sentences which do not include any such nouns. That was, as we are told in *The Development Stages of Concretism*, "the genetic and morphological nucleus of concretism." The ontological thesis presented itself as an explanatory guess: it is probably true because only things exist and we can speak only about things. Hence if we appear to be speaking about properties, relations and other alleged objects which are not things, we are in fact speaking about things.[3]

These two theses, the descriptively semantic and the ontological, have resulted in turn in certain assumptions regarding the properties and structure of language. These are intended, as may be supposed, to bring about a reconstruction of ordinary speech according to certain guidelines adopted in advance. The point is, on the one hand, to be in agreement with the real world, which consists only of things (if only things exist, then only names of things can occur in such a true, and hence also meaningful, utterance which states by implication that designata of such names exist), and on the other, to be in agreement with ordinary speech, in which such statements as 'Seniority is a transitive relation' are accepted as both meaningful and true. Here assistance can be found in the distinction between utterances interpreted in their literal sense and those which are interpreted in a metaphorical sense closely connected with the distinction between genuine names, or just names, and what are called apparent names or onomatoids. Roughly speaking, semantic reism now becomes a viewpoint according to which utterances which include names of properties, relations, etc. (in general, names of alleged objects belonging to ontological categories other than that of things), are meaningful, and hence also true, only in a non-literal interpretation of those names, and the said supposed names of properties, relations, etc., are merely apparent names, or onomatoids. The thesis of semantic reism includes a number of variants. These are the most important, quoted *verbatim* or almost *verbatim*.

(a) All sentences in which something is seemingly said about an object which is not a thing, are treated as substitutive formulations standing for other sentences, the latter understood literally and predicating exclusively about things. In other words, any 'term' which is not a name of a thing, is held to be an onomatoid.[4]

(b) A reist is a person who interprets ordinary speech so that everything which is a name by appearance, and denotes an alleged object of a category other than that of things, is for him an onomatoid.[5]

(c) Any sentence which formally implies that an alleged name of an object which is of a category other than that of things, is a name of a thing, can be true only on condition that it is interpreted so that it cannot be proved by reference to it that the name in question is a name of a thing. For the sake of brevity, this can be reformulated as follows: any alleged name of an object which is not a thing ontologically is an apparent name (an onomatoid).[6]

All these formulations are focused, above all, on the concept of genuine name, or simply name, on the one hand, and that of apparent name, or onomatoid, on the other. What a name is, can be seen from the following definition: "to be a term is to be usable as a predicate (in a later text: as a subject or a predicate — J. K.) in any sentence '*A* is *B*' with the primary understanding of the copula 'is'. A given saying is usable as a predicate not only if, when substituted for *B*, it makes the whole sentence a true one, but even if it makes the sentence a meaningful one." Since onomatoids can also be meaningfully substituted for *B* in the sentences of the type '*A* is *B*' (for instance, 'Seniority is a transitive relation'), the reservation with the primary understanding of the copula 'is' is intended to make sure that the words usable as subjective complements are genuine names.[7]

The interpretation of this reservation is, however, a source of additional problems. The examples and comments given by Kotarbiński seem to indicate that the copula 'is' is used in its primary sense if and only if it stands between names and if, moreover, it satisfies the formal conditions resulting from Leśniewski's axiom. Yet, as has been noted by Ajdukiewicz, that combination of criteria appears to be a vicious circle (Ajdukiewicz's original formulation was "All that system of definitions looks like a circle, though not a vicious one."),[8] and at any rate it does not provide a method which would make it possible to distinguish, in concrete cases, expressions which are (genuine) names from those which are not. This is why, in practice, when the reist has to decide whether a given expression is, or is not, a (genuine) name, he makes use of criteria which do not refer to the syntactic properties of the expressions in question, but to their semantic properties. In general terms, he will class *N* as a name only (and probably always) if it follows from the very meaning of that *N* that *N* is such and such a thing, or simply that *N* is a thing.[9] It is perhaps worth mentioning here that for such an interpretation the primary thesis of ontological reism is no longer a hypothesis, but an analytic statement which is true on the strength of the rules of language themselves, since in accordance with these rules it is only about a thing that we can meaningfully say, given a primary understanding of the copula 'is', that an object is precisely that thing, or, that that thing does exist.

As regards an apparent name we are told in turn that it is "any word (or phrase) which may meaningfully stand for *B* in a structure of the type '*A* is *B*', but only if that structure plays not its primary role but the role of a substitute and an abbreviation." We are further told that an apparent name, when inserted in a singular sentence, yields a meaningless whole if the fundamental, non-metaphorical sense of the copula 'is' is retained, and that those sentences in which apparent names occur are substitutive expressions to which the structural criteria of truth, used for sentences with genuine names, cannot be applied; in particular, we cannot infer from them sentences of the type 'A exists'. Finally, it is said that "apparent terms (onomatoids) (...) are certain nouns, adjectives, noun phrases and adjectival phrases which because of their appearance are taken to be terms, but are not terms; they can be used meaningfully only in substitutive formulations, and can always be eliminated in favour of genuine terms alone."[10]

When all this is taken into consideration, we have the possibility of the following interpretation of the terms in question, which allows us to avoid the objection of a vicious circle (we offer these formulations with no claim to maximum precision).

(1) An expression *N* is a name if and only if it is assigned, by the ostensive method, to a specified thing or things of a specified kind (briefly: if it is an ostensive term), or if it is definable, completely or partially, by means of ostensive terms as the only ostensive terms occurring in a given definition.[11]

(2) The copula 'is' occurs in its primary role if and only if it occurs in a sentence of the type '*A* is *B*', if that sentence has structural criteria of truth fixed by the axiom of ontology, and if both *A* and *B* are names.

(3) An expression of the type '*A* is *B*' is meaningful in a literal interpretation if and only if the copula 'is' occurs in it in its primary role.

(4) An expression of the type '*A* is *B*' is meaningful in a metaphorical interpretation if and only if *A* and *B* are not names and if it is translatable into a sentence which is meaningful in a literal interpretation.

(5) If an expression '*W*' occurs in a formulation *Z* of the type '*A* is *B*' as the subject or the predicate, then '*W*' is an apparent name if and only if it is not a (genuine — Tr.) name and if *Z* is meaningful in a metaphorical interpretation.[12]

(6) An expression which has the form of a sentence is meaningful if and only if it is meaningful either in a literal or a metaphorical interpretation.

In the light of these explanations the thesis advanced by semantic reism, which states that any alleged names of objects which are not in the ontological category of things are apparent names, affirms, first, that utterances in the form of sentences in which alleged names occur are not meaningful in a literal interpretation, and, second, that they are meaningful only on condition that they are translatable into formulations that are meaningful in a literal interpretation.[13]

In the reistic system, the above statement is accompanied by the additional assumption that utterances which include apparent names and which are in current use in ordinary speech and in the language of science are nothing other than substitutive abbreviations of sentences formulated so that they are meaningful in a literal interpretation. This additional assumption, we are told, "is naively intuitive and based on common induction."[14] It is therefore a hypothetical assumption, uncertain and only partially substantiated. It must also be emphasized that the range of observations on which it is based does not go beyond the simplest cases; all endeavours to find adequate translations into a reistic language of more complicated sentences — such as sentences about the properties of properties, about relations betwen properties, etc., in general and in another terminology: sentences about sets of higher orders than sets of individuals — have so far ended in failure. Let us add that in these circumstances the feasibility of this task with regard to such more complicated sentences is highly problematic; hence the meaningfulness of such sentences is problematic, too, since it has been made dependent by definition on their translatability into sentences liable of a literal interpretation, as both a necessary and a sufficient condition.

At this point we have reached the main source of difficulties with which reists have to cope. This becomes quite clear if we realize that the aforementioned more complicated sentences include mathematical theorems as well as theorems of all sciences which make use of the conceptual apparatus of mathematics, and that the meaningfulness of these sentences cannot be affirmed, even with a merely tolerably sufficient substantiation. The conflict with the exact sciences thus becomes quite clear.

The question arises whether reists can avoid this conflict without abandoning their most important goals. To answer this we must take a further step in our analysis. To do this we shall go back to the assumptions of semantic reism quoted earlier, and try to comprehend as clearly as possible the underlying intentions of that standpoint.

4. We can start from the statement (c) above, especially the formulation that "any sentence which formally implies that an alleged name of an object which is of a category other than that of things, is a name of a thing, can be true only on condition that it is interpreted so that it cannot be proved by reference to it that the name in question is a name of a thing."

A commentary on this formulation, expounding in greater detail the ideas expressed here, would be to the point. We shall quote it therefore, almost *in extenso*, even though it is fairly long. It goes as follows:

"A critical reader who sees that such words as 'property', 'relation', etc., are called pseudo-names or appaıent names, will be right if he asks which linguistic system is meant here: is it the terminology of a certain scientist, ordinary literary language, or the language which the present author has imposed upon himself?

"But it is only apparently the case that what one is aiming at here is to describe the usages and conventions of such and such a linguistic system, and it is also only apparently the case that what one is aiming at here is to suggest such and such linguistic conventions. In my innermost intentions, when I call the words 'relation,' 'property', etc., apparent names I mean a certain condition of the truth of sentences in any language in which, according to the intention of the speaker, those words do not denote things. Now in any such language any utterance which formally implies the existence of designata of such words can be true only in so far as it is substitutive or non-literal in nature — if it has a secondary interpretation in which such a proof of existence would not be possible. Consider, for instance, the sentence 'Whiteness is an attribute of snow'. Formally, the existence of whiteness follows from it, on the strength of the formula: 'A est $B \rightarrow$ ex A'. Now we say that in this sentence the word 'whiteness' is an apparent name. By saying this (...) we claim that if this sentence is to be true in the speaker's language, it must be interpreted in a substitutive, secondary sense (e.g., as a substitute of the sentence 'Snow is white'), for in its literal interpretation we could prove the existence of whiteness on the strength of this sentence."[15]

When we consider this and similar statements we can hardly avoid the impression that the motive by which the reist has been guided while formulating his semantic doctrine has been an endeavour to ensure to sentences with abstract terms as subjects (briefly: sentences about abstract entities) an interpretation for which those sentences would have existential consequences incompatible with the main principle of ontological reism. This motive seems to have been decisive for the imposition on the concepts of name, the primary understanding of the copula 'is', a meaningful sentence, etc., of definitional conditions which have resulted in difficulties which are already familiar to us.

We may ask, however, whether — from the point of view of the said intentions — all these conditions are really necessary, in particular, whether it is necessary to refer to assumptions which are mainly responsible for such untoward consequences.

The procedure adopted in the reistic system is basically as follows. Assumptions are adopted as a result of which:

(a) a sentence of the type '*A* is *B*', with *A* standing for an abstract term, is meaningful only on condition that the copula 'is' occurs in it in its non-primary, metaphorical sense;

(b) a sentence of the type '*A* exists' follows logically from a sentence of the type '*A* is *B*', with *A* standing for the same subject in both sentences, if and only if in the latter sentence the copula 'is' is used in its primary sense.

As a result:

(c) consequences of the type '*A* exists', which would affirm the existence of abstract entities, cannot be deduced from the sentences described under (a).

It can be clearly seen now that, first, the meaninglessness of sentences of the type in question, given a primary understanding of the copula 'is', suffices to prevent existential consequences from being deducible from such sentences, but this is by no means necessary: the fact that the copula 'is' does not occur in these sentences in its primary sense is by itself sufficient. Secondly, the sentences in question are here deprived of all existential consequences, whereas the ontological thesis of the reistic system requires merely that these sentences have no consequences of the type '*A* exists', given a fundamental interpretation of the verb 'exist' (see pp. 211–12 above). In both cases the requirements exceed what is really indispensable for the attainment of the desired result.

It might therefore be suggested that this result could perhaps be attained by reference to weaker means, which would not require such excessively severe restrictions, and that this would perhaps be the proper method of avoiding difficulties which tax the reists so much.

The answer will be easier to find if we become aware in greater detail how the verb 'exist' ought to be interpreted when used in its fundamental meaning. Some light is shed upon this issue by a statement which we have already quoted on another occasion (see pp. 211–12 above). According to that explanation: "In that fundamental sense: '*A* exists' is the same as: 'a certain object is *A*'," and an object is defined as anything, and only that thing, about which we can meaningfully formulate singular sentences with the copula 'is' interpreted as in singular empirical statements, while the latter can be meaningfully made only about things. Such a standpoint has numerous consequences. First of all, it follows that sentences of the type '*A* exists', given a fundamental interpretation of the verb 'exist', are empirical statements; secondly, sentences of the type '*A* is *B*', if they contain existential sentences interpreted in their fundamental sense, are also empirical statements; thirdly, the fundamental sense of the verbs 'is' and 'exists' (verb forms of 'be' and 'exist', treated as technical terms — Tr.) is made to depend not only on the syntactic and semantic criteria alone, as was done previously (see p. 215 above), but also on the methodological nature of the sentences in which those terms occur.[16] To distinguish these two meanings, and for lack of better terms, we shall use the term *fundamental meaning* to denote only that narrower, methodological sense, and the term *primary meaning*, to denote only the broader sense, even though in the reistic system the terms *fundamental meaning* and *primary meaning* are used interchangeably.

Having made these explanations we can revert to the issue which has become the starting point of the present considerations. The question was to satisfy two requirements which appear to be discordant. One of them calls for agreement with the guiding ideas of ontological reism, and the other, for avoiding conflicts with statements which are accepted in science. Now it seems that, contrary to appearances, a solution can be found, at the price, however, of a number of radical changes in the present form of the semantic doctrine of the reistic system. Such changes would result in a far-reaching liberalization of the criteria of meaningfulness so far adopted.

The basic method would be to extend the concept of name to cover both those expressions which the reists accept as genuine names, i.e., "names of things," and those which they class as apparent names (onomatoids); to make the primary understanding of the terms 'is' and 'exists' depend, as a necessary condition, on their occurrence together with names, and the fundamental understanding of these terms, on their occurrence with names of things; to make the meaningfulness of statements about abstract entities independent of their translatability into reistic language.

This can be done in two ways. They differ from one another above all by the fact that in one of them the same syntactic functions — usable both as the subject and as the predicative word in any sentence of the type '*A* is *B*' — are ascribed to all names, and hence all names are classed in the same syntactic category; whereas in the other, names are treated as expressions in the various syntactic categories, since it is held that every name is usable as the subject or predicate only in such a sentence of the type '*A* is *B*' as satisfies certain additional conditions, which are different for different kinds of names. The variety of these conditions accounts for the variety of categories of names.

Without going into details, we shall briefly describe each of these methods separately.

The multicategorial concept of names was developed by Ajdukiewicz (by referring to the suggestions of Aristotle and Johnson) for use in the controversy over the universals[17] His point, however, was not consistently to carry into effect the requirements of the reistic system, but, on the contrary, to demonstrate that it is possible consistently to defend the realistic standpoint against reistic criticism. This idea came close to Russell's theory of logical types. In accordance with the assumptions adopted here we can single out among names an infinite number of separate syntactic categories which form a certain hierarchy: the lowest category — that of the zero order — includes names which are usable only as subjects in sentences of the type '*A* is *B*' (names of individuals); names of the first order are those which are usable as predicates in those sentences of the type '*A* is *B*' which have names of individuals as their subjects; names of the second order are those which are usable as predicates in those sentences of the type '*A* is *B*' which have names of the first order as their subjects; etc. The general principle is: names of the k-th order (where k is a positive integer) are those which are usable as predicates in those sentences of the type '*A* is *B*' which have names of the $k-1$-th order as their subjects. The multicategoriality of names implies, obviously, the multicategoriality, and thus the systematic polysemy, of all functors of name (term) arguments. For instance, the copula 'is' may have the category of a sentence-forming functor of two name arguments of zero and first order, respectively; of first and second order, respectively; of second and third order, respectively; etc. To use Ajdukiewicz's notation, its categories could be symbolized thus:

$$\frac{z}{n_0\, n_1}, \qquad \frac{z}{n_1\, n_2}, \qquad \frac{z}{n_2\, n_3}, \quad \ldots, \text{etc.}$$

A language which has such a syntax (let it be called language *L*) is considered by Ajdukiewicz, apparently with justice, to be an admissible intepretation of everyday language. It is, as we can see, a multilevel language, consisting of a number of partial languages which differ from one another by the category of names and the category of functors of name arguments. Moreover, each such language includes names of two kinds: those of the *k*—1-th order, which are usable in that language as subjects in singular sentences (and only as subjects), and names of the *k*-th order, which are usable as predicates (and only as predicates) in such sentences. Each such language also includes name - argument functors of the appropriate categories. The system of the calculus of names (terms), on which such a language is based, is also multilevel. Under additional assumptions, each of its partial systems is constructed according to the pattern of Leśniewski's ontology with the same shape of functors and the same structure of axioms, definitions, and theorems. Hence sentences of the type '*A* is *B*', '*A* exists', '*A* is an object', etc., occur at all levels of language *L*; they have at all levels the same structural conditions of truth, but at each level they have a different meaning. Let us suppose now, that the lowest-level language satisfies the conditions which, in the reistic system, are imposed upon the language about things, whereas the remaining partial languages are not about things, but about abstract entities. The copula 'is' therefore occurs at all levels in its primary meaning, but it occurs in its fundamental meaning only in the language about things, that is, in the lowest-level language.[18]

Now it is obvious that given such assumptions neither the meaningfulness nor the truth of statements which have the structure of elementary sentences with abstract names as their subjects depends on the translatability of those statements into a language about things. It is also obvious that the acceptance of statements about abstract entities is now not at variance with the principal guidelines of the reistic system, since none of these statements assumes the existence of abstract entities in a fundamental interpretation of the verb 'exist'.

The situation is the same in this respect with regard to the second of the concepts of name previously mentioned, that, is, the unicategorial approach. We are then dealing with a single-level language, in which 'is' and 'exists' (like all other functors of name arguments), when used in their primary role, occur in the same meaning regardless of whether their arguments are names of concrete objects or names of abstract entities. In that language we can therefore truly assert, given a primary understanding of the verb 'exists', both that such and such individuals exist, and that the class of those individuals, the class of classes of those individuals, etc., exist. On the other hand, given a fundamental interpretation, as singled out in the by now familiar way (see pp. 218–19 above), we can truly predicate existence about things alone.

The final result is, of course, the same as before. The reists can accept the existence of abstract entities of the most diverse kinds without running the risk of being at variance with their own doctrine, regardless of whether given existential statements can be interpreted as statements about things. This eliminates their conflict with mathematics.

It could be argued, however, that this approach makes the reistic doctrine so closely resemble the opposite standpoint that the difference between the two is almost completely blurred. Now, this would in fact be true if the opposition between those two standpoints were reduced to an opposition of theses.[19] However, in addition to their theses, the reists also advance a programme. They demand that despite partial failures we should not stop searching for methods that will make it possible, on an increasing scale, to reduce statements about abstract entities to those about concrete objects, and to strive stubbornly for the elimination of apparent names.[20] This programme has not lost its validity and continues to be a specific trait of the reistic doctrine.

The point is, however, that to substantiate this programme one can no longer refer to the argument that has been used so far as the principal one. It is not possible to claim that the meaningfulness of statements about abstract entities is ensured only if we are able to translate them into a language about concrete objects. The other arguments, however, retain their validity. They emphasize above all (a) the agreement of the reistic interpretation with ordinary linguistic usage;

whenever, in current speech, we use abstract terms we only seem to speak about abstract entities, and in fact our genuine intentions are to speak about things; (b) the advantages of such an interpretation in philosophical reflections: problems which have been the subject of endless sterile discussions often vanish immediately as being ill-posed if we stick to the principle that in our ultimate formulation we should use no names which are not names of concrete objects; (c) the psychological naturalness of the reistic guideline: we grasp the intuitive sense of statements about abstract entities only when we are able to translate them into statements about concrete objects.[21]

The strongest emphasis is usually placed on this last argument. This stress largely refers to the sometimes glaringly non-intuitive nature of the concepts of set theory and of the disciplines derived from it. There is nothing strange in this fact, for it would be difficult to treat as intuitive such concepts as that of existence, for which in at least some cases existence depends exclusively on the syntactic properties of language;[22] or that of the distributive interpretation of sets, since in this interpretation existence is ascribed to sets precisely because of the somewhat peculiar criteria of applicability of that term; or the concept of reality, which identifies reality with the totality of not only all concrete objects, but all abstract objects as well (the latter being sometimes termed ideal or "non-real"), i.e., sets; or the concept of semantics as that sphere of research which, while intended to link language with extra-linguistic reality, links it, *inter alia*, with that "non-real reality" which consists of sets of all possible kinds, including — let us note — sets of non-existent objects. The non-intuitive character of set theory is well known, being cited even by those who cannot be classed as supporters of the reistic approach.[23]

It would seem that the arguments to which reists refer are a sufficient justification of their programme. Even if it were to turn out that this programme can be implemented only on a limited scale, it would be reasonable to put it into effect wherever possible. To turn something which is less comprehensible into something more comprehensible and to eliminate at least some clashes with intuition is always an intellectually profitable undertaking.

In conclusion we might say that unlike Berkeley, who once recommended that we should think like scholars and talk like the common people,[24] the reists strive to talk like scholars and to think like the common people, whereas the realists want to think like scholars and to talk like scholars. The remaining possibility is to think like the common people and to talk like the common people. To complete the picture it is worth adding that the last formulation corresponds almost exactly to the trends represented by Wittgenstein in the last period of his activity and to the standpoint of the Oxford School.

Notes

[1] S. Leśniewski, 'O podstawach matematyki' (On the Foundations of Mathematics), *Przegląd Filozoficzny*, Vol. 30, No. 2–3, 1927 (especially pp. 166–9 and 186–202), and Vol. 31, No. 3, 1928 (especially pp. 261–3). Similar ideas are to be found in S. Leonard and N. Goodman, 'The Calculus of Individuals and Its Uses', *Journal of Symbolic Logic*, Vol. 5, 1940; in N. Goodman, *The Structure of Appearance*, Cambridge, Mass., 1951; and in J. Słupecki, 'Towards a Generalized Mereology of Leśniewski', *Studia Logica*, VIII, 1958. (English-language readers will find a more comprehensive discussion of Leśniewski's ideas in E. C. Luschei, *The Logical Systems of Leśniewski*, North-Holland 1962, — Tr.)

[2] T. Kotarbiński, 'The Reistic, or Concretistic, Approach', p. 45 in the present book. It may be helpful to explain that reism is based on Leśniewski's ontology and uses the concepts characteristic of that system. In that ontology the only primitive term is '*est*' ('*is*'), introduced by the axiom $\Pi A, B\,(A\ est\ B \equiv$ $\equiv (\Pi x\,(x\ est\ A \rightarrow x\ est\ B) \wedge \Sigma x\,(x\ est\ A) \wedge \Pi x, y\,(x\ est\ A \wedge y\ est\ A \rightarrow x\ est\ y)))$. The term *ex* ("exists") is defined thus: $\Pi A\,(ex\ A \equiv \Sigma x\,(x\ est\ A))$. This obviously yields the theorem: $\Pi A, B\,(A\ est\ B \rightarrow ex\ A)$, by which a statement of the type *A est B* has en existential consequence *ex A* which applies to the designatum of A.

[3] T. Kotarbiński, *The Development Stages of Concretism*, in the Supplement to *Gnosiology*, Pergamon Press, 1966, pp. 429–37.

[4] T. Kotarbiński, *Gnosiology*, ed. cit., p. 51.

[5] T. Kotarbiński, *Wybór pism* (Selected Works), Vol. II, Warszawa 1958, p. 75.

[6] *Ut supra*, pp. 109–10. Both (c) and (a) have two formulations each: both cases formulations which form a given pair are treated as synonymous. These will be marked (a_1), (a_2), (c_1), (c_2).

[7] T. Kotarbiński, *Gnosiology*, ed. cit., p. 7. See also his *Selected Works* (cf. footnote[5]), Vol. II, pp. 232–3. (It is to be noted that the 'term' used in *Gnosiology* corresponds to the term 'name' often used in the present book. It would appear that neither is a completely satisfactory rendering of the Polish term '*nazwa*', simply because of a certain vagueness of the latter. — Tr.)

[8] From K. Ajdukiewicz's review of the first Polish edition of *Gnosiology*, included in the Supplement to the English-language version. The formulation mentioned here is on p. 519 of the latter version. (Tr.)

[9] T. Kotarbiński, *Selected Works*, Vol. II, pp. 232–3.

[10] T. Kotarbiński, *Gnosiology*, pp. 9 and 406; see also his *Selected Works*, Vol. II, pp. 74 and 232–3.

[11] This interpretation is partly based on the formulations found in *Gnosiology*, p. 9, and in the paper 'The Reistic, or Concretistic, Approach', also in the present book. It is to be noted that at the time when the Polish version of *Gnosiology* first appeared partial definitions were not yet known. Yet partial definitions should also be taken into consideration; this is justified by the present-day approach and at the same time would not appear to be at variance with the intentions of the reists.

[12] Formulations (3), (4) and (5) are, for the sake of simplicity, given only for statements of the type '*A* is *B*'. However, since in ontology all simple statements are definitionally reducible to statements of that type, and since all compound statements are combinations of simple ones, the same formulations can easily be made to cover all other statements.

[13] Note that, if this analysis is correct, then of the two statements, (a_1) and (c_1), which in the reistic theory are considered to be synonymous with (a_2) or (c_2) (cf. footnote[6] above), (a_2) alone evokes no doubts or objections. In the case of (c_2), it seems to differ in content from (a_1) and (c_1); in particular, it does not state — not directly anyway — that translatability into the language of things is a necessary condition of the meaningfulness of the statements under consideration. It points rather to the motives which account for the adoption of that assumption.

[14] T. Kotarbiński, *Gnosiology*, ed. cit., p. 434.

[15] T. Kotarbiński, 'Uwagi na temat reizmu' (Comments on Reism), *Ruch Filozoficzny*, Nos 1–10, 1930–1 (quoted from the text reprinted in *Selected Works*, Vol. II, pp. 108–9).

[16] The last issue is certainly the most controversial of all. It does seem, however, that making the semantic role of the terms 'is' and 'exists' depend on the methodological properties of the statements in which those terms are used is fairly natural. Moreover, this is not specifically linked to the reistic standpoint. For instance, this assumption is adopted by K. Ajdukiewicz in his paper 'W sprawie pojęcia istnienia' (On the Concept of Existence), 1951, reprinted in *Język i poznanie* (Language and Cognition), Vol. II, 1965. It is also in harmony with the tendencies discussed by R. Carnap in 'Empiricism, Semantics and Ontology', *Revue Internationale de Philosophie*, II, 1950, and with H. Reichenbach's views as formulated in *Experience and Prediction*, 5th ed., Chicago 1957, Sec. 24 *et passim*.

[17] Cf. K. Ajdukiewicz, 'W sprawie uniwersaliów' (Concerning Universals), *Przegląd Filozoficzny* 1934 (reprinted in *Język i poznanie* (Language and Cognition), Vol. I, Warsaw 1960.

[18] Cf. footnote[16] above. To avoid misunderstandings it should be said that the primary meaning of the copula 'is' is interpreted here as in Sec. 3, formulation (2), above, but with the understanding that the term 'name' is interpreted more broadly.

[19] In 'The Development Stages of Concretism' (first published in Polish in 1958; for an English-language version see the Supplement to *Gnosiology*, ed. cit.), T. Kotarbiński even shifts his main emphasis from theorems to a programme. He maintains that the reistic doctrine in its earlier stages was hypothetical in nature and concludes that "in its mature form concretism absolutely insists on its programme only." (Cf. *Gnosiology*, p. 435) This is certainly a much more cautious formulation, and far less open to objections and criticism than the previous one. It does not, however, reject the previous concept of meaningfulness (see the paper under consideration *in fine*) and thus does not remove the essential obstacle, i.e., the conflict with mathematics and the exact sciences.

[20] *The Development Stages of Concretism*, loc. cit.

[21] Cf. the two papers on concretism in the Supplement to *Gnosiology*, ed. cit.

[22] If a language L assumes set theory, and if an expression M is a predicate in L, then as we know, we may rightly conclude in L that there is an infinite hierarchy of sets: the set of M's, the set of the sets of the sets of M's, etc., *ad infinitum*.

[23] Cf. A. Mostowski, 'Thirty Years of Foundational Mathematics', *Acta Philosophica Fennica*, XVII, 1965, p. 7, and H. Mehlberg's paper in *Logic and Language*, Dordrecht 1963, pp. 79 and 84.

[24] Cf. G. Berkeley, *A Treatise concerning the Principles of Human Knowledge*, Dublin 1710.

VAGUE WORDS

by

Adam Schaff

(Fragment)

Both in the case of everyday language and in the case of the specialized language of science we are always faced with the fundamental problem: what must we do so as not to be misled by an improper use of language? When we speak of being misled by a certain use of language we have two cases in mind: first, when language performs its communicative function in a defective manner so that the speaker is unable to convey his ideas to the listener; and second, when the language in which we think imposes upon us false ideas about the real world (cf. the problem of hypostases), and does so through its structure and forms which, in the case of natural languages, have been fixed by tradition. Such and similar problems are due to various factors. It is these problems which have suggested the idea that language must be not only an instrument, but an object, of research — an idea which has enriched 20th century philosophy through ontological, gnosiological, and methodological studies of languages. These defects of language have become the target of semantic analysis, understood as a semantic analysis of words. Such operations are intended to eliminate factual and logical errors which burden the process of thinking and which make interhuman communication and also one's communication with oneself (in the process of thinking) more difficult. Such errors are, among other things, due to the ambiguity and vagueness of words.

I am not raising here the issue of the obscurity of statements which express certain ideas in an incomprehensible manner. It is irrelevant whether a statement is obscure because the speaker has intentionally tried to make it incomprehensible (like certain philosophers who used to take professional pride in the esoteric form of their statements) or because he has not been competent enough to formulate his ideas correctly. In any case, we are dealing here with a mistake which is of a subjective nature and therefore less interesting.

The situation is completely different when it comes to statements which include ambiguous and vague words.

An ambiguous word links different meanings with one and the same sound or combination of sounds. It is an objective linguistic fact that can normally be explained by reference to the history of a given language. The risk of a confusion of different meanings and the resulting risk of misunderstandings and logical errors in reasoning is reduced by the fact that ambiguous words are usually disambiguated by the context formed by the sentence or group of sentences in which they occur. This is the case because the context indicates the meaning in which a given word is used in a given case. Moreover, the simple operation of explicitly specifiying the meaning in which a given ambiguous word is used in a given case eliminates the danger of misunderstanding. That is why this case may be disregarded in further analyses.

We shall concentrate here on the issue of vague words. In this connection we shall be interested in the ontological aspect of the issue, that is, in the problem of the origin of vague words and in the subjective or objective nature of their vagueness. The methods of measuring the degree of vagueness and the precise use of vague words[1] are important issues, but are nevertheless derivative with respect to the ontological analysis of the origin of the vagueness of words, for unless this problem is solved we are not in a position to answer the question of how to counteract or reduce that vagueness (or, at the very least, such an answer will be difficult to formulate).

Is the vagueness of words subjective in nature?

To answer this question we must first of all state precisely what we mean by the term 'vague word'. I shall refer in my analysis to two essays, both bearing the same title (one by Betrand Russell and the other by Max Black), which I believe to be of particular importance in the literature of the subject.[2]

The literature of the subject — and this applies to the two essays just mentioned as well — is mainly, if not exclusively, concerned with the problem of vague terms. I consider this to be a unjustified narrowing of the subject matter of research, which should cover all lexical items. For if a vague term is one which divides the universe into its extension (the class of its designata) and the complement of that extension (the class of objects which are not its designata), so that there are objects which we are unable to class in either category, then such a trait is also shared by words which are not terms. In this sense, not only terms, i.e., nouns and adjectives, but verbs, adverbs and conjunctions are also vague. If we find it difficult (not for subjective reasons, as we shall see) to answer the question "is this a river?", because the objects which may be designata of the term 'river' include those about which we are unable to decide whether or not they are still rivers (since they are rivulets, streams, etc.), then we may face a similar difficulty when confronted with the questions 'is this a red object?', 'does it walk?' (or, e.g., creep, crawl, etc.), 'was that a noble act?', 'did he behave bravely?', etc. Thus it is not only in the case of terms whose designata are individuals or classes of individuals, but also in the case of other lexical items which state something about objects, their actions, the mode of their actions, etc., that we are dealing with vagueness in the sense discussed above (i.e., inasmuch as there are actions, modes of action, relationships, etc., about which we cannot say — and not for subjective reasons — whether given lexical items fit them or not). As we shall see later, we may also have doubts (as Russell obviously has) as to the non-vagueness of such words as 'or', 'if', etc., which occur as logical constants.

Before we proceed to define the vagueness of words in a more rigorous manner and to investigate the origin of that vagueness, we must make a clear statement on the following issue: of what is vagueness an attribute? Is it an attribute of linguistic formulations or of the facts to which such formulations refer? This is very important for a study of the problem of the subjective *versus* the objective nature of the vagueness of words.

We must make a clear distinction between the question of the origin of the vagueness of words and that *of what* is vagueness an attribute. Both issues are somehow connected with the intriguing nature of vagueness, but these connections differ in aspects and approaches.

In his essay mentioned above, Russell writes that vagueness is characteristic of "that which represents things (language being a case of such a representation), and not of things as such." Apart from representation, whether cognitive or mechanical, there can be no such thing as vagueness or precision; things are what they are, and there is an end of it" (p. 85). I am in full solidarity with his statement, and I consider it very important for an understanding of the problem. Things are neither vague nor precise, just as they are neither true nor false. Things are always things and nothing more. What is vague is our knowledge of things and the linguistic formulations which reflect this knowledge of ours, just as it is our knowledge and the corresponding linguistic formulations which are true or false. In the case of both vagueness and truth, we are dealing with the characteristics of the relation between the cognizing subject and the real world (a relation which is always a combined mental and linguistic operation), and not with characteristics of the real world alone.

The answer to the question *why* cognition or the corresponding linguistic formulation is vague is, however, a quite different issue. Are the causes of this fact objective or subjective in nature? To answer this we have to go back to the definition of vagueness.

Let us begin with the problem of terms (in interpreting the meaning of the word 'term' I am following Kotarbiński: to be a term is to be usable as the predicative word in any sentence of the type '*A* is *B*' given a primary understanding of the copula 'is').

We shall also make a distinction between singular and general terms, that is those which denote only one object each (the relation between a term and its designatum being one-one) and those which denote many objects each (the corresponding relation being one-many). We shall analyse in some detail the case of a term which we can characterize as being vague beyond all doubt.

Now, water is flowing in a bed which it has channelled in the soil. Let us suppose we are dealing with the Vistula near Warsaw, the Seine near Paris, and the Thames near London. We would not hesitate to say that the Vistula, the Seine, and the Thames are rivers. Thus, the term 'river' is a general name (the corresponding relation being one-many) of such objects as the Vistula, the Seine, and the Thames. In a great number of other cases as well we encounter no difficulty in answering the question: 'is that a river?', for instance, when we are referring to the Volga, the Danube, or the Mississippi. The answer is determined mainly by the length and width of the bed through which the water flows. But our globe contains a large number of such objects, which are a mass of water flowing in a bed channelled through the soil, even though the number of such objects is finite. Moreover, the beds differ in width, depth, and length. Suppose, now, that we have arranged all the objects into a single sequence, following the three dimensions of their beds as indicators — from the largest to the smallest. Let us now try to label these objects with names like 'river', 'rivulet', 'stream', 'creek', 'brook', 'spring', etc. We immediately see that at the ends of the sequence we are dealing with objects which we can class without any difficulty as far as the range of the said terms is concerned, but the further we move toward the centre of the sequence the greater the difficulty we face. We can easily mention cases in which we are at a loss whether a given object is to be called a river or a brook. Note that this difficulty is not due to our ignorance, but to the fact that in the limiting cases the characteristics which determine the use of such terms become blurred (in this particular case we mean the dimensions of the bed). In such cases we often say 'rivulet' to indicate, by the use of a diminutive, that what we mean is neither a river nor a brook. Or else we may say 'small river' or 'large brook'. However, this does not eliminate the problem, but merely shifts it elsewhere: for how are we to draw a distinction between a river and a rivulet, or a brook and large brook? If between the extension of a given term and the complement of that extension there is a border area (which Russell picturesquely calls *penumbra*) which for objective reasons (a lack of clear criteria) cannot simply be classed in one of the two, then inventing a new name to denote that border area or leaving it unnamed does not solve the problem. There is still the question of how to delimitate that no man's land from the areas covered, beyond any controversy, by the extension of the term and its complement, respectively. This complement is usually formed through the aid of a negation (let it be a 'non'), and this term-forming functor 'non', as Kubiński has pertinently pointed out, is used here to construct a term which is not contradictory to a given term, but is its opposite in a narrower sense: the extensions of both terms are included in the universe of discourse U, they are disjunctive, and their union is a proper subset of U, which leaves our problem unsolved, even though by making new distinctions and inventing new terms we can succeed in shifting the boundaries of the extension of a given vague term. This has already been emphasized by Russell. Note, too, that the device most often used in such cases in science, namely making terms more precise by the adoption of certain conventions, in no way settles the issue (although it may have important consequences in practice), but merely shifts the boundaries. For if, for instance, we adopt the convention that by 'brook' we mean an object which consists of a mass of water flowing in a bed that is no wider than five metres, and that an analogous object whose bed is wider than that is termed a 'river', then we merely shift the problem, since, as we know, establishing whether something has a width of no more than five metres involves problems of boundary analogous to, although subtler than, the distinction between river and brook.

What we have seen in the foregoing analysis of the term 'river' recurs in the analysis of such terms as 'red', 'bald', 'heap', 'heap of stones', 'noble', 'brave', etc., that is, in all cases in which terms are used to establish qualitative differences among objects on the basis of sense data, moral or aesthetic valuation, etc., whenever such differences are subject to gradation. The problem boils down to this: when we are dealing with series of objects denoted by common terms and revealing qualitative differences in certain properties, with sequences of colours, attitudes, types of conduct, etc., which share common terms, but also reveal quantitative differences in some respect — differences which are measurable according to a specified scale — are we dealing then with a *continuum* or with distinguishable *quanta* of changes? It is only in the latter case that we could speak, if only theoretically, about a specified point which fully divides the universe into the extension of a given term and the complement of that extension. As long as we know no methods of finding such *quanta* — and we do not in fact know any — all attempts to define precisely the boundaries of the extension of such a term, and thus eliminate its vagueness, are doomed to failure.

This gives rise to a new problem for analysis from the standpoint of Marxist dialectic, which — by following Hegel in the view that quantitative changes at one moment yield a qualitative one, which takes the form of a leap — seems to suggest that Marxists believe in such a *quantum*-like interpretation of all changes. Obviously, such an interpretation — which is, by the way, not supported by any empirical proof provided by the specialized sciences — is imprudent, to say the least, and quite unnecessary, too. Dialectic need not link that law either with any belief in such a *quantuum*-like nature of change or with the belief that terms which fit well must be ideally precise. The transition of quantitative changes into qualitative ones can be perfectly well interpreted, not as an ideal point on the time axis, but as a time interval. This applies not only to social revolutions, which have long been interpreted, from the dialectical point of view, as periods, but to other "leaps" in development as well.

The same may be said about an object which is continuously changing in a certain respect, and which may be termed young at one time and old at another. The sequence to which a general term applies in this case is not a class of different objects which share a common characteristic subject to gradation, but a set of development stages of one and the same object, stages which differ by the degree of a certain characteristic. The problem is the same as before: is this sequence a *continuum*, or is it *discrete*? The answer is also the same as before.

We have already touched upon the problem of the origin of the vagueness of terms, but we are still concerned only with a definition of vague terms and the relationship between vague terms, on the one hand, and singular and general terms, on the other. Such a definition may be formulated on the basis of the examples discussed above: in all of them the focal point is the relation of a term to the set of objects that it denotes, given the proviso that the extension of that term is ill-defined (its boundary is not clearly marked). Our definition could be worded as follows: a term is vague if it denotes a number of objects whose class is not strictly defined. This is also the line of reasoning of M. Black,[3] who says that "vagueness is defined by a finite extension and the indefiniteness of its boundaries" (p. 31). Black criticizes Russell for having confused the general character of terms with their vagueness, which in fact can be found in Russell ("*Per contra*, a re presentation is vague when the relation of the representing system to the represented system is not one-one, but one-many" — p. 89), but in the same essay Russell makes a correct distinction between the general and the vague character of statements, and Black's definition clearly follows Russell's reasoning ("It follows that every proposition that can be framed in practice has a certain degree of vagueness; that is to say, there is not one definite fact necessary and sufficient for its truth, but a certain region of possible facts, any one of which would make it true. And this region is itself ill-defined: we cannot assign to it a definite boundary. This is the difference between vagueness and generality." — p. 88).

When we discuss the applicability of this definition of vagueness to the categories of terms listed earlier, the situation is obvious with respect to general terms. General terms may be vague, but they need not be so. One can say that the generality of a term is a necessary, but not a sufficient, condition of its being vague. For instance, the term 'planet' (restricted to the solar system) is general, but not vague as to its extension: the class of planets of the solar system has nine elements. If, however, a term is general and if, in addition, the boundary of the class of the objects it denotes is indefinite (as in the case of 'river'), then that term is vague.

The fact, however, that a term is vague in some respect does not determine its being vague in another respect. An example is offered by singular terms: such a term is precise as to its extension since it denotes a single specified object, but it is vague when treated as a process (e.g., the life of a person from birth to death), since the same term is used to denote the different stages of that process (including, in particular birth and death, which are themselves processes, so that one may ask whether these terms are already applicable to a person at his birth, and still applicable at his death). The problem of the vagueness of singular terms thus arises when a singular term is a crypto-general one as it denotes consecutive stages in the development of a given object.

A discussion of the grammatical categories of verbs and adverbs will not contribute anything new when it comes to the principles of deciding whether a given term is vague or not, the more so as in both cases we can easily pass to the category of nouns (from 'to walk' to 'the walk', 'the walking', from 'bravely' to 'bravery', etc.); and although we shall be dealing with apparent terms (as is always the case when it comes to abstract terms), the problem with which we are concerned here can be formulated as above.

Such words as 'slowly', 'rapidly', 'little', 'much', etc., are vague through the very intention of the speaker, since they are meant to convey about objects something which either cannot be precisely defined or may be not precisely known. This is why we cannot go beyond a vague statement which, while dividing the universe of discourse by means of disjunctive terms, does not divide it completely: it leaves a subset of that universe about which we cannot decide whether either of the two terms can be used properly.

An interesting problem is posed by the logical constants, such as the sentential connectives 'or', 'not', etc. According to Russell, these are vague, too, since their respective meanings in logic are defined in a way which implies the truth or falsehood of statements, so that the issue of vagueness is involved indirectly. Russell defines the connective 'or' thus: '*p* or *q*' is true if it is true that *p*, if it is true that *q*, and it is false if and only if it is false that *p* and false that *q*. This is why Russell is justified in his skepticism about the precision of logical constants, even though they are assumed to be precise. "All traditional logic usually assumes that precise symbols are being used. Hence that logic does not apply to earthly life, but only to an imaginary heavenly existence. Yet that heavenly existence differs from ours in what is of interest to logic, not the nature of the objects of cognition, but merely the perfection of our cognition. (...) Here I agree with Plato. Yet those who are not fond of logic will — I am afraid — be disappointed with my heavens" (pp. 88–9).

Only a step separates this formulation from the standpoint adopted in fact by Russell, and later by Black, that *all* linguistic signs are vague. I see no necessity for arriving at such extreme and radical conclusions. Of course, it is always possible to demonstrate that a linguistic sign is vague in some respects. But linguistic signs are precise in certain other respects, and there is no need to deny this. It is enough to state that words may be vague, that such cases are frequent (or at least more frequent than is usually assumed), and that vagueness is objective in character.

Our subsequent analysis will concentrate on this issue. But we must first conclude the foregoing discussions by suggesting a definition of vague words. By paraphrasing the definitions of vague terms we shall say that a word is vague if it has a "fringe". This directly occurs if a word states something about certain objects (it does not necessarily have to name them, this being a function of terms) so that the sum of its extension and its complement (i.e., the cases to which a given word, be it 'walks', 'slowly', etc., applies, and those to which it does not clearly apply) does not cover the whole universe. It occurs indirectly if the definition of a word (e.g., a logical constant) is made up of statements which include vague words.

What is the origin of vagueness? Is vagueness objective or subjective in nature? Please note the practical aspect of these questions. Answers to them will affect our possibility and methods of handling the problem of vagueness — a vagueness which not only hampers our communication processes, but also gives rise to specific paradoxes threatening the logical foundations of our thinking. Consider the fact that the principle of the excluded middle is valid only if well-defined words are involved. And wherever the validity of the principle of the excluded middle is undermined, logical contradiction emerges as a natural threat (this follows from De Morgan's laws). This has been convincingly demonstrated by Black (p. 36), who suggests, as a way out, an interpretation different to the usual interpretation of the word 'not' with reference to vague words (Kubiński has taken up this idea with reference to vague terms). If the issue could be reduced to a subjective imperfection in the handling of words, then the situation would differ from one in which objective causes were involved. But what is the actual situation?

Black is positively in favour of the objective nature of the vagueness of words and uses the following criterion to distinguish objective nature from a subjective one: are the manifestations of the vagueness of words facts related to human behaviour (and hence psychological facts) or facts related to the physical world? His resolution of the problem is correct, even though the criterion adopted is not a good one. Vagueness is not an attribute of things, but rather of words. The vagueness of words is not, however, merely subjective — something which is due only to incompetence or error. It originates from the relation between words and the objective world about which these words convey certain information. From what these words are and what that objective world is, it follows that words cannot precisely map all the wealth of the facts which they describe. The importance of this problem requires an analysis in depth. As we shall see later, Russell treats it peripherally and, to make matters worse, inconsistently. On the other hand, Black is interested mainly in the methodological aspect of counteracting or eliminating the vagueness of words, and only just mentions the related ontological problem, a problem which is not a trifling one.

Verbalized cognition, and hence the usual scientific cognition or ordinary cognition in the broad sense of the term, has long been criticized by various philosophical schools. The motif runs from Plato to Bergson. If we reject the irrationalistic consequences of that criticism, which in most cases makes the critics groundlessly believe in some direct, non-verbal "true cognition", then we arrive at the rational element of that criticism, which consists in stressing the imperfection of linguistic means as instruments of mapping the real world.

Verbal signs, which generalize, are products of the process of abstraction. In the meaning of a verbal sign — and it is in meaning that we are interested — we always encounter a result of an abstraction which is an element of the process of classification; classification is based in turn on a given characteristic and dismisses all other characteristics as irrelevant from a given point of view. This is why verbal signs are — as befits logical entities — static, rigid, and non-flexible. By this I mean that a verbal sign immobilizes in its meaning the image of that to which it refers, even if it refers to motion and change, because it refers in a classificatory sense to those phenomena, too, namely by bringing out characteristics common to them and by generalizing them in the form of a specified category. By this I mean furthermore that a verbal sign maps in its meaning certain elements of the world by encasing that mapping in a rigid framework, and by separating — through classification — given things, their properties, actions, etc., from the rest of the world around us. The more precise a term is, and the more rigorous it is from a logical point of view, the more clearly these characteristics of verbal signs are marked.

And yet every element and/or aspect of the real world is changing and in constant movement, and linked with other elements and/or aspects of that world by an infinite number of connections and interdependences. If we disregard these changes, interconnections and interdependences we obtain — to put it figuratively — a cross-section or an anatomical specimen of the real world. And what else could we obtain if we place the changing real world in the Procrustean bed of categories that disregard these changes, and if we try to map an element of the real world, with all its interconnections, with the aid of verbal signs — which have rigid semantic boundaries — verbal signs which become less and less flexible the more precise they are made? If we view the relation between linguistic signs and the real world — a relation which consists in the real world as mapped or reflected by linguistic categories — from such a standpoint, then it becomes clear that the instrument, which is language, does not fit the object to be mapped, and hence that the relation of reflection (between language and the real world) is not one-one. This is a fact which we have to bear in mind when analyzing linguisic facts. Of course, this by no means leads to a metaphysics of "true cognition" which would be non-verbal and non-intellectual. There is no logical connection between empirically established linguistic facts and metaphysical speculation. But we can learn something about the vagueness of words and about its origin.

Verbal signs are vague not because they are imperfect (which would suggest that there are technical means to remove their imperfection), but because there is a relation of being ill-fitted which exists between the rigid classification, in certain respects, of the elements of the real world and that real world itself, which by its changes and transitions from one state to another evades all rigid classification. Of course, this evasion of a classification based on a given verbal sign takes place at the peripheries of the area covered by a given classification and can be reduced by making that sign more precise. Yet the boundary of ill-fittedness can be shifted (in the sense that by making a given verbal sign more precise we reduce the fringe of its extension), but it cannot be removed, for precisely the reason indicated above: something which is variable and in manifold ways interconnected with other elements of the real world cannot be fully mapped in terms of categories which grasp as motionless something that is variable, which squeeze into rigid classifications something which crosses every artificial barrier by the wealth of its shades, gradations, and interconnections with other phenomena.

Is this basic defect of language — a defect which prevents us from knowing the real world — as is claimed by the irration alistically-minded believers in "true cognition"? Not in the least. Every measurement made with an instrument is loaded with error, but it is enough to know the limit of that error to be able to judge its consequences for our reasoning. This applies to language to some extent, too. It is enough to know the nature of error in the reflection, by words of a language, of the real world to be able to neutralize that error in such and such a way and in such and such a degree by resorting to other words of that language.

In any case, we can shift the limits of error and reduce it until idealization, i.e., a deliberately adopted false assumption that error has vanished, is admissible and justifiable not only in practice, but from a theoretical point of view as well.

The ill-fittedness of words to the real world is therefore objective in nature. This fact does not prevent us from acquiring knowledge of the real world with the aid of a language consisting of such ill-fitted words. The point is only that we have to know the nature and the degree of that ill-fittedness.

Both Russell and Black agree with this in principle, though Russell takes a double approach to the problem. On the one hand, he is positive about every verbal sign being vague, and that any logic which would use perfectly precise signs would be good for discussing Platonic entities (and hence ideal entities, which are changeless), and not real entities (this implies an admission that verbal signs are ill-fitted to the changing real world). On the other hand, however (and this certainly results from the metaphysical assumptions of logical atomism), he does not reject the idea of a perfect language in which there is no place for vagueness. Such a language would have to be constructed on the principle of one-one relations between verbal signs and (atomic) facts. Apart from all other objections, such a language would have catastrophic consequences. This, however, naturally leads us to the issue of a perfect language as an instrument for counteracting the vagueness of words.

NOTES

[1] In the Polish literature on the subject the issue has been discussed by T. Kubiński, 'Wyrazy nieostre' (Vague Terms), *Studia Logica*, VII, 1958.

[2] B. Russell, 'Vagueness', *Australian Journal of Psychology and Philosophy*, 1923, and M. Black, *Vagueness, Language and Philosophy*, New York 1949.

[3] M. Black, op. cit.

NAMES AND PREDICABLES

by

Peter Thomas Geach

If we are concentrating on the logic of predication, we shall easily come to the conclusion that this part of the history of logic is largely a story of error. For Aristotle, the ancestor of all logicians, began the development of this thought by formulating some masterly insights, but later rejected these insights and took a wrong direction. The consequences of his mistake were almost as fatal for his successors as the consequences of the first sin were for the heirs of Adam.

Adam's state before the Fall is a question of theological speculation, but we know pretty well what views Aristotle originally held on the nature of predication. He took over from Plato the thesis that there is a fundamental difference between two kinds of expression, names (*onomata*) and predicables (*rhemata*). In Plato's *Sophist* we read that the simplest sentence (*logos anankaiotatos*) consists of two heterogeneous parts, a name and predicable, e.g. 'Soctrates sings', '(A) man runs'. We get no proposition (*logos*) from a string just of names ('man lion...' or just of predicables ('sings runs...'). Aristotle gave these Platonic views clear expression and further development. In the few examples Plato gives all the *rhemata* are in fact verbs; but I prefer to render '*rhemata*' by 'predicables'; this better fits Aristotle's view as expressed in his masterpiece *De interpretatione*, where he says expressly that *rhemata* 'always are signs of what is said of something else' — that is, they are essentially predicative.

By the Platonic-Aristotelian theory these two classes of expressions — names and predicables — were to be mutually exclusive. Aristotle asserts that tense differences apply to predicables but not to names. We may indeed doubt whether tense differences make sense for all predicables; for if we predicate something of an arithmetical object like the number 142857 or a geometrical object like the equilateral triangle, time-restrictions appear senseless. But if the object X does exist in time, then undoubtedly we may ask whether a given predication about *X is* true, *was* true, or *will* in the future be true. Nothing of the sort applies to names. Suppose a schoolmaster asks a boy the question when Augustus Caesar was born; it would be gross impudence if the boy ventured to reply 'He wasn't called Augustus then, sir' The name 'Augustus' now relates to that emperor at every stage of his career, and regardless of his now having been dead long since.

A more fundamental difference is bound up with the nature of negation. We sometimes get the negation of a sentence by replacing a predicable with a negated predicable — but never by replacing the subject with a negated subject. Uncritical reliance upon surface grammar might make the matter look different; for the negation of the sentence '*petetai anthropos*', 'a man flies', is the sentence '*petetai oudeis anthropos*' 'not a man flies' or 'no man flies'. We have here, it seems, a negative subject '*oudeis anthropos*', and here in Greek '*oudeis*' is as it were an adjective grammatically agreeing with the name 'man'. But Aristotle was not deceived by this grammatical mirage, anybody who thinks 'no man' is a negative subject is commiting a gross logical error — an error into which some of my logic pupils in England have fallen. The corresponding Polish sentence, 'Żaden człowiek nie lata', has a negating word at the predicate end, and thus gives no ground for the idea that an unchanged predicate is attached to a negated subject.

On the Platonian-Aristotelian analysis, the class of names includes not only proper names like 'Socrates' and 'Theodoros' but also common nouns answering to genera and species, like 'lion', 'animal' 'man'.

Aristotle states that a name must be syntactically simple — must not have any parts with a significance of their own. This requirement seems entirely sound. The role of a name is just to refer to its bearer, and for this role separate significant parts are needless. If an inn is called 'The White Eagle', we are not speaking of eagles when we mention the inn. The three words evoke certain associations, but these play no logical role; from the point of view of logical syntax 'The White Eagle' is a *simple* name. This thesis had been pretty well forgotten for more than two thousand years when Russell and Wittgenstein restated the doctrine of names being logically non-compound. Wittgenstein used the argument that a complex sign relates to reality only by the mediation of its constituent signs, whereas a name relates to its bearer directly. We therefore need not and must not make the meaning of a name depend on its syntactical make-up. (Russell and Wittgenstein had indeed quite a different view of the world from Aristotle, but these differences of ontology are not of present concern.)

The theory of names and predicables stated in masterly fashion by Aristotle in the *De interpretatione* seems to me to be true, and I have several times written in defence of it. I disagree with Aristotle, however, when he holds that predicables cannot be complex. This view is to my mind quite unjustified. To name a thing we need only a single word; but if we need to say something complicated about a thing, then this will require a complex predicable or (as Aristotle would say) a complex *rhema*. I can only conjecture why Aristotle thought otherwise; when he wrote the *De Interpretatione* he may have aimed at reducing all sentences to simple two-membered sentences. These simple 'atomic' sentences would consist of a simple name and a simple predicable; and a complex sentence would be a combination (*syndesmos*) of the simple sentences.

If Aristotle ever really had such a programme, he must have convinced himself that it could not be carried through. For Aristotle must have been aware of certain kinds of sentences that are irreducible to any *syndesmos* of two-word sentences, such as 'Socrates loves Theodoros' or 'Every man either wakes or sleeps'. No combination of simple sentences like 'Socrates wakes' will be logically equivalent to such a complex sentence. Anyhow, Aristotle did undoubtedly conclude, at a later stage of his logical enquiries, that the predicable occurring in a predicative sentence need not necessarily be syntactically simple. He was persuaded of this before he wrote his greatest logical work, the *Prior Analytics*.

Unfortunately, Aristotle abandoned also his position that any predicative sentence consists of two heterogeneous parts, a name and a predicable, and adopted the view that a sentence has two homogeneous parts, the so-called 'terms' (*horoi*). According to the earlier theory, a predicable, *rhema*, is necessarily predicative, since it is 'always a sign of what is said of something else'. But according to the later theory no term is necessarily predicative; the same term is predicate in one proposition and subject in another. I shall call this thesis of Aristotle's the *interchangeability thesis* (as regards terms). By accepting this thesis Aristotle replaced the older name-and-predicable theory with a *two-term* theory. And since a term in the role of predicate may be syntactically complex, it follows from the interchangeability thesis that a logical subject may also be syntactically complex.

To my mind the devising of the two-term theory was a disaster for logic comparable only to the Fall of Adam. It is not hard to see why Aristotle did not realize his mistake. His syllogistic was a great success, and it was easy to believe that its validity depended on the possibility of interchanging terms. In any Aristotelian syllogism the premises are two categoricals, that is, predicative propositions; and from these premises we get a conclusion containing the two terms that remain, after eliminating the term common to the premises, the so-called middle term. It is easy to show that in any argument of this structure, at least one of the three terms will play the part of subject in one proposition and of predicate in another. If we do not consider any parts of logic outside syllogistic, we shall find nothing obviously wrong with this theory; and I believe Aristotle was so much captivated by the beauty and power of the theory he had created that he never noticed how shaky its foundations were.

The foundations were indeed shaky. This does not mean that in concrete examples syllogistic reasoning that is supposedly sound is really unsound; only that Aristotle's own analysis of such reasoning is confused and defective. The change of a term's logical role carries with it of necessity a change of sense, in medieval terminology, a change of *modus significandi*,

Only names are logical subjects; when a name becomes a logical predicate, it *ipso facto* ceases to be a name. For in a predicate we express what holds good or does not hold good of a given object, but a name just *designates* the object. (Of course a name can be *part* of a predicable e.g. the name 'Socrates' enters into the predicable 'thought by Socrates, or 'older than Socrates'.) This is not at all a matter of degree. Some Oxford philosophers think there is a scale with essentially descriptive expressions at one end and purely demonstrative words at the other end. I agree with Wittgenstein: any logical difference is a *big* difference; in logic we are not making *subtle* distinctions, like the distinctions of odours or tastes.

Aristotle himself held that at any rate proper names cannot be predicated, not really (*haplós*) predicated; and he felt some doubt about the interchangeability thesis even as applied to common names. Is it really true, he asks, that 'This timber is white' and 'This white is timber' are equivalent ? There are also other signs that Aristotle's Fall did not put his logical conscience quite out of kilter. Although grammatically speaking it is only nouns or noun-phrases that can be alike subject or predicate, it would be wrong to assume that Aristotle recognised only such expressions as possible terms. In fact Aristotle generally avoids such a logical schema as '*A* is *B*', his preferred form is rather '*B* applies (*hyparchei*) to *A*'. Moreover he gives some concrete interpretations of this schema in which the grammatical form '*A* is *B*' does not occur; e.g. in the sentence 'There is a *single* science concerned with opposites' he puts *A* = opposites (e.g. health and disease) *B* = there being a single science concerned with them; and he expressly spells it out that this does not mean that opposites *are* there being a single science of them but that *it is true to say* of opposites that there is a single science of them. Splendid! But in this way Aristotle himself has overthrown his interchangeability thesis. In this example '*B* applies to *A*' makes sense, but not '*A* applies to *B*'; it would be nonsense to say that it is true of there being one science of them that it is (they are?) opposites!

When a common noun like 'philosopher' occurs in a proposition as a name, the name relates directly to this or that philosopher. We can thus state the truth-conditions for such a sentence in terms of other sentences in which names of individual philosophers explicitly occur. Thus, the sentence 'Socrates taught a philosopher' is true just in case we get a true sentence from the formula 'Socrates taught ...' by inserting the (proper) name of some philosopher. The case is quite altered when common nouns are predicative. The truth-conditions for the sentences 'Socrates was a philosopher' or 'Alcibiades became a philosopher' nowise depend on the possibility of replacing the common noun 'philosopher' by the proper name of some philosopher.

Let us not be misled by the fact that 'Alcibiades became a philosopher' is a false proposition. For take the true proposition 'Lord Home became prime minister'; obviously this has nothing to do with any individual prime minister, say Harold Macmillan or Harold Wilson. To be sure Lord Home 'became' Sir Alec Douglas-Home, but in affirming that I am not answering the nonsense question *which* British prime minister Lord Home 'became'. Certainly he did not become *himself* just at the time when he became prime minister; and neither did he become another prime minister, say Churchill!

It is not only in predicative contexts that common nouns like 'philosopher' have a part to play other than that of naming; there are many other cases, e.g. the sentence:

'Philosopher' means by etymology 'lover of wisdom'.

In both sorts of proposition the term 'philosopher' is used in a special way and does not relate to particular philosophers. Careful logicians use quotes when they are concerned with the word itself and not with its designation. It would be useful to employ a special logical sign in order to mark out the places where a letter in logical formulas that corresponds to a common noun is playing a predicative role. It would be absurd to protest that in ordinary language we get on without such a sign. Indeed there actually *is* such a sign in Polish; the Polish sign of the predicative sense is the case-ending that a noun must have when it is the complement of such verbs as 'to be', 'to become', 'to look'. Anyhow logical subjects and logical predicates just *have* a different *modus significandi*; the change from one role to the other has the same importance for a logician as a gear change for a motorist, and if a faulty logical training makes us deaf to such changes of gear, we shall jam the works of the machine.

Aristotle's Fall had many disastrous consequences, but in our time there has been an important repair of the damage. Our gratitude is due chiefly to Frege and Russell. To Frege we owe it that modern logicians almost universally recognise an absolute difference of logical category between names and predicates (what I have been calling predicables); different founts of type are chosen for the schematic letters and variables answering to these two categories. Russell re-emphasized this difference; and moreover he rejected the view that logically complex expressions can play the part of names, whereas Frege still regarded some of these as names.

This was the point of Russell's famous Theory of Descriptions. Take an example adapted from Quine. From the premise:

The broker who hired Joseph hired no negroes

we easily infer that Joseph is not a negro. But if 'the broker who hired Joseph' were a name, then it would be logically one and indivisible, and then the conclusion that Joseph is not a negro could not be get out of the premise: it would no more follow from the premise we were given than from the premise:

Theodore hired no negroes,

even supposing 'Theodore' to be the broker's name.

All that I regret is that neither Russell nor Frege recognised a naming role for common nouns like 'man' or 'philosopher'. It would be highly desirable if someone succeded in formulating a logical theory which satisfied this Aristotelian requirement but did not blur the distinction between names and predicables. To my mind such a theory would be an integral restoration of that was lost in Aristotle's Fall: it would be like Paradise Regained.

ON THE ANTINOMY OF THE LIAR AND THE SEMANTICS OF NATURAL LANGUAGE

by

Roman Suszko

The antinomy of the liar has been discussed many times in formal logic. It is associated with remarkable advances in logic: the formulation of the semantic theory of truth [4][1] and the discovery of undecidable statements and the impossibility of proofs of consistency under specified conditions ([2]; see also [3], Vol. II, pp. 256ff).

All those results make fundamental use of *self-referential expressions*, which were first used, in the history of logic, in the antinomy of the liar. The aim of this paper is to demonstrate, by quite elementary methods, something that has been known since the birth of semantics, namely, that the concept of truth and other semantic concepts are relative in nature [5] and that using relative semantic concepts, including the construction of self-referential expressions, does not result in antinomies in natural language.

Semantics, and in particular the semantic theory of truth, presupposes syntax. The wealth of semantic analysis thus depends on the wealth of syntactic information about those expressions to which semantic analyses refer. Since in this paper no systematic syntactic studies on the structure of expressions are made, except for the construction of self-referential expressions, the set of concepts used in the semantic theory of truth discussed here is very modest.

1. An analysis of the antinomy of the liar will be preceded by some comments on the construction of self-referential expressions.[2]

For every expression, we can form and use its *quotation*; to do so it suffices to write that expression in quotation marks.[3]

We can perform various operations on expressions. In particular, we can, in a given expression, substitute some other expression for an expression contained in the original one (e.g., for one of the variables used here: 'x', 'y', 'z', etc.). Hereafter we shall need only the operation of substitution for one fixed variable, e.g., the variable 'x'.

We therefore consider the following two formulations to be completely comprehensible:

(1) The expression obtained by the substitution in an expression y (for the variable 'x') of the quotation of an expression z.

(2) The expression obtained by the substitution in an expression y (for the variable 'x') of the quotation of the expression y.

Formulation (2), which is a special case of (1), will be abbreviated as:

(3) $w(y)$.

Now, (2) and (3) contain the variable 'y', for which any other variable may, of course, be used. Let us now write down the following two expressions:

(4) $w(x)$.

(5) w('$w(x)$').

Expression (5) is a name of an expression, namely a name for w('$w(x)$'), i.e., the expression obtained from '$w(x)$', i.e., expression (4), by substitution in it, for the variable 'x', of its own quotation. But this substitution yields (5), which is therefore a self-referential name.

Self-referential sentences (statements) are constructed in a similar manner. We shall describe their construction schematically. Let us write for the sake of brevity:

$x \in M$

for any of the following formulations:

x is an M, x is M-y, x is in the set of M's.

We shall write down two schematic statements:

(6) $w(x) \in M$.

(7) w('$w(x) \in M$') $\in M$.

Now, (7) states that w('$w(x) \in M$') is M-y which means that in the set of M's is the expression obtained from '$w(x) \in M$', i.e., from (6), by the substitution in it (for the variable 'x') of its own quotation. But this substitution yields (7), which is thus a self-referential statement, which states with reference to itself that it is M-y.

The self-referential nature of (5) and (7) is expressed by the following equalities:

(8) w('$w(x)$') = 'w('$w(x)$')'.

(9) w('$w(x) \in M$') = 'w('$w(x) \in M$') $\in M$'.

2. The schema of the classical definition of truth (and falsehood) takes on the form of the set of the following two schemata, which are in the form of equivalences:[4]

(10.1) $`a' \in Ver \leftrightarrow a$.

(10.2) $`a' \in Fls \leftrightarrow$ *non* a.

These schemata contain the semantic concept of truth *Ver* and the semantic concept of falsehood *Fls*. These schemata are abbreviated notations of all statements which can be obtained from them by the substitution in them, for the letter *a*, of any statements.

There is no need to explain in what sense schemata (10.1) and (10.2) express the interpretations of truth and falsehood as agreement and disagreement, respectively, of a given statement with that to which that statement refers.

We shall also adopt the principle of two-valued logic, under which for any expression x

(11) $(x \in S$ (i.e., x is a statement)$) \leftrightarrow (x \in Ver \veebar x \in Fls)$.

The antinomy of the liar develops within the framework of the semantics based on (10.1) and (11). Let us use, to wit, the predicate 'false' in the schema (7) of self-referential statements. This yields a self-referential statement, termed an *Eubulidean expression*—*eub* for short — which has the form:

(12) $w\,(`w\,(x) \in Fls') \in Fls$.

In accordance with the symbolism adopted and the example of (9) we have the (triple) equality:

(13) $eub = `w'\,(`w\,(x) \in Fls') \in Fls' = w\,(`w\,(x) \in Fls')$

and on the strength of (13) we rewrite an Eubulidean expression thus:

(14) $eub \in Fls$.

The antinomy of the liar may be treated as proof of the fact that an Eubulidean expression thus constructed is not a statement. In antinomial reasoning we apply to an Eubulidean expression the schema of the classical definition of truth, which is possible only if an Eubulidean expression is a statement. The interpretation that follows stresses this aspect of the antinomy of the liar.

If an Eubulidean expression is a statement, then by virtue of (11)

(15) $eub \in Ver \veebar eub \in Fls$.

Now the application of (10.1) yields

(16) $`w\,(`w\,(x) \in Fls') \in Fls' \in Ver \leftrightarrow w\,(`w\,(x) \in Fls') \in Fls$

which by virtue of (13) yields

(17) $eub \in Ver \leftrightarrow eub \in Fls$,

which contradicts (15). Hence the assumption that an Eubulidean expression is a statement is not acceptable, so that

(18) *non* $(eub \in S)$.

This conclusion, however, is at variance with our intuition, which makes us adopt the contrary opinion, namely that

(19) $eub \in S$.

This intuition of ours is based on the syntactic correctness (well-formedness) of the structure of self-referential statements. Note that the results obtained by Gödel [2] are based on the structure of self-referential statements. Some of his results can be partially reconstructed by recourse to the means adopted in this paper. If in the schema of the structure of self-referential statements we use the predicate 'unprovable' in one case, and the predicate 'refutable' in the other, and if we assume that every provable statement is true (but not conversely)[5], and that every refutable statement is false (but not conversely)[5], then elementary reasoning, analogical to the antinomial reasoning carried out above, does not result in a contradiction, but proves that of the two self-referential statements constructed above, one is true but unprovable, and the other is false, but irrefutable.[6]

3. It appears that responsibility for the antinomy of the liar is borne by the concept of statement, or rather the range of those statements to which we apply semantic concepts. To clarify this matter we shall consider an Eubulidean expression in a somewhat different situation.

Suppose a set $S(L)$ of statements is given which includes statements of a fixed and defined type L, but such that L is not described in any greater detail. The concepts of truth and falsehood are applicable to statements of type L. The set of true statements of type L and that of false statements of type L will be denoted, respectively, by

$$Ver(L) \text{ and } Fls(L).$$

It is obvious that

(20) if $x \in Ver(L) \vee x \in Fls(L)$, then $x \in S(L)$.

We have thus restricted the concepts discussed earlier by introducing the concepts of a *statement of the type* (L), a *true statement of the type* (L), and a *false statement of the type* (L), as restricted to the type L of expressions. These concepts will now be used to reformulate (10.1), (10.2) and (11):

(21) $x \in S(L) \leftrightarrow (x \in Ver(L) \veebar x \in Fls(L))$.

(22) 'a' $\in Ver(L) \leftrightarrow a$.

(23) 'a' $\in Fls(L) \leftrightarrow$ *non* a.

Schemata (22) and (23) will, of course, henceforth be applied to statements of type L.

Schema (7) for self-referential statements will now be applied to the predicate 'false of type L'. This yields a self-referential statement which we shall term an Eubulidean expression with respect to the type L of statements — *eub* (L); for short this expression has the form:

(24) w ('$w(x) \in Fls(L)$') $\in Fls(L)$.

We note the triple equality:

(25) *eub* (L) = 'w ('$w(x) \in Fls(L)$') $\in Fls(L)$' = w ('$w(x) \in Fls(L)$'),

and, accordingly, rewrite the equivalent expression (24) thus:

(26) *eub* $(L) \in Fls(L)$.

We continue our line of reasoning along the path which has led us to the antinomy of the liar. If (24), i.e., an Eubulidean expression with respect to the type L of statements, were a statement of type L, then by virtue of (21) we would have the disjunction

(27) *eub* $(L) \in Ver(L) \veebar$ *eub* $(L) \in Fls(L)$.

By virtue of (22), we would have

(28) 'w ('$w(x) \in Fls(L)$') $\in Fls(L)$' $\in Ver(L) \leftrightarrow w$ ('$w(x) \in Fls(L)$') $\in Fls(L)$,

which, in accordance with (25), we can transform into

(29) *eub* $(L) \in Ver(L) \leftrightarrow$ *eub* $(L) \in Fls(L)$.

Now, (29) and (27) yield a contradiction. Hence we conclude that

(30) *non* (*eub* $(L) \in S(L)$).

Antinomial reasoning which makes use of the concepts restricted to the type L of statements leads us to conclude that we do not arrive at the antinomy of the liar (by constructing self-referential statements) if we apply our semantic analyses to statements of a type that does not include statements constructed of those semantic concepts which we use in our semantic analyses, in particular such as does not include statement (24).

In semantic analyses restricted to the type L of statements there occurs a "new" well-formed statement, which cannot, under penalty of contradiction, be of type L. This new statement is an Eubulidean expression restricted to the type L of statements. The statement that

(31) $eub\,(L) \in S$

does not now result in a contradiction.

4. Restriction of semantic concepts to a definite set of statements satisfying the condition that bars the antinomy of the liar can be carried out with respect to various types of statements. In particular, we can apply it in semantic analysis of statements which include (24). Let us consider statements of type L, to which we have restricted semantic concepts, so as to avoid antinomies while engaging in semantic analysis of statements of type L. Let us now consider a more comprehensive type L' of statements; such statements will include all statements of type L as well as statements of the semantic system restricted to L. Let us now carry out a certain brief reasoning within the semantic system restricted to L'. We shall accordingly, use the restricted concepts

$$S\,(L'), \qquad Ver\,(L'), \qquad Fls\,(L').$$

The semantic principles on which we now base our procedure assume the following forms:

(32) $x \in S\,(L') \leftrightarrow (x \in Ver\,(L') \vee x \in Fls\,(L'))$.

(33) $`a' \in Ver\,(L') \leftrightarrow a$.

(34) $`a' \in Fls\,(L') \leftrightarrow non\ a$.

Schemata (33) and (34) are applied exclusively to statements of type L'.

Statements of type L' include Eubulidean statements restricted to L, so that

(35) $eub\,(L) \in S\,(L')$,

and, by virtue of (32).

(36) $eub\,(L) \in Ver\,(L') \vee eub\,(L) \in Fls\,(L')$.

On the other hand, schema (33) yields

(37) $`w\,(`w\,(x) \in Fls\,(L)') \in Fls\,(L)' \in Ver\,(L')$
$\leftrightarrow w\,(`w\,(x) \in Fls\,(L)') \in Fls\,(L)$.

which by virtue of (25) yields

(38) $eub\,(L) \in Ver\,(L') \leftrightarrow eub\,(L) \in Fls\,(L)$.

But, by virtue of (20), it follows from (30) that

(39) $non\,(eub\,(L) \in Fls\,(L))$.

Now, (38) and (39) yield

(40) $non\,(eub\,(L) \in Ver\,(L'))$.

In view of (38) we conclude that

(41) $eub(L) \in Fls(L')$.

Thus, a Eubulidean statement of type L is a false statement of type L'.

A Eubulidean statement $eub(L)$ is, of course, not to be confused with $eub(L')$, i.e., with a Eubulidean statement of type L': $eub(L')$ is a self-referential statement constructed with the aid of the predicate 'false of type L'' and, as can easily be demonstrated, is not a statement of type L.

5. The semantic theory of truth does not result in the antinomy of the liar if we use concepts restricted to a set of statements which does not include statements from the theory of truth which we are studying in a given case.

It can be shown that the same applies to other parts of semantics, namely those in which the other semantic concepts (denoting, satisfying, etc.) are used [4], [5], [6].

To do this it suffices to analyse other antinomies constructed with the aid of semantic concepts, and to modify them in a manner analogical to that applied above in the case of the antinomy of the liar.

The semantic concepts which we can use in semantic research without being involved in antinomies are relative (restricted). They have a certain reference to a type L of expressions, which includes neither those semantic terms which have a reference to L, nor statements containing those semantic terms. Within those semantic analyses in which we use semantic concepts restricted to type L of expressions we can construct, in accordance with general syntactic rules, an expression which can be proved not to be of type L. The proof consists in a reasoning which changes into an appropriate antinomy if the restrictive reference to L, applied to the semantic concepts used in that case, is disregarded.

BIBLIOGRAPHY

[1] Carnap, R., 'Die Antinomien und die Unvollständigkeit der Mathematik', *Monatshefte für Mathematik und Physik*, 41, 1934, pp. 263–84.

[2] Gödel, K., Über formal unentscheidbare Sätze der "Principia Mathematica" und verwandter Systeme I', *Monatshefte für Mathematik und Physik* 38, 1931, pp. 173–98.

[3] Hilbert, D., Bernays, P., *Grundlagen der Mathematik*, Berlin 1934, 1939.

[4] Tarski, A., 'The Concept of Truth in Formalized Languages', in: Tarski, A., *Logic, Semantics, Metamathematics*, Oxford 1956.

[5] Tarski, A., 'The Establishment of Scientific Semantics', *ibid.*

[6] Tarski, A., 'On the Concept of Logical Consequence', *ibid.*

Notes

[1] Numbers in brackets refer to the bibliography at the end of the paper.

[2] This construction comes from K. Gödel [2], who applies it through the arithmetization of linguistic expressions.

[3] Note that the existence of what is termed the quotational function is not assumed here.

[4] The symbol $\leftrightarrow$ stands for equivalence (if and only if). Use will also be made of the symbol $\vee$ of non-exclusive disjunction and the symbol $\veebar$ of exclusive disjunction (aut ... aut ...).

[5] Otherwise we come back to the antinomy of the liar.

[6] The reader is asked to work out the details of this reasoning. See also [1].

NORMAL AND NON-NORMAL CLASSES IN CURRENT LANGUAGE[1]

STUDIES IN THE CONCEPT OF CLASS (I)

by

Zdzisław Kraszewski and Roman Suszko

Russell's antinomy of the class of normal classes, i.e., the class of those classes which are not their own elements, emerged when the current concept of class was being given more precision. It is this current concept of class which is blamed for Russell's antinomy.

The task of the present paper is to offer a fairly precise definition of the current concept of class, which has subsequently come to be split into the collective (concretistic) concept of class and the distributive (mathematical) concept of class or set. S. Leśniewski's mereology, to which T. Kotarbiński's concretism refers, is a theory of classes in the collective sense. The theory of classes in the distributive sense has taken the form of mathematical set theory, which originated with E. Zermelo; other versions of the theory of classes in the distributive sense are provided by B. Russell's type theory and S. Leśniewski's ontology.

After making the current concept of class more precise, which will consist in a systematization of the assumptions concerning that concept, we shall define normal and non-normal classes as well as the class of normal classes and the class of non-normal classes. Several variations of these definitions are possible, and Russell's antinomy can be reconstructed in each case. We shall see, however, that his antinomy cannot be reconstructed in current language, since the corresponding reasonings do not yield a contradiction. The thesis of this paper is that the current concept of class, as described below, is not self-contradictory.

1. *Formalization of current language*

In our interpretation, current language coincides with an elementary logical formalism, namely the logic of statements and the logic of terms. This formalism will not be described rigorously,[2] and attention will be paid to major points only.

We assume that current language, in our interpretation, includes statements of the following kinds.

(1) Compound statements formed by means of ordinary sentential connectives, i.e., symbols of negation, conjunction, disjunction, implication, and equivalence, which are denoted, respectively, thus: $\sim$, $\wedge$, $\vee$, $\rightarrow$, $\leftrightarrow$.

(2) Statements occurring in the square of opposition: 'every A is B, some A are not B', etc. The conjunction: 'every A is $B \wedge$ every B is A' will be abbreviated into '$A = B$', and the negation '$\sim A = B$' will be replaced by: '$A \neq B$'.

(3) Singular statements: 'a is A', 'a is not A', 'a is non-A', with singular terms as subjects. The conjunction: 'a is $b \wedge b$ is a' will be abbreviated into: '$a = b$', and the negation: '$\sim a = b$' will be replaced by: '$a \neq b$'.

(4) Existential statements: '$\vee A$, '$\dot{\vee} A$·', which are to be read as: 'there is an A', 'there is exactly one A', respectively.

We are adopting the rules of inference of two-valued sentential calculus and the elementary calculus of terms (the laws of the square of opposition, with singular and existential statements, as above, the laws concerning negative terms in the form of '*non-A*' and compound terms in the form of 'A and B', the laws of obversion, etc.), including the laws of extensionality for co-extensional terms. We mention, by way of example, that the singular statements 'a is not A', 'a is *non-A*' are logically equivalent to the statement '$\sim a$ is A'. We also have the tautologies: '$\dot{\vee} A \rightarrow A$ is A', 'A is $B \rightarrow \vee A$'.

The lack of apparent variables bound by quantifiers is an essential feature of the formalism which corresponds to natural language. This is why some of our analyses will go beyond the limits of that formalism: in some cases we shall make use of existential quantifiers. The expression in the form '$\vee x \Phi(x)$', where the variable x stands for singular (individual) terms, will be read: 'there is an x such that $\Phi(x)$'. In the special case when the sentential function $\Phi(x)$ has the form 'x is A' the statement '$\bigvee x$ x is A' is equivalent to the statement '$\bigvee A$'.

In some cases, mainly those of definitions of normal and non-normal classes, we shall make use of existential quantifiers to bind variables that stand for predicates (general terms). The expression in the form '$\bigvee A\, \Phi(A)$' means 'for some A, $\Phi(A)$'.

2. *Specific terms and rules*

Our language includes certain specific terms with which we are concerned here. They are:

(1) *The class of A's, denoted by* $K(A)$, where A stands for any term.[3]

(2) *A class* (*in general*), denoted by KL.

(3) *A normal class*, denoted by KN.

(4) *A non-normal class*, denoted by KNN.

These terms are predicates. Some of them will occur in several variations, which will be marked by indices. We shall assume below that the predicate $K(A)$ is either empty or singular. In the latter case, $K(A)$ may also occur as subject in a singular statement. It is also assumed that compound terms of the type $K(A)$ are extensional with respect to their term arguments. It is therefore assumed that

(2.1) $\quad A = B \rightarrow K(A) = K(B).$

In accordance with the observation made above we shall adopt the rule

(2.2) $\quad x$ is $K(A) \wedge y$ is $K(A) \rightarrow x = y.$

We shall also make the obvious assumption concerning the predicate KL that

(2.3) $\quad x$ is $K(A) \rightarrow x$ is KL.

Under these assumptions the statements

$$\bigvee K(A), \quad \dot{\bigvee} K(A), \quad K(A) \text{ is } K(A),\ K(A) \text{ is } KL$$

are equivalent to one another.

The concept of element is linked with that of class. It occurs in our language as the relative term *element*, denoted by El. This relative term, when taken together with any singular term a, forms the predicate $El(a)$, which is read as *an element of a*. In our formalism, relative terms may also occur in certain combinations with any predicates. For instance, for any predicate A we can form the predicate: *an element of an A*.

The only assumptions about the concept of element are as follows: (i) every class is the class of its elements, (ii) every A is an element of the class of A's, (iii) only classes have elements. This means that the class of A's is the class of all A's, and that class somehow consists exactly of its elements. In symbolic notation, these three rules are:

(2.4) x is $KL \rightarrow x$ is $K(El(x))$,

(2.5) x is $K(A) \rightarrow$ every A is $El(x)$,

(2.6) x is $El(y) \rightarrow y$ is KL.

Now, (2.5) may be reformulated thus:

(2.7) $\vee K(A) \rightarrow$ every A is $El(K(A))$.

The following two equivalences, which are easily obtainable from (2.3) and (2.4), are more important:

(2.8) x is $KL \leftrightarrow x$ is $K(El(x))$,

(2.9) x is $KL \leftrightarrow \bigvee A \quad x$ is $K(A)$.

We shall also perform reasoning which leads to the conclusion that the relative term El is non-empty if for some predicate A we have: $\vee A$ and $\vee K(A)$. We shall resort for that purpose to the universal predicate 'object', to be denoted by P, and for which the following tautologies hold: (i) $\vee P$, and (ii) *every A is P*.

Suppose that $\vee K(A)$. By virtue of (2.7), every A is $El(K(A))$. Hence, *every A is an El of some P*. If also $\vee A$, then *some P is an El of some P*. We have thus demonstrated that

(2.10) $\vee A \wedge \vee K(A) \rightarrow$ some P is an El of some P.

Hence

(2.11) $\vee K(P) \rightarrow$ some P is an El of some P.

3. *Existential assumptions*

The rules (2.1), (2.2), (2.3), (2.4), (2.5), (2.6) adopted above are not existential in nature. In current language, the problem of the existence of classes, is, so it would seem, settled in the sense that all classes exist. There is a tendency in current language to state the existence of a class of A's, regardless of whether A's exist or not. Current language explicitly admits a discussion of a class of A's as something unique, and hence as something which exists, in those cases in which it is not known whether A's exist or not. If it turns out that A's do not exist, then we simply cease to be interested in the class of A's, and it may be supposed that current language in no way inclines us to the verdict that the class of A's does not exist.

These comments are an argument in favour of including in the set of rules which describe the current concepts of class and element, and alongside rules (2.1) – (2.6), the following universal existential rule: for any predicate A

(3.1) $\qquad \vee K(A)$.

It follows therefore that, for any predicate A: $\dot{\vee} K(A)$, $K(A)$ *is* $K(A)$, and $K(A)$ *is* KL.

The non-emptiness of the predicate KL and of the relative term El follows from assumption (3.1) concerning the non-emptiness of the predicates in the form of $K(A)$ and from (2.3) and (2.11), respectively. In other words, we now have:

(3.2) $\qquad \vee KL$,

(3.3) $\qquad$ Some P is an El of some P.

The rules (2.1) – (2.6) and (3.1) form the axiom system of the current concept of class and the current concept of element. This axiom system is contradiction-free, which means that no pair of contradictory statements (Z, $\sim$ Z) follows from it logically. To prove this it suffices to consider the domain which consists of only one object p. If we assume that p is the only designatum of the predicate KL and of all predicates in the form of $K(A)$, and that the relation denoted by the relative term El exists between p and p, then the axiom system under consideration is satisfied, i.e., it consists of statements which are true relative to that domain of subjects. The logical conclusions which can be drawn from that axiom system must also be true in that sense, and hence they do not contain any pair of contradictory statements.

The existential assumption (3.1) differs from the other rules that describe the current concept of class and the current concept of element.

The rules (2.1) – (2.6) are called the non-existential axiom system for the current concepts of class and element. Note that the concept of element does not occur in (2.1), (2.2), (2.3). Nor does it occur in formula (2.9), which is characteristic of the current concept of class. If we make a point of singling out the axiom system for the current concept of class without making use of the corresponding concept of element, then in our axiom system we have to leave out (2.4), (2.5) and (2.6), and make (2.3) stronger, as it were, by turning it into an equivalence, i.e., by replacing it by (2.9).

4. *Other rules*

In mereology, i.e., the theory of collective classes (and the corresponding relation of elementhood) and in set theory, i.e., the theory of distributive classes (and the corresponding relation of elementhood) we find a number of other, fairly intuitive, rules which we are not adopting here. We shall discuss some of them and draw attention to certain logical relations that hold between them.

For instance, here we are not adopting the rule

$$x \text{ is } K(A) \rightarrow \text{every } El(x) \text{ is an } A,$$

which in a sense is a conversion of (2.5). In view of (2.5) this formula is equivalent to

$$(4.1) \qquad x \text{ is } K(A) \rightarrow El(x) = A,$$

which we are obviously not adopting, either. Nor are we adopting the rule

$$(4.2) \qquad x \text{ is } K(A) \wedge x \text{ is } K(B) \rightarrow A = B,$$

which in a sense is a conversion of (2.1).

Consequence relation and equivalence will — unless explicitly mentioned otherwise — be interpreted as consequence relation and equivalence in the non-existential axiom system for the currentconcepts of class and element.

It will now be proved that

(4.3) rules (4.1) and (4.2) are equivalent.

It can be seen immediately that (4.2) follows from (4.1). To prove the converse relationship we assume that (4.2) holds and that x is $K(A)$. Then, by virtue of (2.3), x is KL. Hence, by virtue of (2.4), x is $K(El(x))$. In accordance with (4.2) we conclude that $El(x) = A$, q.e.d.

We assume here neither the reflexivity nor the anti-reflexivity of the relation of elementhood, i.e., the formulas

(4.4) $\quad x \text{ is } KL \rightarrow x \text{ is } El(x),$

(4.5) $\quad x \text{ is not } El(x),$

nor any of the following rules concerning the relation of elementhood (asymmetry, antisymmetry, transitivity):

(4.6) $\quad x \text{ is } El(y) \rightarrow y \text{ is not } El(x),$

(4.7) $\quad x \text{ is } El(y) \wedge y \text{ is } El(x) \rightarrow x = y,$

(4.8) $\quad x \text{ is } El(y) \wedge y \text{ is } El(z) \rightarrow x \text{ is } El(z).$

5. *Normal and non-normal classes in current language*

The distinction between normal and non-normal classes made by B. Russell is introduced in current language by discussing problems in the form of:

is $K(A)$ an A, or is it not?

where the predicate A is given. We say that $K(A)$ is a *KNN* if $K(A)$ is an A, and we say that $K(A)$ is a *KN* if $K(A)$ is not A, i.e., if $K(A)$ is *non-A*.

If A is a common predicate, such as 'town', 'human being', 'planet', 'angel', then $K(A)$ is a *KN*.

If A is a universal predicate, such as 'object', 'class', then we obtain a result which states that '$K(A)$ is a *KNN*'. The same holds if A is a negative predicate in the form of '*non-B*', where 'B' is a common predicate in the sense used above.

Non-normal classes are also obtained in a special case: if a term F is a relative predicate, such as 'component', 'part', 'fragment', 'element', which satisfies the conditions of reflexivity antisymmetry and transitivity,[4] namely:

$$x \text{ is a } KL \rightarrow x \text{ is } F(x),$$
$$x \text{ is } F(y) \wedge y \text{ is } F(x) \rightarrow x = y,$$
$$x \text{ is } F(y) \wedge y \text{ is } F(z) \rightarrow x \text{ is } F(z),$$

then by substituting, for the predicate A, terms in the form of '$F(a)$' and 'F of some B', where a is a singular term, and B is any non-empty predicate, we obtain: $K(A)$ is KNN.

It may be supposed that the concepts of normal class and non-normal class in current language are described by the following rules:

(5.1) $K(A)$ is a $KN \leftrightarrow K(A)$ is *non-A*,

(5.2) $K(A)$ is a $KNN \leftrightarrow K(A)$ is A,

(5.3) every KN is KL,

(5.4) every KNN is KL.

Note that these rules do not refer to the concept of element.

Rules (5.1) and (5.2) can be reformulated thus (taking into consideration the existential axiom (3.1)): if x is $K(A)$, then the following two formulas hold:

(5.5) x is $KN \leftrightarrow x$ is *non-A*,

(5.6) x is $KNN \leftrightarrow x$ is A.

This yields the following conclusions:

(5.7) x is $K(A) \rightarrow (x$ is $KN \vee x$ is $KNN)$,

(5.8) x is $KNN \rightarrow x$ is *non-KN*.

Formula (5.8) states that no normal class is a non-normal class; we call this the *principle of exclusion*.

It is obvious that: x is $K(A) \wedge x$ is *non-A* $\rightarrow x$ is KN. On the other hand, if x is KN, then, by virtue of (5.3) and (2.9), $\bigvee A$ (x is $K(A) \wedge x$ is KN), i.e., $\bigvee A$ (x is $K(A) \wedge x$ is *non-A*). This shows that the following equivalence holds:

(DN) x is $KN \leftrightarrow \bigvee A$ (x is $K(A) \wedge x$ is *non-A*),

(DNN) x is $KNN \leftrightarrow \bigvee A$ (x is $K(A) \wedge x$ is A).

The following equivalence is obtained analogously:

It turns out — we shall not give the proofs here — that (DN), (DNN) and (5.8) yield, under consideration of (3.1), rules (5.1) – (5.4), which means that those rules, when taken together and in consideration of (3.1) are equivalent to the system of formulas (DN), (DNN), (5.8).

Our standpoint at this moment is as follows: we are adopting (DN) and (DNN) and joining them to the axiom system described in Secs. 2 and 3 as certain possible definitions of the concepts of normal and non-normal class. The principle of exclusion (5.8) will be discussed elsewhere, and at this moment we are not adopting it. Accordingly we are not adopting the system of rules (5.1) – (5.4) as a whole.

Now (DN) and (DNN) easily yield (5.3), (5.4) and (5.7). On the other hand, (5.8) does not follow from them, which means that these definitions admit the existence of classes which are both normal and non-normal. Other definitions of normal and non-normal classes are possible, too. They will be marked by subscripts, as follows: KN_1, KNN_1, KN_2, KNN_2.

We shall adopt the following definitions:[5]

(DN_1) $KN_1 = KN$.

(DNN_1) $KNN_1 = KL$ and *non-KN*,

(DNN_2) $KNN_2 = KNN$,

(DN_2) $KN_2 = KL$ and *non-KNN*.

These definitions can also be equivalently written thus:

(5.9) x is $KN_1 \leftrightarrow \bigvee A$ (x is $K(A) \wedge x$ is *non-A*),

(5.10) x is $KNN_1 \leftrightarrow x$ is $KL \wedge x$ is *non-*KN_1,

(5.11) x is $KNN_2 \leftrightarrow \bigvee A$ (x is $K(A) \wedge x$ is A).

(5.12) x is $KN_2 \leftrightarrow x$ is $KL \wedge x$ is *non-*KNN_2.

These definitions yield not only rules analogous to (5.7), i.e. (for $i = 1, 2$) x is $K(A) \rightarrow$ (x is $KN_1 \vee x$ is KNN_1) but also rules analogous to (5.8), i.e. (for $i = 1, 2$):

(5.13) x is $KNN_i \rightarrow x$ is *non-*KN_i.

Thus definitions (DN_i) and (DNN_i), i.e., (5.9) and (5.10) and (DN_2) and (DNN_2), i.e., (5.11) and (5.12), guarantee, unlike (DN) and (DNN), that the concept of normal class and the concept of non-normal class are mutually exclusive.

6. *The class of normal classes*

In the preceding Section we analysed possible definitions of the concepts of normal and non-normal class in current language, namely (DN, DNN), (DN_1, DNN_1), (DN_2, DNN_2). When joined to a contradiction-free axiom system for the current concept of class (and the current concept of element) these definitions obviously do not result in any contradiction. In particular, the analysis of the class of normal classes, which we can now carry out, does not result in a contradiction. Thus, if we start from the assumptions which we have adopted and the definitions which we have formulated, we are not in a position to reconstruct B. Russell's antinomy concerning the class of normal classes.

The schema of Russell's antinomy is as follows. We assume that the class of normal classes is normal, and by inference we conclude that that class is non-normal. On the other hand, from the assumption that the class of normal classes is non-normal we arrive at the conclusion that that class is normal.

It is self-evident that some of those reasonings cannot be carried out once the axiom system described in Secs. 2 and 3 has been adopted. Those which can be carried out as a result of the adoption of that axiom system yield certain positive results.

Let us begin with definitions (DN_1, DNN_1), i.e., (5.9) and (5.10). We set

$$k_1 = K(KN_1)$$

and suppose that k_1 is KNN, i.e., k_1 is *non*-KN_1. In view of the symbolism adopted above, k_1 is $K(KN_1)$. Hence $\bigvee A\,(k_1 \text{ is } K(A) \wedge k_1$ is *non*-$A)$. By virtue of (5.9), k_1 is KN_1. This contradicts our assumption, and hence

(6.1) k_1 is KN_1.

The class of normal classes in the sense of (DN_1) is thus a normal class in that sense.

We now consider definitions (DN_2, DNN_2), i.e., (5.11) and (5.12), and set

$$k_2 = K(KN_2).$$

Naturally, k_2 is $K(KN_2)$. If we suppose that k_2 is KN_2, this yields $\bigvee A\,(k_2 \text{ is } K(A) \wedge k_2 \text{ is } A)$. Hence, by virtue of (5.11), k_2 is KNN_2, however, this contradicts our assumption, and hence we conclude that

(6.2) k_2 is KNN_2,

i.e., that the class of normal classes in the sense of (DN_2, DNN_2) is a non-normal class in that sense.

Other reasonings resulting from the schema of Russell's antinomy and pertaining to the class of normal classes (k_1, k_2) in the sense of the two said definitions cannot be reconstructed. Analogous reasonings pertaining to the class of non-normal classes in the sense of (DN_1, DNN_1) and (DN_2, DNN_2) fail, too.

In the case of (DN, DNN), from which the principle of exclusion (5.8) does not follow, the situation is more complicated. We must here make the distinction between normal classes and non-normal non-classes, and also between non-normal classes and normal non-classes. Therefore, we adopt the following symbolism:

$$q = K(KN), \qquad p = K(KNN),$$
$$q^* = K(non\text{-}KNN), \qquad p^* = K(non\text{-}KN).$$

We shall now prove that q and q^* are both normal and non-normal in the sense of (*DN*, *DNN*).

If q is *KN*, then, of course, q is K (*KN*) and q is *KN*. Hence, $\bigvee A\,(q \text{ is } K(A) \wedge q \text{ is } A)$, which by (DNN) yields q is *KNN*. If q is not *KN*, then q is *KNN* by virtue of (5.7.) In both cases we have the same conclusion that

(6.3) q is *KNN*,

which is thus proved. In a similar way we obtain

(6.4) q is *KN*.

For if q is not *KN*, then q is K (*KN*) and q is *non-KN*. Hence, $\bigvee A\,(q \text{ is } K(A) \wedge q \text{ is } non\text{-}A)$. By virtue of (DN) it follows that q is *KN*. Thus, we have arrived at a contradiction of the assumption made in the proof of (6.4). These reasonings fall under part one of the schema of Russell's antinomy. The reasonings which fall under part two of that schema cannot be reconstructed here. Neither q is *KN* nor q is not *KNN* can be inferred from the assumption that q is *KNN*.

The reasonings which pertain to q^* are analogical. If q^* is not *KNN*, then q^* is K (*non-KNN*) and q^* is *non-KNN*. Hence $\bigvee A\,(q^* \text{ is } K(A) \wedge$ $\wedge\, q^* \text{ is } A)$. This, by virtue of (DNN), yields q^* is *KNN*, which contradicts our assumption. Thus, we have proved that

(6.5) q^* is *KNN*.

We shall also see that

(6.6) q^* is *KN*.

For if q^* is *KNN*, then obviously q^* is not *non-KNN* and q^* is *K* (*non-KNN*) by assumption. In other words, q^* is *K* (*non-KNN*) and q^* is *non-non-KNN*. Hence, $\bigvee A$ (q^* is $K(A) \wedge q^*$ is *non-A*). In view of (DN) this means that q^* is *KN*. If q^* is not *KNN*, then, by virtue of (5.7), we infer that q^* is *KN*. In both cases (6.6) holds, q.e.d.

In the case of q^*, as in that of q, the other reasonings which fall under the schema of Russell's antinomy cannot be reconstructed.

No results are obtainable concerning the normal or non-normal character of p and p^*.

7. *The relative concepts of normal and non-normal class*

The three possible definitions of normal and non-normal classes, as analysed above, make use of quantifiers that bind predicate variables.[6] Those quantifiers are often treated metatheoretically, according to the interpretation given in Sec. 1, and not existentially. That interpretation treats a statement in the form of $\bigvee A\, \Phi(A)$ as an abbreviaton of $\Phi(Q)$, where Q stands for a given predicate about which little has been said. The metatheoretical interpretation is not sufficiently clear, and we shall not analyse it here. It does, however, suggest the following point of view. It may be supposed that in current language there are relative concepts of normal and non-normal classes, which means that the corresponding terms are not predicates, but operators of a certain kind which form predicates when combined with predicates. The situation is the same as in the case of the concept of class in general. The term $K(A)$ is a predicate formed of an operator and a predicate. On the other hand the term *KL* is simply a predicate. When applied to (2.9), the metatheoretical interpretation of the quantifiers that bind predicate variables makes it possible to treat *KL* as an abbreviation of $K(Q)$, where Q is a given predicate, but one about which little has been said.

In passing now to the relative concepts of normal and non-normal class, we turn our attention to definitions (DN) and (DNN). We shall assume that *KN* and *KNN* are abbreviations of

(7.1) $KN(Q)$,

(7.2) $KNN(Q)$,

where Q is a given predicate, but one about which little has been said. These terms are read as: *normal class relative to Q's, non-normal class relative to Q's*, respectively. Now, (7.1) and (7.2) are formed of the predicate Q and certain operators which, without risk of confusion, can be symbolized (KN, KNN), in the same way as the predicates defined in (DN, DNN).

We refer to an analogy to the concept of class in general and to formula (2.9) when we adopt the following definitions:

(DN*) x is $KN(A) \leftrightarrow x$ is $K(A) \wedge x$ is *non-A*,

(DNN*) x is $KNN(A) \leftrightarrow x$ is $K(A) \wedge x$ is A.

This immediately yields:

(7.3) x is $KN \leftrightarrow \bigvee A\, x$ is $KN(A)$.

(7.4) x is $KNN \leftrightarrow \bigvee A\, x$ is $KNN(A)$.

In view of (2.9) we can easily infer that

(7.5) x is $KL \rightarrow (x$ is $KN(A) \vee x$ is $KNN(A))$.

(7.6) x is $KNN(A) \rightarrow x$ is *non-KN* (A).

(7.7) $A = B \rightarrow (KN(A) = KN(B) \vee KNN(A) = KNN(B))$.

It is not, however, excluded that the following may hold:

(7.8) a is $KN(A) \wedge a$ is $KNN(B)$,

where a is a singular term, and A and B are certain predicates, where naturally $A \neq B$.

The following may serve as an example: $a =$ the class of human beings, $A =$ human being, $B =$ a group of human beings. We therefore have: a is $K(A)$ and a is $K(B)$, and a is *non-A* and a is B. Obviously, both $A \neq B$ and (7.8) hold in this case.

Formulas (7.3) and (7.4) reflect the close link between two pairs of concepts of normal and non-normal classes, where the concepts in one pair are absolute, and those in the other are relative in nature. In current language these concepts are not distinguished with any precision. Such confusion of the various categories of concepts is fairly frequent in natural language.[7] For instance, the term 'uncle' occurs as a relative predicate in the sentence 'John is Paul's uncle'; it also occurs as an absolute predicate, and in two different ways. In a sentence uttered by Paul, 'Uncle says that ...', or 'Uncle John says that ...', 'uncle' occurs as a pseudo-absolute predicate because it is in fact an abbreviaton for 'My uncle' or 'My uncle John' (which indicates the relation between John and Paul). The term 'uncle' may also occur (though presumably this does not occur frequently) as an absolute predicate, namely as a general term whose designata include all human males whose brothers and/or sisters have children.

In reasonings in current language which refer to normal and non-normal classes, these terms may also occur in the two roles defined by (DN, DNN), and (DN*, DNN*). This distinction between the relative and the absolute interpretation of the terms *KN* and *KNN*, in accordance with these definitions, does not, however, cover cases in which a relative predicate occurs as pseudo-absolute, i.e., when its argument is somehow implied, but does not occur in the sentence. Description of such cases evades formal procedures.

From the standpoint solely of definitions (DN, DNN) and (DN*, DNN*) we can analyse certain structures in which the terms '*KN*' and '*KNN*' occur in various roles. We may consider $K(KN)$ and $K(KNN)$, and ask whether those classes are $KN(A)$ or $KNN(A)$ for a given A. By way of example we shall consider classes q and q^*, which were discussed in Sec. 6. It has been proved (cf. (6.3) and (6.5)) that (7.9) q is *KN* and q^* is *KNN*, i.e., q^* is *non-non-KNN*. In view of the symbolism adopted there q is $K(KN)$ and q^* is $K(non\text{-}KNN)$, hence, by virtue of (*DN**, *DNN**), we infer that

(7.10) q is $KNN(KN)$,

and

(7.11) q^* is $KN(non\text{-}KNN)$.

Of course, (7.10) and (7.11) are not contradictory in the current interpretation of the concept of class. This is closely related to the non-identification of normal classes with non-normal classes. Such an identification may easily take place in imprecise current reasonings. It results in a contradiction, for if we assume that

(7.12) $KN = non\text{-}KNN$,

then $q = q^*$ and (7.10) and (7.11) become a pair of contradictory statements.

8. *Normal and non-normal classes and the concept of element*

The concept of element has not been used in Secs. 5, 6 and 7. Nor do we use there rules (2.4), (2.5) and (2.6). We refer there only to the axiom system of a current interpretation of the concept of class, without the concept of element, and to rule (2.9) as somewhat stronger than (2.3). Existential axiom (3.1) is also referred to in certain formulations.

Normal and non-normal classes are often termed, respectively,

(i) classes which are not their own elements,

(ii) classes which are their own elements.

Our intention now is to grasp the relationship which this terminology expresses. To do so we refer to rules (2.4) and (2.5), which pertain to the current concept of element.

We shall first prove that

(8.1) $\quad x \text{ is } KL \wedge x \text{ is not } El(x) \rightarrow x \text{ is } KN.$

If x is KL, then, by virtue of (2.4), x is $K(El(x))$. If at the same time x is not $El(x)$, then $\bigvee A$ (x is $K(A) \wedge x$ is not A), so that x is KN, q.e.d.

We shall now show that

(8.2) $\quad x \text{ is } KNN \rightarrow x \text{ is } KL \wedge x \text{ is } El(x).$

For if x is KNN, then $\bigvee A$ (x is $K(A) \wedge x$ is A). From the fact that x is $K(A)$ it follows, by virtue of (2.5), that every A is $El(x)$. But since x is A, x is $El(x)$, q.e.d.

A reasoning analogical to the proof of (8.1) yields an implication converse to (8.2). Hence we have the equivalence

(8.3) $\quad x \text{ is } KNN \leftrightarrow x \text{ is } KL \wedge x \text{ is } El(x).$

Implication (8.1) is not subject to conversion: it can easily be seen that its conversion, by yielding the equivalence

$$x \text{ is } KN \leftrightarrow x \text{ is } KL \wedge x \text{ is } \textit{not } El(x),$$

would mean, in view of (8.3), the same as the principle of exclusion (5.8), the validity of which we have left in suspension.

A similar asymmetry is observed when (DN, DNN) is replaced by (DN*, DNN*). By virtue of (2.8), we easily obtain the following two equivalences:

(8.4) x is $KL \wedge x$ is not $El(x) \leftrightarrow x$ is $KN(El(x))$,
(8.5) x is $KL \wedge x$ is $El(x) \leftrightarrow x$ is $KNN(El(x))$.

We also obtain, for any predicate A:

(8.6) x is $KNN(A) \rightarrow x$ is $El(x)$.

Proof. If x is $K(A)$ and x is A, then by virtue of (2.5), *every A* is $El(x)$. Hence x is $El(x)$, q.e.d.

On the other hand, the analogical implication for the normal classes:

$$x \text{ is } KN(A) \rightarrow x \text{ is } \textit{not } El(x),$$

cannot be proved.

We still have definitions (DN_i, DNN_i) for $i = 1, 2$, i.e., formulas (5.9) and (5.10), and (5.11) and (5.12). It will be proved that with some additional assumption the following equivalences hold (for $i = 1, 2$):

(8.7) x is $KN_i \leftrightarrow x$ is $KL \wedge x$ is *not* $El(x)$,
(8.8) x is $KNN_i \leftrightarrow x$ is $KL \wedge x$ is $El(x)$.

The assumptions mentioned above are certain principles of anti-reflexivity (4.5) and reflexivity (4.4), formulated in Sec. 4, but not included by us in the axiom system for the current interpretation of the concept of element.

It turns out that

(8.9) Equivalences (8.7) and (8.8) hold for $i = 1$ if anti-reflexivity (4.5) is assumed.

In view of (5.13) it suffices to prove only (8.7). Implication from left to right requires merely our realizing that *every* KN_1 is KL. And if x is KL, then, by virtue of (2.4), x is $KL(El(x))$. But since x is *non-*$El(x)$, by virtue of (5.9) x is KN_1.

We likewise conclude that

(8.10) Equivalences (8.7) and (8.8) hold for $i = 2$ if reflexivity (4.4) is assumed.

In view of (5.13) it suffices to prove only (8.8). We obtain implication from left to right, since every KNN_2 is KL. And if x is KL, then, by virtue of (2.4), x is $K(El(x))$. But in this case x is $El(x)$, and hence, by virtue of (5.11), x is KNN_2.

Relationships (8.9) and (8.10) make us adopt the following terminology. Definitions (DN_1, DNN_1), i.e., formulas (5.9) and (5.10), will be termed the *set-theoretical version* of the current definitions of the concepts of normal and non-normal class, and definitions (DN_2, DNN_2), i.e., formulas (5.11) and (5.12), will be termed the *mereological version* of the current definitions of the concepts of normal and non-normal class.

The reflexivity of the relation of elementhood is essentially connected with mereology, whereas the anti-reflexivity of that relation corresponds to the intuitions of mathematical set theory. Furthermore, it is a mereological thesis that normal classes, i.e., those which are not their own elements, do not exist, whereas the non-existence of non-normal classes, i.e., those which are their own elements, seems to be essentially connected with the mathematical concepts of class (set) and element. This is also an argument in favour of the terminology adopted above. For the above proofs of (8.9) and (8.10) show clearly that

(8.11) on the assumption of anti-reflexivity (4.5) *every KL* is KN_1, i.e., $\sim \bigvee KNN_1$,

and

(8.12) on the assumption of reflexivity (4.4) *every KL* is KNN_2, i.e., $\sim \bigvee KN_2$.

Notes

[1] The problems to be discussed and the main theses of the present paper come from Z. Kraszewski. R. Suszko's contribution is confined to the proposal that everyday language should be identified with the formalism of the logic of statements and the logic of terms, and to certain suggestions concerning the arrangement of the material.

[2] For a limited version of a formalism of the logic of terms see A. Morawiec, 'Podstawy logiki nazw' (Foundations of the Theory of Names), *Studia Logica*, XII, 1961.

[3] It may be assumed that the argument A in $K(A)$ is always a subjective complement (a general term). While adopting this position we also admit terms in the form of $K(a)$, where a is a proper name (an individual term): we can consider $K(a)$ to be an abbreviation for $K(Id(a))$, where Id is the relative predicate 'identical with'; in that case the term '$Id(a)$' is a subjective complement.

[4] The relative terms mentioned previously are thus interpreted in a weak sense, in which a whole is its own part (fragment, component, element).

[5] It is obvious, by virtue of (2.3), that every *KN* is *KL* and that every *KNN* is *KL*.

[6] We have disregarded here the fourth, combinatorily possible, definition of normal and non-normal classes, namely $KNN_3 = KL$ and non-KN, $KN_3 = KL$ and non-KNN. This case seems to be totally unintuitive.

[7] Certain examples in this paragraph have had to be modified, because the use of articles in English renders some of the original Polish formulations pointless. (Tr.)

NORMAL AND NON-NORMAL CLASSES VERSUS THE SET-THEORETICAL AND THE MEREOLOGICAL CONCEPT OF CLASS[1]

STUDIES IN THE CONCEPT OF CLASS (II)

by

Zdzisław Kraszewski and Roman Suszko

We shall concern ourselves here with the transition from the current concept of class to the distributive (set-theoretical) and the collective (mereological) concept of class. This transition is linked to the concepts of normal and non-normal class. Preliminary remarks on that issue have already been made in Sec. 8.

We assume here a non-existential axiom system for the current concepts of class and element, as described in Secs. 2 and 3. Consequence and equivalence are interpreted, as before, as consequence and equivalence in the light of that axiom system.

9. *Certain absolute concepts of normal and non-normal class*

The non-relative concepts *KN* and *KNN* are connected with the relative concepts of normal and non-normal class (cf. DN* and DNN* in Sec. 7), respectively (cf. (7.3) and (7.4)). For the non-relative concepts, however, the principle of exclusion (5.8) does not hold. On the other hand, it holds (5.13) for KN_i and KNN_i for $i = 1, 2$.

We shall now discuss other absolute concepts of normal and non-normal class, AKN_i, $AKNN_i$, for $i = 1, 2$. Definitions of these concepts do not include quantifiers binding predicate variables, and are obtained from the definitions $A/El(x)$ and A/x (i.e., $A/Id\ (x)$) of the corresponding relative concepts by appropriate substitutions:
We shall adopt the following definitions:

(DAN$_1$) x is $AKN_1 \leftrightarrow x$ is $KN(El(x))$,

(DANN$_1$) x is $AKNN_1 \leftrightarrow x$ is $KNN(El(x))$,

(DAN$_2$) x is $AKN_2 \leftrightarrow x$ is $KN(x)$,
(DANN$_2$) x is $AKNN_2 \leftrightarrow x$ is $KNN(x)$.

We shall also refer to the set-theoretical ($i = 1$) and the mereological ($i = 2$) version of the concepts AKN_i and $AKNN_i$.

It could be assumed that classes which are not their own elements and classes which are their own elements are yet another version of the normal and non-normal classes. However, this version coincides with the set-theoretical version for $i = 1$, as formulated above.

In fact, by breaking down (DAN$_1$, DANN$_1$) and making use of (2.8) we obtain the following equivalences:

(9.1) x is $AKN_1 \leftrightarrow x$ is $KL \wedge x$ is not $El(x)$,
(9.2) x is $AKNN_1 \leftrightarrow x$ is $KL \wedge x$ is $El(x)$.

In the case of the mereological interpretation, on breaking down (DAN$_2$, DANN$_2$) we obtain different equivalences:

(9.3) x is $AKN_2 \leftrightarrow x$ is $K(x) \wedge x$ is not x,
(9.4) x is $AKNN_2 \leftrightarrow x$ is $K(x) \wedge x$ is x,

which are equivalent to the following formulas, respectively:

(9.5) $\sim \bigvee AKN_2$, i.e., $AKN_2 = non\text{-}P$,

where P is the universal predicate that stands for 'object', and

(9.6) x is $AKNN_2 \leftrightarrow x$ is $K(x)$.

Now, (9.6) will be written thus:

(9.7) x is $AKNN_2 \leftrightarrow x = K(x)$,

as it is obvious, by virtue of (2.2), that: x is $K(x) \rightarrow x = K(x)$.

On comparing, for $i = 1, 2$, the concepts AKN_i and $AKNN_i$ with KN and KNN we easily find the relationships:

(9.8) every AKN_i is KN,
(9.9) every $AKNN_i$ is KNN.

Note also that

(9.10) every KL is either AKN_1 or $AKNN_1$.

The analogical relationship for the mereological version cannot be proved, which means, in view of (9.5), precisely that we cannot prove that every *KL* is $AKNN_2$.

10. *The class of normal classes and the principle of exclusion*

The following symbolism (for $i = 1, 2$) will be adopted:

(10.1) $h_i = K(AKN_i)$.

It will be proved that

(10.2) h_1 is *KN*, and h_1 is *KNN*.

Proof. There are two possibilities: either h_1 is AKN_1, or h_1 is $AKNN_1$. In the former case we infer, in view of DN in Sec. 5, that h_1 is *KNN*. On the other hand, however, h_1 is *KN* by virtue of (9.8). In the latter case h_1 is *non-*AKN_1, and hence in view of DNN in Sec. 5, we infer that h_1 is *KN*. But on applying (9.9) we obtain the result that h_1 is *KNN*. Thus, in both cases we arrive at (10.2).

In the mereological version, i.e., for $i = 2$, the analogical reasoning cannot be carried out (see the final remark in Sec. 9).

Now, (10.2) would be a contradiction if we were to adopt the principle of exclusion (5.8). In other words, joining the principle of exclusion to the axiom system adopted in this paper yields a contradiction, and the reasoning carried out above, which results in (10.2), is a form of Russell's antinomy. That reasoning clearly shows that the principle of exclusion (5.8) results in the negation of that special case of the existential axiom (3.1) which is implied by (10.1).[2] This is so because

(10.3) it follows from (5.8) that $\sim \bigvee K(AKN_1)$.

The joining of the principle of exclusion (5.8), or any other principle from which (5.8) follows, to our axiom system for the current interpretation of the concept of class makes us restrict the general existential axiom (3.1).

Note also that (5.8) is equivalent to the rejection of the possibility of (7.8), i.e., to the formula

(10.4) x is $KN(A) \rightarrow x$ is not $KNN(B)$.

Now formula (10.4) can be written thus:

(10.5) $(x$ is $K(A) \wedge x$ is not $A) \rightarrow (x$ is $K(B) \wedge x$ is not $B)$.

On the other hand, by breaking down the principle of exclusion in accordance with (DN, DNN) we obtain the formula

(10.6) $\bigvee A\,(x \text{ is } K(A) \wedge x \text{ is not } A) \rightarrow \sim \bigvee B\,(x \text{ is } K(B) \wedge x \text{ is } B).$

which, under the rules of joining and dropping quantifiers, is equivalent to (10.5).

It can also be easily seen that identification of all concepts of normal and non-normal classes, as discussed in Sec. 5, with the corresponding absolute concepts in the set-theoretical version is another equivalent of the principle of exclusion (5.8). Such identification takes on the form of the following equations:

(10.7) $KN = KN_1 = KN_2 = AKN_1,$

(10.8) $KNN = KNN_1 = KNN_2 = AKNN_1.$

11. *The set-theoretical concept of class*

Adoption of rule (4.1), i.e., the implication

$$x \text{ is } K(A) \rightarrow El(x) = A,$$

is an essential step toward making the current concept of class correspond to the distributive (set-theoretical) concept of class. Note that (4.1) is equivalent to (4.2), i.e., the implication

$$x \text{ is } K(A) \wedge x \text{ is } K(B) \rightarrow A = B.$$

A more important point, however, is that (4.1) is equivalent — in view of (2.8) — to the following definition of the class of A's:

(11.1) $x \text{ is } K(A) \leftrightarrow x \text{ is } KL \wedge El(x) = A.$

KL here remains a primitive concept.

Note now that

(11.2) (5.8) follows from (4.1).

This is so because (4.2) immediately yields.

(11.3) $(x \text{ is } K(A) \wedge x \text{ is } K(B)) \rightarrow (x \text{ is } B \rightarrow x \text{ is } A),$

which is equivalent to (10.5).

Adoption of (4.1) results in the adoption of (5.8), (10.4), (10.7) and (10.8), and thus yields a contradiction. Hence, the adoption of (4.1) makes us restrict the general existential axiom (3.1) if we want to avoid the contradiction resulting from the adoption of (4.1). We also see that in view of (10.7) and (10.8) the antinomial reasoning concerning $K(AKN_1)$ applies to the remaining classes, i.e., $K(KN)$, $K(KN_1)$, $K(KN_2)$.

12. *Elimination of quantifiers binding predicate variables*

Adoption of (4.1) or (11.1) enables us to eliminate existential quantifiers binding predicate variables from contexts which occur in definitions of the concepts of normal and non-normal class, as discussed in Sec. 5. It can easily be seen that (11.1) and (2.8) yield

$$(12.1) \qquad \bigvee A\,(x \text{ is } K(A) \wedge F(A)) \leftrightarrow x \text{ is } KL \wedge F(El\,(x)),$$

where $F(A)$ has the free predicate variable A, and $F(El\,(x))$ is obtained from $F(A)$ by the substitution $A/El\,(x)$. In particular, the statements

$$\bigvee A\,(x \text{ is } K(A) \wedge x \text{ is } non\text{-}A),$$
$$\bigvee A\,(x \text{ is } K(A) \wedge x \text{ is } A),$$

are equivalent, respectively, to

$$x \text{ is } KL \wedge x \text{ is not } El\,(x),$$
$$x \text{ is } KL \wedge x \text{ is } El\,(x).$$

These immediately yield (10.7) and (10.8).

13. *The set-theoretical version of Russell's antinomy*

It has been shown in Sec. 10 that Russell's antinomial reasoning can be reconstructed when the principle of exclusion (5.8) is joined to the axiom system for the current concept of class. That reasoning will now be repeated here in its usual form.

To the axiom system for the current concept of class we join (4.1) or (11.1), so that (10.7) and (10.8) hold. We are considering normal classes in the sense of (DAN_1). Now, on applying the existential axiom (3.1) we obtain

$$(1) \qquad \bigvee K(AKN_1).$$

Hence $K(AKN_1)$ is $K(AKN_1)$, i.e., by virtue of (10.1),

$$(2) \qquad h_1 \text{ is } K(AKN_1).$$

Now, by virtue of (11.1), (2) is equivalent to the following two formulas:

(3) h_1 is KL,

(4) $El(h_1) = AKN_1$.

We now make use of (DAN_1) transformed into (9.1), in order to write

(5) h_1 is $AKN_1 \leftrightarrow h_1$ is $KL \wedge h_1$ is not $El(h_1)$.

Formulas (5) and (4) yield the equivalence

(6) h_1 is $AKN_1 \leftrightarrow h_1$ is $KL \wedge h_1$ is not AKN_1,

from which it immediately follows that

(7) h_1 is not KL.

We have thus obtained a contradiction between (3) and (7). The starting point of our reasoning must therefore have been false, and for this the existential axiom (3.1) is to be blamed.

Beginning with (3), the foregoing antinomial reasoning is based on the following logical relationship, which has been remarked on by S. Leśniewski (cf. T. Kotarbiński, *Leçons sur l'Histoire de la Logique*, Warszawa 1965, pp. 288–9). The very definition of a predicate D in the form:

$$x \text{ is } D \leftrightarrow x \text{ is } C \wedge x \text{ is not } R(x),$$

where C is any given predicate, and R is any given relative predicate, yields the implication

$$R(y) = D \rightarrow y \text{ is not } C.$$

The application of this logical relationship to antinomial reasoning consists in making the substitutions: D/AKN_1, C/KL, R/El, x/h_1, y/h_1.

The set-theoretical (distributive) approach to the concept of class, which consists in the adoption of (4.1), i.e., definition (11.1), forces us to modify the existential axiom (3.1). It therefore poses the question: for which A does $K(A)$ exist? This question is the focal issue in the axiomatization of set theory.

14. *The set-theoretical concept of element*

We shall not concern ourselves here with the axiom system(s) of set theory.[3] We shall, however, touch upon a minor issue, which concerns the set-theoretical concept of element.

It is often said that a class of A's interpreted in the distributive (set-theoretical) sense is a certain property shared by all A's and by A's only. From this viewpoint the relation of elementhood in the distributive sense is converse to the relation that holds between a property and an object of which that property is an attribute. This relation of a property as an attribute of an object will be denoted by the relative predicate $\breve{El}$, to be read as "converse element", which can be defined (thus independently, by the way, of whether our approach is distributive or collective):

$$(14.1) \qquad x \text{ is } \breve{El}\,(y) \leftrightarrow y \text{ is } El\,(x).$$

It seems natural to think that the relation of a property as an attribute of an object is asymmetrical and intransitive. This would mean that the principle of transitivity (4.8) does not hold for the set-theoretical concept of elementhood; on the other hand, the principle of asymmetry (4.6) would hold for it, with the principle of anti-reflexivity (4.5) following immediately from the latter. Compare also (8.9).

If we accordingly adopt (4.6) alongside of (4.1), then — in view of anti-reflexivity — all predicates in (10.7) coincide with KL, and all predicates in (10.8) are empty:

$$(14.2) \qquad AKN_1 = KL \wedge AKNN_1 = non\text{-}P.$$

The antinomial class of normal classes is thus identical with the class of all classes:

$$(14.3) \qquad K\,(AKN_1) = K\,(KL).$$

15. *The mereological concepts of class and element*

We shall now discuss the absolute concepts of normal and non-normal classes in their mereological version (DAN_2, $DANN_2$). We know in view of (9.5) that AKN_2 do not exist, but we are unable to prove that every KL is $AKNN_2$. Hence, it cannot be ruled out that for a singular term a we could have:

$$(15.1) \qquad a \text{ is } KL \wedge a \text{ is not } K\,(a).$$

Yet such a case is excluded in mereology, where the following general principle is satisfied:

(15.2) $K(x) = x$,

i.e., in other words, x is $P \rightarrow K(x) = x$.

Now, (15.2) first yields, in view of (2.3), the statement that every object is a class,

(15.3) $KL = P$.

The term KL is here a universal predicate and as such is superfluous.

But (15.2) also yields, by virtue of (9.7),

(15.4) $AKNN_2 = KL$,

which excludes the case described by (15.1). From this, in view of (9.9) for $i = 2$, we obtain

(15.5) $AKNN_2 = KNN$.

Now, in view of (15.2), we have the following equations:

(15.6) $AKNN_2 = KNN = KNN_2 = KL = P$,

(15.7) $AKN_2 = non\text{-}KNN = KN_2 = non\text{-}KL = non\text{-}P$.

Finally, it becomes evident that the universal reflexivity of the mereological relation of elementhood follows from (15.2). We shall demonstrate that

(15.8) x is $El(x)$,

which is stronger than the principle of reflexivity (4.4). To prove (15.8) we take any object x. Now x is KL by virtue of (15.3), and hence, by virtue of (2.4), we have (1) x is $K(El(x))$. But then $\dot{\vee}\, K(El(x))$, and by virtue of (15.2) we have: $K(El(x)) = K(K(El(x)))$.

From this and from (1) we conclude that x is $K(K(El(x)))$. From that it follows by virtue of (2.5), that (2) every $K(El(x))$ is $El(x)$. From (1) and (2) we obtain (15.8).

Note, too, that the mereological relation of elementhood is not only reflexive, but antisymmetric and transitive as well. Mereology accepts rules (4.7) and (4.8).

16. *The class of normal classes in mereology*

The mereological interpretation of the concept of class is free from antinomies, but it modifies the existential axiom (3.1). It is assumed in mereology that a class of A's does not exist if the predicate A is empty, so that

(16.1) $\sim \bigvee A \rightarrow \sim \bigvee K(A)$.

Axiom (3.1) is therefore reduced to a weaker form

(16.2) $\bigvee A \rightarrow \bigvee K(A)$.

The mereological modification of (3.1) bears upon the class of normal classes and the principle of exclusion.

First of all, it follows from (16.1) and (15.7) that $K(KN_2)$ and $K(non\text{-}KNN)$ do not exist. Hence the reasonings carried out in Sec. 6 concerning those classes (k_2, q^*) are plainly pointless.

In the case of $K(KN)$ and $K(KN_1)$ the situation is somewhat different. By definition, $KN = KN_1$, and hence, by virtue of (2.1), $K(KN) = K(KN_1)$. If these classes exist, they are (q, k_1) — as has been demonstrated in Sec. 6 — and this applies to KN as well as to KNN. Since, however, now $KNN = KL$, the result obtained in Sec. 6 concerning $K(KN)$ and $K(KN_1)$ now reduces to

(16.3) $\bigvee K(KN) \rightarrow K(KN)$ is KN.

In view of (16.1) and (16.2) we rewrite this implication as

(16.4) $\bigvee KN \rightarrow K(KN)$ is KN.

It can easily be seen that since now $KNN = KL$, the existence of normal classes in the sense of (DN) or (DN_1) is equivalent to the rejection of the principle of exclusion (5.8). In other words,

(16.5) $(5.8) \leftrightarrow \sim \bigvee KN$.

Mereology does not settle the issue of the existence of normal classes, nor does it settle the validity of the principle of exclusion. It may be said, however, that non-existence of normal classes and the validity of the principle of exclusion are basically at variance with mereology: we shall prove below (cf. (16.7)) that if at least two objects exist, then normal classes exist and the principle of exclusion (5.8) does not hold.

Note first that

(16.6) $El(x) = El(y) \leftrightarrow x = y$

follows easily from the principles of reflexivity (15.8) and antisymmetry (4.7).

It will now be demonstrated that

(16.7) $a \neq b \rightarrow \bigvee KN$.

Assume that $a \neq b$ and let the predicate C be defined thus:[4]

(1) $C = a$ or b.

Obviously, $\bigvee C$, and hence, by virtue of (16.2), $\bigvee K(C)$. Let

(2) $c = K(C)$.

By virtue of (2.5), every C is $El(c)$, hence

(3) a is $El(c) \wedge b$ is $El(c)$.

Since $a \neq b$, $El(a) \neq El(b)$ in view of (16.6). Let, for instance (the converse case may be disregarded),

(4) x is $El(a) \wedge x$ is not $El(b)$.

Transitivity (4.8) and (3) and (4) lead us to the conclusion that x is $El(c)$. Hence, by virtue of (4) and (16.6), it follows that

(5) $c \neq b$.

We conclude analogically that

(6) $c \neq a$.

It follows from (1), (5) and (6) that c is not C. But c is $K(C)$. Hence c is KN, q.e.d.[5]

17. *The axiom system of mereology*

Set theory and mereology are not rival theories. They use different concepts of class and element. These concepts, both distributive (set-theoretical) and collective (mereological), originate from current concepts. Axiomatization of set theory led to the emergence of Russell's antinomy of the class of normal classes, which in turn resulted in a modification of the general existential axiom (3.1) by its appropriate restriction. The birth of S. Leśniewski's mereology is historically linked with Russell's antinomy, but this antinomy does not play any logical role in the axiomatization of mereology. The mereological modification of (3.1) is not due to any contradictions connected with the class of normal classes. That modification is just a more precise formulation of the current interpretation of the concept of class and can be outlined as follows.

The concept *KL* is made universal (cf. (15.3)) and hence becomes superfluous. Next, it is assumed that the relation *El* (the relation of being an element, a fragment, a component, (a) part) is reflexive (15.8), antisymmetric (4.7) and transitive (4.8). The concept *El* is the only primitive concept, since the term *the class of A's* is defined as follows (see the auxiliary definition (14.1))

$$x\ K\ (A) \leftrightarrow A \text{ every } A \text{ is } El\ (x) \wedge \text{ every } El\ (x) \text{ is } \breve{El} \text{ of some } El \text{ of some } A.$$

Finally, axiom (3.1) is reduced to its weaker form (16.2).

The above definition of the class of *A*'s and the rules (15.8) (4.7), (4.8), and (2.2) form, in principle, the first axiom system of mereology, as given by S. Leśniewski.[6] It leads to the mereological principles (15.2) and (16.1) analysed above, and to the whole non-existential axiom system for the current concepts of class and element.

NOTES

[1] This is a continuation of the preceding paper by the same authors. It also assumes a knowledge of that paper on the part of the reader.

[2] Note that the tentative reconstructions of Russell's antinomy, as carried out in Sec. 6, refer to the existential axiom (3.1). The symbolisms adopted there ($k_1 = K(KN_1)$, $k_2 = K(KN_2)$, $q = K(KN)$, $q^* = K(\textit{non-KNN})$) include implicit assumptions of the existence of the respective classes $K(KN_1)$, $K(KN_2)$, $K(KN)$, $K(\textit{non-KNN})$ in accordance with (3.1).

[3] Note that the axiom of extensionality:

$$x \text{ is } KL \wedge y \text{ is } KL \wedge El(x) = El(y) \rightarrow x = y,$$

follows easily from (2.2) and hence is a theorem for the current interpretation of the concepts of class and element.

[4] We are using abbreviated symbolism here: for *a* and *b* we should write *Id* (*a*) and *Id* (*b*), respectively (cf. footnote[3] to the first paper by the same authors).

[5] This is the situation described by (7.8) in Sec. 7 we have: c is $KN(C) \wedge c$ is $KNN(c)$.

[6] S. Leśniewski, *Podstawy ogólnej teorii mnogości* (The Foundation of General Set Theory), I, Moscow 1916, and 'O podstawach matematyki' (On the Foundations of Mathematics), *Przegląd Filozoficzny*, 30–34, Warsaw 1927–1934.

THE SEMANTICS OF OPEN CONCEPTS

by

Marian Przełęcki

I

1. The main problem to be discussed in this paper has a long philosophical tradition. It emerged for the first time in reflections on vague terms in common language. The term (a) 'youth' (in the sense of a young man) may serve as a classical example. There is practically no doubt that a man under 18 is still a youth, and that no one over 30 is any longer a youth. But what about a person who is, for instance, 25 years old? The meaning assigned to that term in common language is such that we are not in a position to answer the question. What then is the nature of a statement that a person aged 25 is a youth? Is it a true or a false, but essentially undecidable statement, as some claim? Or is it deprived of any truth value, as others would have it? Or is it, perhaps, the case that both the statement and its negation are false, as still others maintain, thereby rejecting the principle of the excluded middle — one of the fundamental laws of logic? These questions lead to further questions. What, in fact, is such a statement about? Does the term 'youth' denote any definite set of objects? And if it does, then what is that set?

These semantic problems, concerned as they are with vague terms in common language, have become topical recently in connection with logical investigations into the language of empirical scientific theories. Theoretical terms, which form one of the basic kinds of terms occurring in empirical theories, resemble vague terms in common language. They, too, occur therefore in statements that give rise to the same problems as those which emerge in connection with vague terms. The source of difficulty seems to be the same in both cases. Vague terms and theoretical terms share the same logical properties. If we simplify matters slightly, then both categories of terms may be described as terms which are definable conditionally, or, to use another terminology, partially. An adequate definition of such a term must be conditional in form. Let a predicate Q be the term to be defined. Its conditional definition has the form

(1) $\bigwedge x\,(\Psi x \rightarrow (Q x \equiv \Phi x)),$

where Ψ and Φ are expressions whose meanings have been fixed previously. As we can see, definition (1) establishes the meaning of predicate Q only partially, namely for those objects which satisfy condition Ψ. Of these, those which are Φ are Q, and those which are not Φ are not Q. On the other hand, definition (1) does not establish any criteria of applicability of predicate Q to those objects which do not satisfy condition Ψ.

Other forms of partial definitions can also be subsumed under the schema of definition (1). For instance, the so-called bilateral reduction sentence for Q by means of predicates P_1 and P_2

(2) $\bigwedge x\,(P_1 x \rightarrow (Q x \equiv P_2 x))$

falls directly under schema (1). But the same applies to the partial definition of Q in its general form:

(3) $\bigwedge x\,((P_1 x \rightarrow Q x) \wedge (P_2 x \rightarrow \sim Q x)).$

Now (3) yields the statement

$$\sim \bigvee x\,(P_1 x \wedge P_2 x),$$

and if this is assumed to be true, then (3) is equivalent to

$$\bigwedge x\,((P_1 x \vee P_2 x) \rightarrow (Q x \equiv P_1 x)),$$

and hence to a statement that directly falls under (1).[1]

The same likewise applies to so-called unilateral reduction sentences for Q:

(4) $\bigwedge x\,(P_1 x \rightarrow Q x)$

and

(5) $\bigwedge x\,(P_2 x \rightarrow \sim Q x).$

The former is logically equivalent to

$$\bigwedge x\,(P_1 x \rightarrow (Q x \equiv P_1 x)),$$

and the latter to

$$\bigwedge x\,(P_2\,x \rightarrow (Q\,x \equiv \sim P_2\,x)).$$

Now the meaning of a vague term appears to be such that its adequate definition, which refers to non-vague terms only, must be a conditional one. Such a definition of the term 'youth' could take the form of the following partial definition of type (3):

$$\bigwedge x\,((x \text{ is less than 18 years old} \rightarrow x \text{ is a youth}) \wedge$$
$$\wedge\, (x \text{ is more than 30 years old} \rightarrow x \text{ is not a youth})).$$

This example also shows in what the simplification of the problem consists. The age limits are to some extent arbitrary and remain such regardless of how we select them. Thus we are dealing here with vagueness of, as it were, the second degree, which we shall completely disregard in our analyses.

In the case of theoretical terms, simplification consists in something else. Logical investigations into the language of empirical scientific theories lead us to conclude that the logical relationships which an empirical theory establishes between theoretical terms, on the one hand, and observation terms, on the other, usually reduce to conditional definitions of theoretical terms by means of observation terms, with such definitions taking one of the forms listed above. Such, for instance, is the nature — in classical genetics — of the definition of the theoretical term 'genotype', which refers exclusively to such observation terms (in the broad sense of the word) as 'phenotype' and 'descendant'. It can be formulated as a partial definition of type (3). It is, however, an open issue whether this kind of logical relationship with observation terms is characteristic of all theoretical terms. It is sometimes claimed that logical links between theoretical and observation terms may be even looser than those established by conditional definitions. It is not clear, however, whether any existing scientific theory has availed itself of such a possibility.[2]

In leaving unsettled the issue of whether our analysis will in fact cover all vague terms or all theoretical terms, we shall confine our discussion to those terms which are definable conditionally. The concepts which correspond to such terms are sometimes called 'open', and they are precisely those which suggest the semantic problems mentioned at the beginning of this paper. Let a_1 be an object which does not satisfy condition Ψ formulated in conditional definition (1) of predicate Q. The statement Qa_1, which assigns predicate Q to object a_1, may serve as an example of statements which are problematic from the semantic point of view. Our task will be, first of all, to define the class of such problematic statements in a general and precise manner. Statement Qa_1 is only one example. On the other hand, it is obvious that such statements do not include all those in which Q occurs. We shall try next, using a possibly precise and consistent conceptual framework, to present both the basic semantic problems suggested by the class of such statements, and the main attempts at solving those problems which can be found in the literature of the subject. The framework in question is provided by contemporary logical semantics, understood as the theory of models of formalized languages. The application of that framework to the problems under consideration follows R. Suszko's use of modern logical semantics in the analysis of philosophical issues.[3]

Our analysis will thus be confined to formalized languages, in particular those which are fairly simple from the syntactic point of view. They will be only elementary languages, based on the first-order predicate calculus with identity. An example is provided by a language L, which besides *individual variables* x, y, ..., and *logical constants* $\sim$, $\wedge$, $\vee$, $\rightarrow$, $\equiv$, $\bigwedge x$, $\bigvee x$, $=$, includes simple extralogical constants of two kinds: *names* a_1, a_2, ..., a_n, and *predicates* (with an arbitrary number of arguments) P_1, P_2, ..., P_m.[4]

The *formation rules* of L, which shall not be listed here, lay down how the expressions of L, in particular the statements, are formed of simple expressions. Language L so conceived is a formal, or rather a semi-formal, construction. Except for the logical constants, concerning which we assume that they have classical semantic properties, the expressions of L are characterized as to their syntactic properties alone. These expressions perform certain semantic functions — i.e., they refer to something and state something — only when alongside of L we have a model of it, which determines some interpretation of L.

The concept of a model of a formalized language cannot be systematically explained here, and hence we have to confine ourselves to a few general comments. A *model* of L is any domain about which we can speak in L. Such a model $\mathfrak{M}$ is identical with a system

$$\langle U, C \rangle$$

consisting of two components U and C. U is a non-empty set of objects called the universe of model $\mathfrak{M}$. It is the range of the variables in L. C, sometimes called the characteristic of $\mathfrak{M}$, includes some elements of U and some subsets of U, or relations holding among U's elements. Every component of C is denoted by a simple extralogical expression in L, and every such expression is denoted by a component of C. There is thus a definite relationship between the syntactic structure of a language and the type of its model. Thus, for instance, every model $\mathfrak{M}$ of the language L described above is a system

$$\langle U, x_1, \ldots, x_n, X_1, \ldots, X_m \rangle$$

which consists of a non-empty set U, n elements $x_1, \ldots, x_n$ of that set, and m relations that hold among its elements: $X_1, \ldots, X_m$ (relation X_i having as many arguments as predicate P_i) .[5] Every such model $\mathfrak{M}$ determines an *interpretation* of L. The variables $x, y, \ldots$ of L range over set U, the names $a_1, \ldots, a_n$ denote, respectively, objects $x_1, \ldots, x_n$, and the predicates $P_1, \ldots, P_m$ denote, respectively, relations $X_1, \ldots, X_m$.

Given a language L and one of its models $\mathfrak{M}$ we can rigorously define the concept of a statement in L which is true in $\mathfrak{M}$. The definition of that concept, which is a fundamental one in modern logical semantics, will not be quoted here. Intuitively speaking, a statement Z in L *is true in* $\mathfrak{M}$ if and only if things are exactly as stated in Z under the interpretation of L determined by $\mathfrak{M}$. Let the following system of objects

$$\langle \boldsymbol{U}, \boldsymbol{a}_1, \ldots, \boldsymbol{a}_n, \boldsymbol{P}_1, \ldots, \boldsymbol{P}_m \rangle$$

be a model $\mathfrak{M}$ of L. A statement $P_1\, a_1$ is true in $\mathfrak{M}$ if and only if the object which in $\mathfrak{M}$ is the denotation of a_1 belongs to the set which in $\mathfrak{M}$ is the denotation of P_1, i.e., if $\boldsymbol{a}_1 \in \boldsymbol{P}_1$. The concept of model of a formalized language and that of a statement which is true in a model are those concepts of logical semantics on which our semantic analysis of open concepts will be based.[6]

2. Let us assume that a language L, as described above, with its extra-

logical constants $a_1, \ldots, a_n, P_1, \ldots, P_m$, is extended by adding a monadic predicate Q, which is introduced by the conditional definition $D(Q)$

$$\bigwedge x (\Psi x \rightarrow (Q x \equiv \Phi x)),$$

where the predicates Ψ and Φ are (simple or compound) expressions in L. We thus obtain a language L', which is an extension of L. Systems of objects of the type

$$\langle U, x_1, \ldots, x_n, X_1, \ldots, X_m, Y \rangle$$

will be models $\mathfrak{M}'$ of L'.

Let $Z(Q)$ stand for any statement in L' in which Q occurs. The following question may be posed concerning any such statement: is that statement equivalent, on the basis of definition $D(Q)$, to any statement in L, i.e., a statement in which Q does not occur? If the definition of Q were a proper (i.e., equivalence) definition, the answer would always be in the affirmative, since such a definition would satisfy the criterion of eliminability, and Q would, on the basis of that definition, be eliminable from every statement $Z(Q)$. To put it more precisely, for every statement $Z(Q)$ there would be a statement Z, in which Q would not occur, and such that

$$D(Q) \rightarrow (Z(Q) \equiv Z)$$

would be a tautology in L', i.e., a statement that would be true in every model $\mathfrak{M}'$ of L'. The definition $D(Q)$ could amount to a proper definition only if $\bigwedge x \Psi x$ were a tautology. We assume here that this is not the case and that $D(Q)$ is not logically equivalent to any proper definition, and as such does not satisfy the criterion of eliminability. There are, however, statements in L' in which Q occurs, and such that Q is eliminable from them on the basis of $D(Q)$. The totality of the statements in L' in which Q occurs can thus be classed into:
(1) statements from which Q is eliminable on the basis of $D(Q)$ and
(2) statements which do not satisfy that condition.

The said condition, to be denoted by (EL), is defined thus:

(EL) Q is eliminable from $Z(Q)$ on the basis of definition $D(Q)$ if and only if there is a statement Z in which Q does not occur, and such that

$$D(Q) \rightarrow (Z(Q) \equiv Z)$$

is true in every model $\mathfrak{M}'$ (of L').[7]

What are the statements $Z(Q)$ which satisfy (EL)? They include, of course, all statements $Z(Q)$ in which Q occurs inessentially, i.e., those which are logically equivalent to statements in which Q does not occur. This applies, above all, to all tautologies, and their negations, such as

$$Q a_1 \vee \sim Q a_1,$$

$$Q a_1 \wedge \sim Q a_1,$$

and also to statements of which an example is provided by

$$P_1 a_1 \wedge (\sim P_1 a_1 \rightarrow Q a_1),$$

a statement which is logically equivalent to $P_1 a_1$. But condition (EL) is also satisfied by such statements $Z(Q)$ in which Q occurs essentially. These are statements which, while not being logically equivalent to statements in which Q does not occur, are equivalent to them on the basis of definition $D(Q)$. Here are some examples:

$$\Psi a_1 \wedge Q a_1,\ \Psi a_1 \rightarrow Q a_1,\ \bigwedge x (\Psi x \rightarrow Q x),$$
$$\bigvee x (\Psi x \wedge Q x),\ \Psi a_1 \rightarrow \sim Q a_1,\ \sim \bigvee x (\Psi x \wedge Q x).$$

It can easily be demonstrated that if the validity of $D(Q)$ is assumed then these statements are equivalent to the following ones in which Q does not occur:

$$\Psi a_1 \wedge \Phi a_1,\ \Psi a_1 \rightarrow \Phi a_1,\ \bigwedge x (\Psi x \rightarrow \Phi x),$$
$$\bigvee x (\Psi x \wedge \Phi x),\ \Psi a_1 \rightarrow \sim \Phi a_1,\ \sim \bigvee x (\Psi x \wedge \Phi x).$$

And here in turn are examples of statements $Z(Q)$ which do not satisfy condition (EL):

$$Q a_1,\ \Psi a_1 \vee Q a_1,\ \Psi a_1 \equiv Q a_1,\ \sim \Psi a_1 \vee Q a_1,$$
$$\sim \Psi a_1 \rightarrow Q a_1,\ \bigwedge x\, Q x,\ \bigvee x\, Q x,\ \bigwedge x (Q x \rightarrow \Psi x).$$

There are no statements in which Q does not occur, to which these statements would be equivalent even if the validity of $D(Q)$ were assumed.

The class of the statements $Z(Q)$ which satisfy condition (EL) can be described in many ways. We mention one of them here, which sheds light upon the type of context in which Q occurs in statements of that class. Let $Z(\Psi \wedge Q)$ stand for a statement which is obtained from $Z(Q)$ by the replacement of every expression of the type $Q x$ by an expression of the type $\Psi x \wedge Q x$. It can easily be demonstrated (we omit the proof here) that

Q is eliminable from $Z(Q)$ on the basis of definition $D(Q)$ if and only if the statement

$$D(Q) \rightarrow (Z(Q) \equiv Z(\Psi \wedge Q)$$

is true in every model $\mathfrak{M}'$.

The statements $Z(Q)$ which satisfy the above condition are, first of all, those which are logically equivalent to statements $Z(\Psi \wedge Q)$. All the examples of statements $Z(Q)$ which satisfy (EL) are like that, e.g., $\Psi a_1 \rightarrow Q a_1$, which is logically equivalent to $\Psi a_1 \rightarrow \Psi a_1 \wedge \wedge Q a_1$. It may be said that they are statements which, if they essentially refer to Q at all, do so only about the Ψ part of Q. Accordingly, they impose, at most, certain conditions on those objects Q which are also Ψ. And those statements which satisfy (EL) and which are not logically equivalent to $Z(\Psi \wedge Q)$, are equivalent to the latter on the assumption $D(Q)$. If this assumption is made, they, too, impose, at most, certain conditions on those objects Q which are also Ψ. The following statements may serve as examples:

$$\bigvee x(\Psi x \wedge \sim Q x \wedge \Phi x) \wedge Q a_1,$$
$$\bigwedge x(\Psi x \rightarrow (Q x \equiv \Phi x)) \vee Q a_1.$$

The relationship now under consideration becomes comprehensible if we realize that definition $D(Q)$ establishes the equivalence of expressions $Q x$ with expressions Φx in which Q does not occur, only for those objects x which are Ψ.[8]

Yet in our analyses a more important role is played by a property which is semantic in nature and which is an attribute of those statements $Z(Q)$ which satisfy (EL). Definition $D(Q)$ admits, for a fixed interpretation of expressions in L, various interpretations of Q. Among all statements $Z(Q)$ we can single out those whose truth value is independent of any admissible interpretation of Q. This condition, to be denoted by (OL), can be defined thus:

(OL) $Z(Q)$ has a definite truth value relative to Q on the basis of definition $D(Q)$ if and only if, for any models $\mathfrak{M}_1'$ and $\mathfrak{M}_2'$ which differ at most in the denotation of Q, the following holds: if $D(Q)$ is true in $\mathfrak{M}_1'$ and in $\mathfrak{M}_2'$, then $Z(Q)$ is true in $\mathfrak{M}_1'$ if and only if $Z(Q)$ is true in $\mathfrak{M}_2'$.

Hence, if $Z(Q)$ is true (false) in a model $\mathfrak{M}'$, then it remains true (false) in every model which differs from $\mathfrak{M}'$ only in the denotation of Q, provided that in both models the denotations of Q satisfy the condition laid down in its definition. Now it turns out that a definite truth value relative to Q is shared by all, and only those, statements $Z(Q)$ from which Q is eliminable. *Conditions* (EL) *and* (OL) *are thus equivalent.*

Here follows an outline of the proof of that proposition. It is obvious that (OL) follows from (EL). Condition (EL) states that there is a statement Z in which Q does not occur and such that, in every model $\mathfrak{M}'$ in which $D(Q)$ holds, $Z(Q)$ has the same truth value as Z. Hence, if we take any models $\mathfrak{M}_1'$ and $\mathfrak{M}_2'$ which differ only in the denotation of Q, and hence such in which Z has the same truth value, and in which $D(Q)$ holds, then $Z(Q)$ must also have the same truth value in both models, namely the same as Z has. And this is precisely what (OL) states. If we want to demonstrate that (EL) follows from (OL) we take as our model the $\mathfrak{M}_2'$ specified in (OL) — a model in which the denotation of Q would be identical with that which in $\mathfrak{M}_1'$ and $\mathfrak{M}_2'$ is the denotation of the predicate Φ. Definition $D(Q)$ must be true in a model $\mathfrak{M}_2'$ so defined, so that the corresponding assumption in the formulation of (OL) may be disregarded. At the same time the statement that $Z(Q)$ is true in a model $\mathfrak{M}_2'$ so defined is equivalent to the statement that $Z(\Phi)$ is true in $\mathfrak{M}_1'$ and may be replaced by the latter.[9] This yields the following result, to be denoted by (EL*):

(EL*) For any model $\mathfrak{M}'$, if $D(Q)$ holds in $\mathfrak{M}'$, then $Z(Q)$ is true in $\mathfrak{M}'$ if and only if $Z(\Phi)$ is true in $\mathfrak{M}'$; or, briefly, $D(Q) \rightarrow (Z(Q) \equiv Z(\Phi))$ is true in every model $\mathfrak{M}'$.

Since Q does not occur in $Z(\Phi)$, (*D*L*) implies (*E*L). Hence all, and only those, statements $Z(Q)$ which satisfy (*D*L), satisfy (OL).

The concept of a statement $Z(Q)$ that satisfies (OL) comes close to the concept of a "determinate" statement introduced into analogous discussions by H. Mehlberg.[10] It appears, therefore, that the latter concept approximately coincides with the concept of a statement $Z(Q)$ from which Q can be eliminated on the basis of $D(Q)$. The proof carried out above also shows that the condition of eliminability (EL) can be replaced by an apparently stronger condition (EL*): not only — — and this is obvious — does (EL*) imply (EL), but, conversely, (EL) implies (EL*), since, as we have just shown, (EL) implies (OL), which in turn implies (EL*). Thus, if as we have assumed, $D(Q)$ has the form

$$\bigwedge x\,(\Psi x \rightarrow (Q x \equiv \Phi x)),$$

then every statement $Z(Q)$ from which Q can be eliminated on the basis of that definition is, in virtue of that definition, equivalent to $Z(\Phi)$, which is obtained from $Z(Q)$ by the replacement of Q by Φ, which is in L.

3. The concepts of eliminability (EL) and definiteness (OL) defined above might be termed absolute. It is also possible to define corresponding relative concepts, with at least two relativizations. First of all, the concepts of eliminability (ET) and definiteness (OT) as relativized to a set T of statements in L. This set T may, and usually is (in the case of such analyses) identified with a theory formulated in L. We shall therefore say that

(ET) Q is eliminable from $Z(Q)$ on the basis of definition $D(Q)$ in a theory T if and only if there is a statement Z in which Q does not occur, and such that the statement

$$D(Q) \rightarrow (Z(Q) \equiv Z)$$

is true in every model $\mathfrak{M}'$ in which T is true.

It is obvious that this condition is, in a sense, a generalization of the preceding one. If T consists solely of tautologies in L, (ET) reduces to (EL). But in a general case, if T includes statements which are not tautologies in L, the class of statements $Z(Q)$ which satisfy (ET) covers not only all those $Z(Q)$ which satisfy (EL), but also some $Z(Q)$ which do not satisfy (EL). Thus, for instance, $Q\,a_1$ satisfies (ET) if Ψa_1 is in T; the same applies to $\bigvee x\,(P_1\,x \wedge Q\,x)$ on the condition that $\bigwedge x\,(P_1\,x \rightarrow \Psi x)$ is a theorem in T. And if T includes the statement $\bigwedge x\,\Psi x$, then Q can be eliminated from every $Z(Q)$ in T, since in T the definition $D(Q)$ becomes equivalent to a proper (equivalence) definition.

The concept of eliminability (ET) has its analogue in the concept of definiteness (OT) relativized in the same manner.

(OT) $Z(Q)$ has a definite truth value relative to Q on the basis of definition $D(Q)$ in theory T if and only if, for any models $\mathfrak{M}'_1$ and $\mathfrak{M}'_2$ which differ at most in the denotation of Q and in which T is true, the following holds: if $D(Q)$ is true in $\mathfrak{M}'_1$ and in $\mathfrak{M}'_2$, then $Z(Q)$ is true in $\mathfrak{M}'_1$ if and only if $Z(Q)$ is true in $\mathfrak{M}'_2$.

As in the case of (EL) and (OL), conditions (ET) and (OT) are mutually equivalent. The proof of their equivalence is analogous to the preceding one.

The concepts of eliminability and definiteness, to be denoted respectively by (EM) and (OM), relativized to a model $\mathfrak{M}$ of L, are more important than those relativized to a set T of statements in L. When defining (EM) and (OM) we shall make use of the concept of an expansion of a given model. If we confine ourselves to the languages L and L' described above, we can say that the model $\mathfrak{M}'$ of L' is an expansion of a model $\mathfrak{M}$ of L if the universes and the denotations of the terms common to L and L' are identical in both models. Thus, if

$$\mathfrak{M} = \langle U, x_1, ..., x_n, X_1, ..., X_m \rangle$$

is a model of L, then the model

$$\mathfrak{M}' = \langle U, x_1, ..., x_n, X_1, ..., X_m, Y \rangle$$

of L' is an expansion of $\mathfrak{M}$. The definition of (EM) now states that

(EM) Q is eliminable from $Z(Q)$ on the basis of definition $D(Q)$ in model $\mathfrak{M}$ if and only if there is a statement Z in which Q does not occur, and such that the statement

$$D(Q) \rightarrow (Z(Q) \equiv Z)$$

is true in every model $\mathfrak{M}'$ which is an expansion of $\mathfrak{M}$.

As in the case of (ET), the class of those $Z(Q)$ which satisfy (EM) covers, in addition to those statements which satisfy (EL), certain statements which do not satisfy that condition. Examples can be given which are analogous to the foregoing ones. Suppose that a given system

$$\langle \boldsymbol{U}, \boldsymbol{a}_1, ..., \boldsymbol{a}_n, \boldsymbol{P}_1, ..., \boldsymbol{P}_m \rangle$$

is the model $\mathfrak{M}$ of L. Now let $\boldsymbol{\Psi}$ stand for the denotation, in $\mathfrak{M}$, of the predicate Ψ. A statement $Q\,a_1$ satisfies (EM) if $\boldsymbol{a}_1 \in \Psi$; the statement $\bigvee x\,(P_1\,x \wedge Q\,x)$ satisfies it if $\bigwedge x \in \boldsymbol{U}\,(x \in \boldsymbol{P}_1 \rightarrow x \in \Psi)$. If it were true that $\bigwedge x \in \boldsymbol{U}\,(x \in \Psi)$, then Q would be eliminable from every $Z(Q)$ in $\mathfrak{M}$, because in that model $D(Q)$ would be equivalent to a proper definition.

The concept (EM) has its coextensional analogue in (OM).

(OM) Z (Q) has a definite truth value relative to Q on the basis of definition D (Q) in model $\mathfrak{M}$ if and only if, for any models $\mathfrak{M}'_1$ and $\mathfrak{M}'_2$ which are expansions of $\mathfrak{M}$, the following holds: if D (Q) is true in $\mathfrak{M}'_1$ and in $\mathfrak{M}'_2$, then Z (Q) is true in $\mathfrak{W}'_1$ if and only if Z (Q) is true in $\mathfrak{M}'_2$.

The mutual equivalence of (EM) and (OM) can be proved analogically to that of (EL) and (OL). Condition (OM) is a concept which exactly coincides with the concept of a determinate statement in Mehlberg. As an example of an indeterminate statement he gives a statement of the type $Q\,a_1$ when it is not true that a_1 is in Ψ , for in such a case that statement becomes true or false according to the interpretation of Q (compatible with $D\,(Q)$). To put it generally, $Z\,(Q)$ is indeterminate when, for a given interpretation of the remaining terms that occur in it, the statement changes its truth value according to the way we interpret Q (in agreement with $D\,(Q)$). And this is just the characteristic trait of those statements $Z\,(Q)$ which do not satisfy (OM).

Attention should be drawn here to the important difference between (EM) and (OM), on the one hand, and all the earlier concepts, on the other. In order to decide whether a given $Z\,(Q)$ satisfies (EL) we have to decide whether the statement

$$(1) \qquad D\,(Q) \rightarrow (Z\,(Q) \equiv Z\,(\Phi))$$

follows logically from the empty class of statements.[11] In the case of, say, $Q\,a_1$ it turns out that this is not so. In order to decide whether a given $Z\,(Q)$ satisfies (ET) we have to decide whether (1) follows logically from statements in T. In the case of $Q\,a_1$ it does so if $\Psi\,a_1$ follows logically from statements in T. Answering these questions requires no reference to empirical data. Hence the statement that a given $Z\,(Q)$ satisfies (EL) or (ET) is always analytic in nature. The situation is otherwise when it comes to (EM). In order to decide whether a given $Z\,(Q)$ satisfies (EM) we have to decide whether (1) is true for any interpretation of Q and for that interpretation of the remaining terms which is determined by a given model $\mathfrak{M}$. In the case of $Q\,a_1$ it is true if the object denoted by a_1 in $\mathfrak{M}$ is in the set which in $\mathfrak{M}$ is denoted by Ψ. Hence the statement that a given $Z\,(Q)$ satisfies (EM) may be synthetic in nature and may require, for its justification or refutation, some reference to empirical data. The same difference obviously holds between (OL) and (OT), on the one hand, and (OM), on the other. We shall turn to it later.

The conditions of eliminability and definiteness, as introduced above, are closely connected with the conditions of the decidability of statements of the type $Z(Q)$. To decide a statement is to justify it or its negation, which, in a loose formulation, means demonstrating that it is true or demonstrating that it is false. But if $Z(Q)$ has no definite truth value, i.e., becomes true or false according to a given admissible interpretation of Q, then that cannot be done without additional assumptions. Thus, if $Z(Q)$ is to be decidable it must satisfy the condition of definiteness (and thereby also that of eliminability). The converse relationship holds on the assumption that statements in L, i.e., statements in which Q does not occur, are decidable. If $Z(Q)$ then satisfies the condition of eliminability, and is thus equivalent to a statement in L, it must be decidable. The various kinds of eliminability (definiteness) have their analogues in the various kinds of decidability, namely D-decidability, T-decidability, M-decidability. To put it more precisely, we are dealing here with various kinds of reducibility of a given statement to decidable ones. Thus, a statement $Z(Q)$ is *D-decidable* if it is reducible to a statement which is decidable on the basis of the definition of Q alone, it is *T-decidable* if it is reducible to a statement which is decidable on the basis of definition $D(Q)$ in theory T; finally, it is *M-decidable* if it is reducible to a statement which is decidable on the basis of definition $D(Q)$ in model $\mathfrak{M}$, and hence if it is equivalent to a decidable statement for any interpretation of Q compatible with $D(Q)$ and for that interpretation of the remaining terms which is determined by the model $\mathfrak{M}$. It is this concept of M-decidability to which we shall chiefly refer below. The indeterminate statements mentioned above are undecidable statements in that sense. As an example of a $Z(Q)$ which is decidable in each of the senses described above we can quote $\Psi^{\cdot} a_1 \wedge Q a_1$. Whereas Qa_1, which is not D-decidable, is T-decidable if $\Psi^{\cdot} a_1$ is a theorem in T, and is M-decidable if $\Psi^{\cdot} a_1$ is true in $\mathfrak{M}$.

II

4. We have thus obtained the answer to the first of our initial questions, namely that concerning the characteristics of the class of semanti-

cally problematic statements, each of which includes a term which has been defined conditionally. If that term is a predicate Q, and if its conditional definition is $D(Q)$, then that class is identical with the class of all statements in L' in which Q occurs and which do not satisfy condition (OM) (or — and this amounts to the same thing — condition (EM)). As we know, the characteristic trait of these statements is that, for a given interpretation of terms of L, as determined by a given model $\mathfrak{M}$, they change their truth value according to the interpretation of Q (compatible with $D(Q)$). The question now arises whether it is legitimate to consider such statements to be true or false, and if so, true or false in what sense? This leads to further questions. To what do such statements actually refer? Does Q have any denotation? And if so, what?

We shall first consider the problem of the truth of those statements $Z(Q)$ which do not satisfy (OM). It must be pointed out here that we have so far not used any absolute concept of truth. The only concept of truth we have used so far has been the relative concept of being true in a model. We have not used the formulation: Z is true, but only the formulation: Z is true in $\mathfrak{M}$.[12] The latter was understood in such a way as to imply the following equivalence:

$$\text{statement } P_1\, a_1 \text{ is true in model}$$
$$\mathfrak{M} = \langle \boldsymbol{U}, \boldsymbol{a}_1, \ldots, \boldsymbol{a}_n, \boldsymbol{P}_1, \ldots, \boldsymbol{P}_m \rangle$$
$$\text{if and only if } \boldsymbol{a}_1 \in \boldsymbol{P}_1,$$

and analogical equivalences for the remaining statements in L. The transition to the absolute concept of truth consists in choosing a fixed model $\mathfrak{M}^*$ of L and in defining a true statement as a statement which is true in $\mathfrak{M}^*$. The family of models $\mathfrak{W}$ of a language L covers all those systems of objects about which it is possible to speak in L. It is assumed that one of them is that about which we are in fact speaking in L. That model, $\mathfrak{M}^*$, is what is termed *the proper model of L*. It is the model which provides the translation of expressions in L into the metalanguage *ML*. The interpretation of expressions in L determined by the proper model $\mathfrak{M}^*$ is the translation of those expressions into the language which we use when describing the language L. The *absolute* concept of truth, as opposed to the *relative* concept of being true in a model, is applicable only to an interpreted language, that is, to such a language L for which its proper model $\mathfrak{M}^*$ is given. If the system mentioned above, that is,

$$\langle \boldsymbol{U}, \boldsymbol{a}_1, \ldots, \boldsymbol{a}_n, \boldsymbol{P}_1, \ldots, \boldsymbol{P}_m \rangle,$$

is given as the proper model $\mathfrak{M}^*$ of the language L here under consideration, then we may say simply that

statement $P_1\, a_1$ is true if and only if $\boldsymbol{a}_1 \in \boldsymbol{P}_1$.

Analogical equivalences hold for the remaining statements in L.[13]

In view of what has been said above, concepts of logical semantics such as the absolute concept of truth, have been used so far only with reference to languages which have a definite and unambiguous interpretation, languages which are fully interpreted (or, in Kemeny's terminology, languages which are semantically determined). What is, in this connection, the nature of languages L and L', which we have been analysing above? Our tentative answer will refer to certain assumptions which are an idealization of the actual state of things. It is assumed, first of all, that L does not deviate from the type of languages which have so far been studied in semantic investigations. It is a fully interpreted language. Hence L has an unambiguously defined proper model $\mathfrak{M}^*$. Let that model be, as before, the system

$$\langle \boldsymbol{U}, \boldsymbol{a}_1, \ldots, \boldsymbol{a}_n, \boldsymbol{P}_1, \ldots, \boldsymbol{P}_m \rangle.$$

It may therefore be assumed that, for any statement Z in L,

Z is true if and only if Z is true in $\mathfrak{M}^*$.

An explanation would not be out of place here. When we previously said simply that a given $Z(Q)$ satisfies (OM) (or (EM)), or does not satisfy it, the model $\mathfrak{M}$ of L, to which that condition was tacitly relativized, was identified with the proper model $\mathfrak{M}^*$ of L.

The problem arises as to how the proper model $\mathfrak{M}^*$ of L is established for that language? And why do we make such an assumption concerning L? Now, as was mentioned at the beginning, the type of the languages we are here concerned with is that of the languages of empirical scientific theories. Specific terms in such theories are usually classed into two kinds: observation terms and theoretical terms. If we identify

L' with the language of an empirical theory, L is its observational part. The terms $a_1, \ldots, a_n, P_1 \ldots, P_m$ are in the class of observation terms, and Q is in the class of theoretical terms. Terms in L which are observation terms, and as such refer to observable objects, admit — as is usually assumed — a direct interpretation. The objects which are to be their respective denotations can be assigned to them directly by indication, i.e., by means of semantic rules which have the nature of ostensive definitions. The assumption that such an assignment is unambiguous is, of course, an idealization of the actual state of things. Here, however, such a simplifying assumption may be adopted, and it may accordingly be assumed that it is possible in this manner to determine uniquely the proper model $\mathfrak{M}^*$ of L.[14]

The situation is otherwise when it comes to the nature of L'. The term Q, being a theoretical term, refers to a non-observable property and hence does not admit a direct interpretation. The interpretation of Q is determined solely by its definition $D(Q)$, which refers to observation terms whose interpretation is determined by the proper model $\mathfrak{M}^*$ of L. Thus the proper model $\mathfrak{M}'^*$ of L' is defined by the following two conditions alone:

(i) $\mathfrak{M}'^*$ is an expansion of $\mathfrak{M}^*$,

(ii) $D(Q)$ is true in $\mathfrak{M}'^*$.

The proper model $\mathfrak{M}'^*$ of L' must therefore be one in which the universe and the denotations of those terms which are in L are the same as in the proper model $\mathfrak{M}^*$ of L, and denotation of Q satisfies the condition formulated in $D(Q)$. Is model $\mathfrak{M}'^*$ established unambiguously in this way? That would be the case only if $D(Q)$:

$$\bigwedge x\,(\Psi x \to (Q x \equiv \Phi x)),$$

were equivalent in $\mathfrak{M}^*$ to a proper equivalence definition. And that in turn would be possible only if the set which in $\mathfrak{M}^*$ is the denotation of Ψ coincided with the set U which is the universe of $\mathfrak{M}^*$. Since we are interested here in the situation characteristic of open concepts — and it is such concepts which correspond to theoretical terms in empirical theories — we assume that that is not the case. Under this assumption model $\mathfrak{M}'^*$ is not determined uniquely. Conditions (i) and (ii) define a family $\mathfrak{R}\mathfrak{M}'$ of models of L', which includes more than one model of that language. The proper model $\mathfrak{M}'^*$ is one of the models in that family:

$$\mathfrak{M}'^* \in \mathfrak{R}\,\mathfrak{M}'.$$

Thus we are dealing with a situation which differs from those so far considered in semantic investigations. L' is a partially interpreted language. What sense then may we associate with the claim that a statement Z in L' is true?

Before we proceed to review various attempts at answering this question, we shall analyse, by way of example, the semantic nature of certain simple statements in L'. To do this we adopt a number of simplifying assumptions, to which we shall refer in further examples, too. Let the definition $D(Q)$ of Q take on the form of a partial definition of type (3), the form which is typical of the definitions of theoretical terms:

$$\wedge\, x\,((P_1\,x \to Q\,x) \wedge (P_2\,x \to \sim Q\,x)).$$

Let us assume that sets $\boldsymbol{P}_1$ and $\boldsymbol{P}_2$ are disjoint and their union does not equal the universe $\boldsymbol{U}$. In such a case the family of models of L', i.e., $\mathfrak{R}\mathfrak{M}'$, includes all, and only those, models

$$\langle \boldsymbol{U}, \boldsymbol{a}_1, \ldots, \boldsymbol{a}_n, \boldsymbol{P}_1, \ldots, \boldsymbol{P}_m, \boldsymbol{Y} \rangle$$

in which

$$\boldsymbol{P}_1 \subset \boldsymbol{Y} \subset -\boldsymbol{P}_2.^{15}$$

If we also assume, to simplify matters even more, that there is only one element of $\boldsymbol{U}$ which is neither in $\boldsymbol{P}_1$ nor in $\boldsymbol{P}_2$, then the family $\mathfrak{R}\mathfrak{M}'$ includes exactly two models:

$$\mathfrak{M}'_1 = \langle \boldsymbol{U}, \boldsymbol{a}_1, \ldots, \boldsymbol{a}_n, \boldsymbol{P}_1, \ldots, \boldsymbol{P}_m, \boldsymbol{P}_1 \rangle,$$

$$\mathfrak{M}'_2 = \langle \boldsymbol{U}, \boldsymbol{a}_1, \ldots, \boldsymbol{a}_n, \boldsymbol{P}_1, \ldots, \boldsymbol{P}_m, -\boldsymbol{P}_2 \rangle.$$

The proper model $\mathfrak{M}'^*$ is then identical with one of them: $\mathfrak{M}'^* = \mathfrak{M}'_1$ or $\mathfrak{M}'^* = \mathfrak{M}'_2$. Let $\boldsymbol{a}_1$ be that object which is neither in $\boldsymbol{P}_1$ nor in $\boldsymbol{P}_2$, and $\boldsymbol{a}_2$ an object other than $\boldsymbol{a}_1$, and hence one which is either in $\boldsymbol{P}_1$ or in $\boldsymbol{P}_2$. These relations are shown by the following diagram:

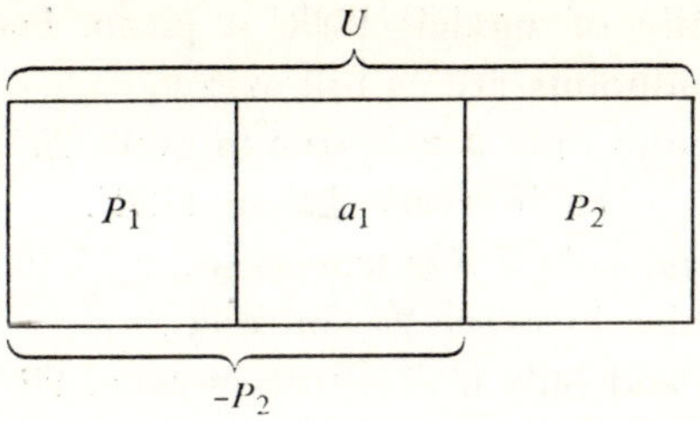

Let us now consider the following three statements in L' in which Q occurs:

(a) $Q\,a_1 \vee \sim Q\,a_1$,

(b) $Q\,a_2$,

(c) $Q\,a_1$,

and let us establish their truth value in the models in $\mathfrak{RM}'$. The first two represent that class of statements $Z(Q)$ which satisfy (OM), and the third that class of $Z(Q)$ which do not satisfy that condition. Further, (a) as a tautology is true in both $\mathfrak{M}'_1$ and $\mathfrak{M}'_2$. Statement (b) has the same truth value in both models regardless of whether Q denotes $\boldsymbol{P}_1$ or $-\boldsymbol{P}_2$. If $\boldsymbol{a}_2 \in \boldsymbol{P}_1$, then (b) is true in both $\mathfrak{M}'_1$ and $\mathfrak{M}'_2$; and if $\boldsymbol{a}_2 \in \boldsymbol{P}_2$, then (b) is false in both $\mathfrak{M}'_1$ and $\mathfrak{M}'_2$. On the other hand, (c) takes on different truth values according to whether Q denotes $\boldsymbol{P}_1$ or $-\boldsymbol{P}_2$: (c) is false in $\mathfrak{M}'_1$ and true in $\mathfrak{M}'_2$.

5. The existence in L' of statements of the last-mentioned type, i.e., those $Z(Q)$ which do not satisfy (OM), accounts for an essential difficulty in attempts to define the absolute concept of truth as applicable to L'. We shall now discuss a number of possible solutions, including the most important ones. We shall be dealing with standpoints which are in fact represented in discussions concerned with the problem, even though the formulations used in such discussions are as a rule rather sketchy and differ from those suggested below. The point is, however, to place these attempts within the precise and uniform framework of contemporary logical semantics. The five standpoints which we list below represent the five main types of answers to the question concerning the nature of the truth and falsehood of statements in L', for

which only a family of models $\mathfrak{N}\mathfrak{M}'$ is given. Let Z be a statement in L'. The said standpoints are as follows:

(I) Z is true if and only if Z is true in every $\mathfrak{M}'$ in $\mathfrak{N}\mathfrak{M}'$; Z is false if and only if Z is false in every $\mathfrak{M}'$ in $\mathfrak{N}\mathfrak{M}'$.

(II) Z is true if and only if Z is true in some $\mathfrak{M}$ in $\mathfrak{N}\mathfrak{M}'$; Z is false if and only if Z is false in every $\mathfrak{M}'$ in $\mathfrak{N}\mathfrak{M}'$.

(III) Z is true if and only if Z is true in every $\mathfrak{M}$ in $\mathfrak{N}\mathfrak{M}'$; Z is false if and only if Z is false in some $\mathfrak{M}'$ in $\mathfrak{N}\mathfrak{M}'$.

(IV) Z is true if and only if Z is true in $\mathfrak{M}'_i$; Z is false if and only if Z is false in $\mathfrak{M}'_i$;
where $\mathfrak{M}'_i$ is a specified model chosen from $\mathfrak{N}\mathfrak{M}'$ on the basis of additional assumptions, to be discussed later.

(V) Z is true if and only if Z is true in $\mathfrak{M}'^*$; Z is false if and only if Z is false in $\mathfrak{M}'^*$;
where $\mathfrak{M}'^*$ is the proper model of L', characterized only by the condition: $\mathfrak{M}'^* \in \mathfrak{N}\mathfrak{M}'$.

As we proceed to discuss these standpoints let us first consider how we are to class statements (a), (b), (c) with regard to their absolute truth value according to each standpoint. They all agree about (a) and (b): (a) is true, (b) is true or false. The differences manifest themselves only when it comes to (c). According to (I), (c) is neither true nor false; according to (II) it is true; according to (III) it is false; according to (IV), it has a definite truth value which depends on which model in $\mathfrak{N}\mathfrak{M}'$ is $\mathfrak{M}_i$ (it will become evident later that in the case under consideration the model is such that (c) is false); according to (V), (c) is true or false, but in view of an ambiguous description of $\mathfrak{M}'^*$ it is impossible to decide which of the two cases holds. This assessment of the absolute truth value of the selected statements in L' suggests certain general conclusions about the nature of each standpoint.

Now according to each of these standpoints all statements in L' in which Q does not occur and those statements Z (Q) which satisfy (OM) have a definite truth value: they are true or false, and in principle they are decidable, too. Likewise, according to each of the above standpoints all tautologies in L', including statements of the type $Z(Q)$, are true. Thus, regardless of differences in interpretation of the truth and falsehood of statements in L', all these standpoints accept all the laws of classical logical calculus, i.e., classical sentential calculus and predicate calculus, as valid. This applies, in particular, to the logical law of the excluded middle for Q: $\bigwedge x\,(Qx \vee \sim Qx)$, and hence to its substitution: $Qa_1 \vee \sim Qa_1$. The same holds, of course, for the logical law of contradiction: $\bigwedge x \sim (Qx \wedge \sim Qx)$ and its substitution: $\sim (Qa_1 \wedge \sim Qa_1)$. On the other hand, all these standpoints differ from one another in their treatment of those $Z\,(Q)$ which do not satisfy (OM), i.e. of the indeterminate statements. We shall now analyze one by one the various solutions and their consequences.

Standpoint (I) differs from the others in denying truth value to the indeterminate statements. These statements, being true in some models in $\mathfrak{RM}'$ and false in others, are neither true nor false in the absolute sense. According to (I), unlike all other standpoints, the set of true and false statements does not cover the totality of statements in L'. How, then, are we to class those formulas which are statements syntactically, but lack truth value? There are two approaches to this issue, (I.1) and (I.2). According to (I.1), these formulas are treated as meaningless. They are considered not to belong to L'. But the consequences of this interpretation are hard to accept. As we have shown, whether a given $Z\,(Q)$ satisfies (OM) or not may depend on empirical data. Hence, whether a given formula is in L' or is meaningless is made to depend on empirical data, and this is what we want to avoid at all costs when constructing a language. Another consequence, which is at variance with our intuitions, is that a conjunction or a disjunction of two meaningless formulas may prove meaningful. And this holds precisely in the case of Qa_1, $\sim Qa_1$. As indeterminate statements, they are both meaningless, but their conjunction and disjunction are meaningful: the former is a false, and the latter is a true statement in L'.[16]

According to (I.2), formulations which are statements syntactically but lack truth value are treated as meaningful statements in L'. Thus (I.2) admits the existence in L' of statements which are neither true nor false. This relieves the advocates of that standpoint of the complications which we have just mentioned. But other complications arise. The very concept of a statement which is meaningful, i.e., states something, but is neither true nor false, seems puzzling. But other consequences are even more embarrassing. Even though (I.2) preserves the classical logical calculus *in toto*, it must reject certain classical metalogical laws. The logical law of the excluded middle holds, but the metalogical law of the excluded middle loses its validity. There are contradictory statements, Qa_1, $\sim Qa_1$, such that none of them is true. And since, at the same time, $Qa_1 \vee \sim Qa_1$ remains true, the classical truth-table for disjunction loses its validity, too: a disjunction of two statements neither of which is true turns out to be true. Similar consequences are obtained in the case of certain other metalogical laws. And these are consequences which cannot be treated lightly.[17]

According to (II) — and all the standpoints listed after (II) — the set of true and false statements in L' covers the totality of statements in that language. Hence all $Z(Q)$ which do not satisfy (OM) are true or false. Moreover, they are not considered to be indeterminate, since their truth value is always defined unambiguously. According to (II), both Qa_1 and $\sim Qa_1$ are classed as true. This immediately shows that in spite of the fact that the classical logical calculus is preserved, certain classical metalogical laws are rejected, and even more radically than in the previous case. Because of the existence of a pair of contradictory statements both of which are true, the metalogical law of contradiction loses its validity. And since $Qa_1 \wedge \sim Qa_1$ is false, the classical truth-table for conjunction is not valid, since the conjunction specified above turns out to be false even though both its components are true.[18]

Standpoint (III), which is dual with respect to (II), has analogous properties. Those $Z(Q)$ which do not satisfy (OM) are true or false, and their truth value is defined unambiguously. In this case, however, both Qa_1 and its negation $\sim Qa_1$ are classed as false. Hence, while the classical logical calculus is preserved, such classical metalogical laws as the metalogical law of the excluded middle and the classical truth-table for disjunction lose their validity, since $Qa_1 \vee \sim Qa_1$ remains true even though both its components are false.[19]

The description of (IV) must remain indefinite until we discuss the model $\mathfrak{M}_i'$ in greater detail, since (IV) refers to that model. In any case it can be said even now that here, too, all $Z(Q)$ which do not satisfy (OM) must be true or false. Moreover, they are not indeterminate here either, and their truth value is defined unambiguously. It will become evident later, when $\mathfrak{M}'_t$ is discussed, that Qa_1 is false and $\sim Qa_1$ is true. But, unlike the previous standpoints, in the case of (IV) both all the laws of classical sentential calculus and classical predicate calculus and all the classical metalogical laws remain valid, even though here, too, there is a certain deviation from the classical logical relations — on an issue to be discussed later.[20]

At this point it is worth-while emphasizing that (II), (III) and (IV) have certain common consequences that may arouse justified doubts or objections. As we have seen, in all these cases the truth values of the indeterminate statements are defined unambiguously. The point is not that these definitions disagree — that one and the same statement is true according to one standpoint, and false according to another. What is at issue is rather the very fact of their being so defined. It appears that this modifies the original semantic nature of those indeterminate statements. Their truth value becomes predetermined, and L' therefore loses to some extent the nature of an open language. A language with open concepts should make it possible for those concepts gradually to be given precision. Such a procedure, which has been discussed by me on another occasion,[21] consists in our providing a term which has been introduced by a conditional definition, with further conditional definitions. Such definitions, by broadening the scope of applicability of a given term, enable us to decide certain statements in which that term occurs and which have been undecidable previously. Now if the truth value of such statements is predetermined, the procedure of making open concepts more and more precise must result in our classing as false some statements which we have previously classed as true, and vice versa. This property of the three standpoints now under consideration seems to show that they do not faithfully reflect the open nature of L'.

Standpoint (V), the last of those listed above, is free from this defect. The concept of truth is defined in the same way as in the case of fully interpreted languages: as being true in the proper model $\mathfrak{M}'^*$. The difference in this case is that this model is not determined uniquely The only assumption is that there is exactly one such model and that it is in the family $\mathfrak{R}\mathfrak{M}'$. Hence all $Z(Q)$ which do not satisfy (OM) are true or false. All the laws of classical logical calculus and all the classical metalogical laws remain fully valid. At the same time, all $Z(Q)$ which do not satisfy (OM), i.e., all indeterminate statements, remain undecidable. They are true or false, but the incomplete description of the proper model of L', i.e., of that to which L' refers, makes it impossible to decide which is the case. As Q is made more and more precise by adding further conditional definitions, some of those statements become decidable. But we never encounter the necessity of classing as false (true) statements which we have previously classed as true (false). It appears therefore that this concept of truth better corresponds to the open nature of L'. On the other hand, however, the acceptance of the fact that there are statements which are true or false, but essentially undecidable, is disquieting from a philosophical point of view. Some of the standpoints discussed earlier have been formulated with the intention of avoiding such a situation, i.e., with the intention of defining such a concept of truth as would deprive it of its character of being an unknowable property.[22]

6. The semantic description of open concepts requires completion. We have to analyse the problem of their denotation, which is a clear analogue of the problem of their truth. This is why the analysis of this issue will be largely analogous to that of the issue of truth. The term Q, on which we have concentrated here, is a predicate, and as such it performs a double semantic function: that of designating and that of denoting. There is a close connection between these two functions, which in classical logical semantics consists precisely in the fact that the set of objects which are designated by a given predicate is its denotation. So far we have used only the relative concepts of designation and denotation in a model. The meanings of these concepts are explained by the following equivalences, which refer to the concept of satisfaction in a model, which is fundamental in contemporary logical semantics:

P designates x in model $\mathfrak{M}$ if and only if x satisfies formula Px in $\mathfrak{M}$,

P denotes X in model $\mathfrak{M}$ if and only if X is the set of objects which satisfy formula Px in $\mathfrak{M}$ (i.e., if X is the set of objects designated by P in $\mathfrak{M}$).[23]

In particular case this yields:

P_1 designates x in model

$$\mathfrak{M} = \langle U, a_1, \ldots, a_n, P_1, \ldots, P_m \rangle$$

if and only if $x \in \boldsymbol{P}_1$;

P_1 denotes $\boldsymbol{P}_1$ in model

$$\mathfrak{M} = \langle\, \boldsymbol{U}, \boldsymbol{a}_1, \ldots, \boldsymbol{a}_n, \boldsymbol{P}_1, \ldots, \boldsymbol{P}_m \,\rangle.$$

But our problem concerns the absolute concepts of designation and denotation, and not the relative ones. Transition to the absolute concepts consists, as in the case of the absolute concept of truth, in choosing a fixed model $\mathfrak{M}^*$ of L as the proper one, capable of providing the translation of expressions in L into those in the metalanguage ML, and in the defining of designation (denotation) as designation (denotation) in $\mathfrak{M}^*$. The absolute concepts of designation and denotation are thus applicable only to an interpreted language, i.e., to an L for which its proper model $\mathfrak{M}^*$ is given. If, for the language L which is being considered here, the system, mentioned above,

$$\langle \boldsymbol{U}, \boldsymbol{a}_1, \ldots, \boldsymbol{a}_n, \boldsymbol{P}_1, \ldots, \boldsymbol{P}_m \rangle$$

is given as its proper model $\mathfrak{M}^*$, then we may simply say that

P_1 designates x if and only if $x \in \boldsymbol{P}_1$;
P_1 denotes $\boldsymbol{P}_1$.

It is assumed here, as before, that this is precisely the case, i.e., that there is an unambiguously defined proper model $\mathfrak{M}^*$ of L, and that the above system is that model. We may therefore assume that, for any predicate P in L:

P designates x if and only if P designates x in $\mathfrak{M}^*$;
P denotes X if and only if P denotes X in $\mathfrak{M}^*$.

We are making the same assumption as before concerning L' as well. Thus we assume that the proper model $\mathfrak{M}'^*$ of L' is defined only as an element of the family $\mathfrak{R}\mathfrak{M}'$, which, as defined earlier, has more than one element. What sense then can be associated here with the concepts: P designates x, and: P denotes X?

Here, too, various solutions are possible, and these correspond to some extent to the standpoints on truth; they are supported, as a rule, by the advocates of the corresponding standpoints. First of all, we can list three concepts of designation:

(A) P designates x if and only if P designates x in every model $\mathfrak{M}'$ in $\mathfrak{R}\mathfrak{M}'$;

(B) P designates x if and only if P designates x in some model $\mathfrak{M}'$ in $\mathfrak{R}\mathfrak{M}'$;

(C) P designates x if and only if P designates x in $\mathfrak{M}'^*$, where $\mathfrak{M}'^*$ is the proper model of L', defined only by the condition: $\mathfrak{M}'^* \in \mathfrak{R}\mathfrak{M}'$.

The corresponding concepts of denotation can be formed in two ways. The first refers to the relative concept of denotation and yields the following formulations:

(a) P denotes X if and only if P denotes X in every model $\mathfrak{M}'$ in $\mathfrak{R}\mathfrak{M}'$;

(b) P denotes X if and only if P denotes X in some model $\mathfrak{M}'$ in $\mathfrak{R}\mathfrak{M}'$;

(c) P denotes X if and only if P denotes X in $\mathfrak{M}'^*$, where $\mathfrak{M}'^*$ is defined as above.

The second refers to the absolute concepts of designation:

P denotes X if and only if X is the set of objects designated by P (in the sense (A), (B), (C)).

In the case of (C) the concept of denotation thus defined is identical with (c). In the case of (A) and (B) we obtain concepts (a*) and (b*) of denotation, which differ from (a) and (b), respectively.

The consequences of the standpoints described above will be illustrated by the term Q in L', concerning which we have made a number of simplifying assumptions, as a result of which $\mathfrak{R}\mathfrak{M}'$ consists solely of two models, $\mathfrak{M}'_1$ and $\mathfrak{M}'_2$. It can easily be seen that, according to (A), Q designates every object which is in $\boldsymbol{P}_1$ and only such objects. According to (B), it designates every object which is in $-\boldsymbol{P}_2$, and only such objects. Finally, according to (C), just one of these two cases holds true, but it is impossible to decide which.

The concepts of denotation presented above lead to the following conclusions. According to (a), Q does not denote any set; according to (b), it denotes both $\boldsymbol{P}_1$ and $-\boldsymbol{P}_2$; according to (a*) it denotes $\boldsymbol{P}_1$; according to (b*), it denotes $-\boldsymbol{P}_2$; finally, according to (c) it denotes $\boldsymbol{P}_1$ or $-\boldsymbol{P}_2$, but we are not in a position to decide which of these two sets it denotes.

These consequences suggest certain general comments on each of the standpoints in question. We are mainly interested here in the standpoints concerning denotation. In contrast to the others, two of them — (a) and (b) — differ from the classical concept of denotation, which, as we know, assumes that every term denotes one and only one object. Yet, according to (a), a term which has been defined conditionally denotes nothing, while according to (b) it denotes more than one object. Standpoints (a) and (b) correspond to the analogous standpoints (I) and (II) on truth, and are in fact represented by the advocates of the latter standpoints.[24] It cannot be denied that (a) and (b) are clearly at variance with our intuitions concerning denotation, since these are linked with the classical concept of it. Furthermore, it seems inconsistent to associate — as indeed happens — standpoints (a) and (b) with the corresponding standpoints (A) and (B) on designation. If according to both (A) and (B) a term which has been defined conditionally designates every object from an unambiguously defined set, and only such objects, then there is no reason to assume that such a term does not denote any set or that it denotes a number of different sets.

This is why concepts (a*) and (b*) of denotation seem to harmonize better with (A) and (B), respectively. These concepts assign to a term which has been defined conditionally an unambiguously defined set of objects as its denotation; in the case of (a*) it is the set of objects designated in accordance with (A), and in the case of (b*), the set of designata in the sense of (B). How is the choice of such sets of designata justified from the point of view of those standpoints? We shall try to answer that question in connection with standpoint (A)–(a*), which, unlike (B)–(b*), is in fact represented by the advocates of standpoints (I.2) and (IV), which we have analysed in connection with the concept of truth.[25]

The intuitions underlying (A) manifest themselves in the following claim, which is explicitly assumed by the advocates of the concept now under consideration:

(D) P designates x if and only if everyone who correctly uses P must, under specified circumstances, predicate P about x.

It is obvious that in the case of objects which are defined conditionally this equivalence implies definition (A) of designation. If at the same time we adopt definition (a*) of denotation, we obtain, for Q in L', the result previously mentioned: Q denotes $\boldsymbol{P}_1$.

This standpoint is also associated with a specific interpretation of term negation — an interpretation which differs from the classical one. Suppose that we know what set is the denotation of P. What then is denoted by its negation P'? All the remaining concepts of denotation lead to the traditional answer:

$$P \text{ denotes } X \text{ if and only if } P' \text{ denotes } -X.$$

The standpoint now under consideration leads to a different answer, however, because it assumes the following proposition, which refers to the same intuitions as (D) does:

(D′) P' designates x if and only if no one who uses P correctly can, under specified circumstances, predicate P about x.

This implies the following definition of designation:

(A′) P' designates x if and only if P does not designate x in any model $\mathfrak{M}'$ in $\mathfrak{RM}'$.

If at the same time we assume for P' definition (a*) of denotation, then we obtain the following conclusion for Q in L': Q' denotes $\boldsymbol{P}_2$. But according to our assumption, $\boldsymbol{P}_2$ is not the complement of $\boldsymbol{P}_1$ (to universe $\boldsymbol{U}$). The object $\boldsymbol{a}_1$ is in neither of these sets, and hence it is in neither the denotation of Q, nor that of Q' (i.e., the negation of Q).

Does the law of the excluded middle then lose its validity for term negation: $\bigwedge x\,(Qx \vee Q'x)$? The answer to this question depends on the particular standpoint on truth, (I.2) or (IV), with which the concepts of designation and denotation, (A)–(A′)–(a*), are linked. If truth is understood in accordance with (I.2), then term negation can be interpreted in the classical sense: $\bigwedge x\,(Q'x \equiv \sim Qx)$. In this case the concept of the denotation of Q and its negation Q' does not directly affect the truth value of the statements in which those terms occur. Hence all the laws of classical calculus of terms including the law of the excluded middle, remain valid. Even though the statements Qa_1 and $Q'a_1$ are devoid of any truth value, the statement $Qa_1 \vee Q'a_1$ remains true. The situation is different, however, in the case of (IV). Here truth is interpreted as truth in $\mathfrak{M}'_i$. Now $\mathfrak{M}'_i$ is meant to be a model in $\mathfrak{RM}'_i$, in which the denotations of Q and Q' are sets defined by (A), (A′), and (a*). If we adopt the assumptions concerning L' formulated earlier, $\mathfrak{M}'$ will be a model of L' in which $\boldsymbol{P}_1$ is the denotation of Q, and $\boldsymbol{P}_2$ is the denotation of Q'. In this case some of the laws of the classical calculus of terms — among them the law of the excluded middle — lose their validity. Both statements, Qa_1 and $Q'a_1$, are false, and their disjunction, $Qa_1 \vee Q'a_1$, is false, too.

Term negation cannot therefore be interpreted in its classical sense. In particular, $\bigwedge x (\sim Qx \rightarrow Q'x)$ does not hold. This is why the classical calculus of terms must be replaced by a non-classical one.[26]

These consequences of the concept of denotation now under consideration argue against that concept, but the latter appears unsatisfactory for other reasons as well. The motives for the adoption of (A), (A′) and (a*), which are manifested in (D) and (D′), point to a change in the meaning of the classical concepts of designation and denotation, a change which takes place in the conception now under discussion. Whether P designates x is here made to depend on whether a person is inclined to accept the statement which predicates P about x, and hence on a certain attitude of the persons using a given language. The concept of designation, and hence, also, that of denotation, become pragmatic rather than semantic in nature. The same comments are applicable to (B) and (b*), for here, too, we can find pragmatic motives for their adoption.[27] But the main defect, which is shared by both conceptions, appears to consist in something else. The point is not that such and such a denotation — and not another — is assigned in them to a term which has been defined conditionally; the point is that such an unambiguous assignment is made at all. The concept which corresponds to such a term thereby loses its character of being an open concept. It is fixed uniquely to what a given term refers. Making that term more precise by adding further conditional definitions results in fixing its denotation in such a way that it does not agree with the previous one. By using this term after it has been made more precise in this way, we begin to speak about something other than what we spoke about before.

This defect is absent from the last concept of designation and denotation: (C)–(c). The concept of denotation is here understood as in the case of fully interpreted languages: as denotation in the proper model $\mathfrak{M}'^*$. But that model is not defined unambiguously here, and hence the denotation of Q, as a term in L' which has been defined conditionally, is not fixed unambiguously, either. The term Q denotes precisely one set in a strictly defined family of sets (in the case of L', the set $\boldsymbol{P}_1$ or the set $-\boldsymbol{P}_2$), but it is impossible to decide which set this is. This standpoint combines some of the advantages of each of the previous solutions while avoiding their shortcomings. On the one hand, we remain within the classical conception of denotation and the classical conception of the negation of terms, and on the other, we preserve the indefiniteness of the function of denotation, which is characteristic of open concepts. Every term, and hence every term which is defined conditionally, denotes one and only one object. But in the case of a term which is defined conditionally, that object is defined only as an element of a class which has more than one element. Suppose now that we make a given term more precise by providing it with further conditional definitons. This results in the narrowing down of the class of possible denotations, but does not involve an inevitable change in the initial denotation of the term in question. We can claim that we are using the term which we have made more precise in such a way as to speak about the same thing as before. It would seem, therefore, that this concept of denotation is in agreement with the open nature of language L'. At the same time it must, however, be admitted that the claim that a certain term denotes something although it cannot be decided what it denotes, sounds somewhat metaphysical, which is exactly the objection raised by some authors.

7. This concludes our analysis of the semantic problems of open concepts. We have singled out the class of statements that raise such problems, and we have formulated both the problems and the possible ways of solving them. Our analyses have been based on a number of assumptions which simplify the existing state of things. Some of them require brief comments. The most important assumption has been the identification of open concepts with terms which are defined conditionally. The simplifying nature of this assumption has already been mentioned. There may exist theoretical terms in empirical theories — and hence terms which correspond to open concepts — which are not defined conditionally. The meaning postulates which introduce such terms into empirical theories may take on various forms which do not fall under the schema of conditional definition. The postulate which introduces Q into L' can be described — in a general, though incomplete way — as a statement in L' in which Q occurs, and such that it admits of at least one interpretation of Q for the interpretation of the remaining terms as determined by the proper model $\mathfrak{M}^*$ of L. Can our analysis be extended to cover terms introduced by any meaning postulates, and hence also by those which are not conditional

definitions? When trying to answer this question we can say in any case that the fundamental results of our analysis are applicable to all terms defined in such a manner. Each such term enables us to define, in the manner described above, the class of indeterminate statements in which that term occurs, and to formulate, as before, standpoints on the truth of such statements and on the denotation of the term in question. The only doubt arises in connection with the description of indeterminate statements as those from which the term involved is eliminable. To put this more precisely, the unsolved problem is whether in the general case under consideration, the condition of definiteness ((OL), (OT), (OM)) logically implies the condition of eliminability ((EL), (ET), (EM)).[28] Or is there perhaps a statement in L' in which Q occurs, which has a definite truth value relative to Q on the basis of the postulate for Q, and which nevertheless cannot be expressed equivalently in L in virtue of that postulate? We must confine ourselves here to a formulation of this problem, and postpone its solution until later analyses.

We should also mention the general assumption which of necessity underlies all semantic investigations that make use of the concepts of modern logical semantics. This assumption pertains to the kind of languages which are the object of semantic analyses and which can be exemplified by languages L and L', which we have been considering here. These are, as we have seen, languages which are formalized in the standard way, and which are idealizations and simplifications of the real language of science. The results of our analyses are therefore applicable, strictly speaking, not to the language actually used by scientists, but to languages which are logical reconstructions of certain fragments of it. This is the price we have to pay for the possibility of presenting certain semantic problems rigorously and consistently. As I have tried to demonstrate, the concepts which enable us to do this in the case of the semantic issues of open languages, are such concepts of modern logical semantics as those of formalized language, model of a formalized language, truth in a model, and denotation in a model. In particular, it does not suffice here to use the current concept of an interpreted language and the absolute concepts of truth and denotation, for open languages are interpreted only in part, and as such they admit of various interpretations. This is why we have to use, on the one hand, the concept of language as a purely syntactic entity, and on the other, the concept of the domain to which such a language may refer. We must also have at our disposal the concept of the truth of statements in a given language, and that of the denotation of its terms — both concepts being relativized to a given domain (a part or aspect of the real world), and hence to a specified interpretation of that language. These concepts are provided by the theory of models of formalized languages. The task I have set myself has been to show — through the use of concepts of that theory — the present controversy in modern logic over the truth and the reference of statements which are characteristic of open languages.

It has not been my task to formulate my own standpoint in this paper. If I have often adduced arguments in favour of the last-named solution, i.e., the concept (V) of truth and the concept (C)-(c) of denotation, I have done so with certain reservations, which would have to accompany my acceptance of that solution. For instance, I do not think that the problems under consideration here can be formulated correctly as questions about "how it really is", that is, questions as to whether indeterminate statements really have a truth value, or whether a term which is defined conditionally in fact denotes an object. I do not think that the accepted meanings of such concepts as truth and denotation can settle these problems definitively. It seems to me that none of the solutions described above manifestly violates those meanings. One may just agree with them more or less, and that is not a decisive argument. The point is, therefore, to choose a consistent, rigorous, and most satisfactory (from a given point of view) mode of speaking. Now according to the point of view we choose, we have to declare ourselves for one solution or another, since each of them has its merits and shortcomings. When declaring myself for concepts of truth and denotation which are extensions, to partially interpreted languages, of that understanding of those concepts which is applied to fully interpreted languages, I have acted above all as a logician, since from the logical point of view this solution seems to be the most satisfactory of all. But if we were to adopt another standpoint, we should have to make another choice. The interpretation which is the least troublesome from the point of view of logic, turns out to be somewhat risky from a philosophical standpoint, and this is why many a positivistic philosopher is reluctant to adopt it. But should not priority be given to the logical criteria of assessment in what is, after all, a logical issue? I have manifested that opinion in declaring myself for the solution I have chosen.

To conclude I will make a few comments on the methodological aspect of our analyses. I have drawn the reader's attention to the fact that languages of empirical scientific theories serve as standards of open languages. In such theories, theoretical terms are definable by postulates which establish the denotations of those terms only in part. The concepts which correspond to those terms are, therefore, open ones. In view of this, the controversy over the truth value and the reference of statements that are characteristic of open languages largely coincides with the controversy over the cognitive value of scientific theories — a controversy which is now topical in methodology of science. For discussion focusses mainly on the issue whether theoretical statements can be treated as true or false. This is why our results are of direct interest for such a discussion. It turns out, first of all, that the doubts may concern only some theoretical statements, for there is a large class of statements in which theoretical terms occur, and whose truth value is beyond dispute, since these terms are eliminable from those statements. This class includes not only tautologies and their negations, but also those statements in which theoretical terms occur essentially. As regards the remaining theoretical statements one may adopt various standpoints, as we have seen. In modern methodology we usually distinguish two principal types of solutions, namely the instrumentalistic and the realistic.[29] The former deny those problematic statements any truth value, while the latter concede such a value to them, even though this truth value is interpreted in various ways. The standpoints discussed above have included examples of both categories. The standpoint I would be inclined to adopt is a typically realistic solution. Statements with ineliminable theoretical terms are treated as fully meaningful ones, with a truth value in the classical sense of the term. The ambiguity of their interpretation results merely in their being undecidable. Moreover, the theoretical terms which occur in them denote, in the traditional sense of the word, certain objects, even though such objects cannot be unambiguously identified. A logically satisfactory formulation of this realistic solution was one of the tasks I set myself in this paper.

Notes

[1] The partial definition of the predicate Q in its general form can be formu lated so as to be a non-creative definition:

$$\wedge x\,((P_1 x \wedge \sim P_2 x \rightarrow Qx) \wedge (P_2 x \wedge \sim P_1 x \rightarrow \sim Qx)).$$

The above definition is logically equivalent to the type (1) statement:

$$\wedge x\,((P_1 x \equiv \sim P_2 x) \rightarrow (Qx \equiv P_1 x)).$$

[2] For a more detailed discussion see M. Przełęcki, 'Pojęcia teoretyczne a do świadczenie' (Theoretical Concepts and Experience), *Studia Logica*, XI, 1961, 'O pojęciu genotypu' (On the Concept of Genotype), *Studia Filozoficzne*, 26, 1961.

[3] R. Suszko, 'Logika formalna a niektóre zagadnienia teorii poznania' (Formal Logic and Some Problems in Epistemology) *Myśl Filozoficzna*, 28, 29, 1957.

[4] To simplify matters we are disregarding here those extra-logical expressions which are function symbols. Likewise, the predicates which occur in the examples will always be treated as monadic predicates. But our analysis can easily be generalized so as to cover predicates of n arguments.

[5] A unary relation is identified with a subset of U.

[6] In piesenting these semantic concepts I have followed R. Suszko's paper, cited above.

[7] The relativization to language will be omitted in the definitions that follow; we do this to simplify our formulations as far as possible. $\mathfrak{M}$ and $\mathfrak{M}'$ will be treated as models of L and L', respectively. The clause in (EL):

$D(Q) \rightarrow (Z(Q) \equiv Z)$ is true in every model $\mathfrak{M}'$

is, of course, equivalent to:

for any model $\mathfrak{M}'$: if $D(Q)$ is true in $\mathfrak{M}'$, then $Z(Q)$ is true in $\mathfrak{M}'$ if and only if Z is true in $\mathfrak{M}'$.

We shall refer to the last formulation from time to time.

[8] An analogous result is obtained if instead of $\Psi \wedge Q$ we adopt the implication $\Psi \rightarrow Q$.

[9] This can be demonstrated more clearly for a case in which Ψ and Q are simple predicates in L, e.g., P_1 and P_2, so that $D(Q)$ becomes $\wedge\, x\,(P_1 x \rightarrow (Q x \equiv P_2 x))$. If $\mathfrak{M}'_1 = \langle U, x_1, \ldots, x_n, X_1, \ldots, X_m, Y\rangle$, then we take $\mathfrak{M}'_2$ to be $\langle U, x_1, \ldots, x_n, X_1, \ldots, X_m, X_2\rangle$. It that case the assumption that $D(Q)$ is true in $\mathfrak{M}'_2$ is equivalent to the tautology $\wedge\, x \in U\,(x \in X_1 \rightarrow (x \in X_2 \equiv x \in X_2))$ and as such may be disregarded. On the other hand, the statement that $Z(Q)$ is true in $\mathfrak{M}'_2$ is equivalent to the statement that $Z(P_2)$ is true in $\mathfrak{M}'_1$, and may be replaced by the latter. By way of example let $Z(Q)$ be $P_1 a_1 \wedge Q a_1$. The latter is true in $\mathfrak{M}'_2$ if and only if $x_1 \in X_1 \wedge x \in X_2$, and hence, if and only if $P_1 a_1 \wedge P_2 a_1$ in true in $\mathfrak{M}'_1$.

[10] See *The Reach of Science*, Toronto 1958. A precise analogue of the concept of a "determinate" statement will be introduced later in this paper.

[11] This formulation is equivalent to the condition (EL) as formulated in the text. An analogous observation applies to (ET) and (EM).

[12] To simplify matters, we omit the necessary relativization to language: a statement Z is true in a language L, or in a model $\mathfrak{M}$ of L.

[13] This interpretation of the concepts of proper model and true statement is to be found in R. Suszko's paper cited above and in J. Kemeny, 'A New Approach to Semantics', *Journal of Symbolic Logic*, 21, 1956.

[14] I discuss the language of empirical theories in my paper *Theoretical Concepts and Experience* (see footnote[2] above), and the ways of interpreting the specific terms of empirical theories, in 'Interpretacja systemów aksjomatycznych' (Interpretation of Axiomatic Systems), *Studia Filozoficzne*, 21, 1960. I have analysed the interpretation of observation terms in 'O definiowaniu terminów spostrzeżeniowych' (On Defining Observation Terms), in: *Rozprawy logiczne* (Papers on Logic). Warsaw 1964.

[15] The set $-P_2$ is the complement of the set P_2 to the universe U.

[16] Standpoint (I.1) roughly corresponds to the standpoint I adopt in my paper 'W sprawie terminów nieostrych' (Concerning Vague Terms), *Studia Logica*, VIII, 1958.

[17] Standpoint (I.2) coincides with that adopted by H. Mehlberg in *The Reach of Science*.

[18] The standpoint (II) approximately corresponds to that adopted by W. Rozeboom in 'The Factual Content of Theoretical Concepts'. *Minnesota Studies* ... Vol. 3, 1962.

[19] I know of no representative of standpoint (III).

[20] The standpoint adopted by T. Kubiński in 'Nazwy nieostre' (Vague Terms), *Studia Logica*, VII, 1958, falls within the general outline of standpoint (IV).

[21] Cf. *Theoretical Concepts and Experience* (see footnote[2] above).

[22] This is explicitly stated by H. Mehlberg in his book cited above. Standpoint (V) is an expanded and more precise version of the standpoint which I adopt in *Theoretical Concepts and Experience* and which is tacitly assumed by many authors.

[23] Both the relative and the absolute concepts of designation and denotation require relativization to a given language, which for simplicity is omitted here.

[24] See footnotes [16] and [18].

[25] See footnotes [17] and [20].

[26] Such a non-classical calculus of terms is constructed by T. Kubiński in his paper cited above.

[27] The following assumption could be a reason for the adoption of definition (B):

P designates *x* if and only if every onewho correctly uses *P* can, under specified cireumstances, predicate *P* about *x*.

[28] The proof given in this paper does not apply to the general case.

[29] Cf. E. Nagel, *The Structure of Science*, New York 1961.

LANGUAGES AND THEORIES ADEQUATE TO THE ONTOLOGY OF THE LANGUAGE OF SCIENCE

by

Henryk Stonert

(Fragments)

I. THE SUBJECT MATTER AND TASKS OF SCIENCE

Philosophers of science differ in their views concerning the subject matter of science, that is, what theorems are about and to what they refer. These problems are of an ontological nature. Some kind of ontology underlies every language of sciences. This discipline is concerned with "the general principles of being", with what exists, the nature of what exists, types of entities, etc. The ontology of the language of science manifests itself in that language, if only in the syntax of that language, in the kinds of expressions used in that language and in the syntactic categories assigned to them. The idea that language determines an ontology has been emphasized in Poland by Roman Suszko.

One of the most fundamental issues in the ontology that underlies the language of science is that of the subject matter of a given discipline, in other words, the objects whose existence is assumed in that discipline. How can we settle this issue, and what are we to adopt as the criterion for establishing that a given discipline is concerned with such and such objects, and that it assumes the existence of such and such entities?

In theorems formulated in the various disciplines we can single out two groups of terms: (a) those which, like terms, predicates, and functors, denote or represent some objects, (b) those which, like quantifiers, certain types of operators, and sentential connectives, neither denote nor represent any objects, but refer to them in some other manner. It is true that some authors are inclined to ascribe to expressions in the latter group — for instance, quantifiers — semantic functions which are proper to the expressions of the former group, but this position does not appear intuitive and will not be adopted here. In answering the question posed above it would seem reasonable to adopt the postulate that a given discipline is concerned with those objects which are denoted or represented by the terms[1] that occur in the statements of that discipline. In adopting this postulate we take account of another assumption as well, namely that which says that scientific statements refer to something, and that that something somehow exists.

Having defined the criterion by which we can decide which objects are referred to in scientific statements we shall now proceed to discuss the next issue: what are those objects, and what are their principal types? To answer this we shall examine several statements chosen at random and diversified enough to make this choice representative.

Let us consider, by way of example, the following statements:

The number π is transcendent.
Mount Everest is the highest mountain on our globe.

In both statements we have expressions which symbolize certain individuals. The name 'π' symbolizes[2] a specified real number which stands for the relation between the circumference of a circle and its diameter. The name 'Mount Everest' denotes a mountain in the Himalayas which is reputed to be the highest mountain on Earth. Finally, the name 'our globe' denotes a specified object, too, namely the planet which is third in the solar system as far as distance from the Sun is concerned.

Let us consider other statements:

The set of all cuts of the set of real numbers is non-denumerable.

As we move to the left of the axis of symmetry in Mendeleyev's Table the intensity of the metallic and basic properties of the various elements increases, and as we move to the right of the axis of symmetry their non-metallic and acidic properties increase in intensity.

These statements include names of sets. In the first, the expression 'the set of real numbers' denotes the familiar set of those mathematical objects which are real numbers. The expression 'the set of the cuts of the set of real numbers' also denotes an object which is the set of a certain type of sets of real numbers. In the second, the expression 'the various elements' represents specified properties of atoms. In both cases we have expressions which denote or represent certain objects which are sets of properties.

We can supply further examples:

If several forces act upon a rigid body, then that body behaves as if one force, which is the resultant of the said forces, has acted upon it.

Hydrogen peroxide, when heated, decomposes very easily and emits a large quantity of heat in the process.

A rigid body upon which no force acts is at rest or is moving uniformly in a straight line.

Let us now consider some of the terms occurring in these statements. The term 'force', which occurs in the first of the three, denotes a four-place relation, in which we have singled out four domains: the set of those rigid bodies which are the "source" of the force, the set of the bodies upon which that force acts, the set of the values of the intensity of that force, and the set of the directions in which the force acts. In the second, too, reference is made to a chemical relation denoted by the expression 'decomposes'. In the third, we also encounter a term which denotes a relation, namely the term 'motion'. Motion can be interpreted as a five-place relation: the first domain would be the set of those bodies relative to which motion takes place (bodies relatively at rest); the second domain would be that of bodies in motion; the third, that of the values of the intensity of motion (velocity); the fourth, that of directions; the fifth, that of the times in which motion takes place.

Analysis of the statements quoted above, as well as an examination of other scientific statements, leads us to conclude that the spheres of investigation of the various disciplines consist of entities which in logical ontology are called individuals, sets, properties, and relations. These kinds of entities, singled out from the point of view of logical ontology — if that term may be used — are the subject matter of investigations in the various disciplines.

To avoid possible misunderstandings we now have to explain the connections between the various entities, and to classify these entities in some way. In the ontology adopted in this paper we assume the following kinds of entities: 'elementary objects', 'properties', 'relations'. The first requires some explanation. We shall use the term 'elementary objects' with reference to those entities which are neither properties nor relations. They will, accordingly, be individuals first and foremost, but possibly sets as well. To avoid being criticized for inconsistency, let us explain that it is fairly common to identify sets with properties, and that it serves a purpose in the case of certain problems; but in the light of recent research in logic and set theory, sets cannot be equated to properties. It must also be emphasized that all these terms are relative, and when they are used to label objects they are largely restricted to the aspects in which we are interested. For instance, numbers are treated as individuals in some theories, and as sets in others. In view of what has been said above we shall be concerned with the following kinds of entities:

1) elementary objects,
2) relations
   a) properties, i.e., one-place relations,
   b) two-place relations.

Note that our approach is to treat in the same manner all the entities denoted by the terms which occur in scientific statements, and to accept all of them as the subject matter of research, without making any distinctions among them in this respect — unlike many other approaches, where certain kinds of entities are singled out and ascribed the property of being the subject matter of research while other kinds of entities are denied that property, whereby their very existence is questioned (this is the approach of T. Kotarbiński's theory of reism). According to still other approaches it is not questioned that research is concerned with objects denoted by the terms which occur in scientific statements, but such objects are subject to differentiation: some are assigned the function of being predicated about, while others, that of predicating. The ontological conception presented in this paper does not differentiate among entities and can therefore claim to be the most "democratic" approach in the sphere of ontology. This issue will be discussed later.

Thus we have agreed to treat the objects denoted or represented by the terms that occur in the statements of a given discipline as the objects with which that discipline is concerned. This approach is not universal. Quite often what is taken to be the subject matter of science is not the individual entities, but certain constructions, namely events. This standpoint would not seem to be justified. In the opinion of the present writer events are not objects in themselves, whatever the nature of such objects might be, but rather configurations of co-existing entities — in other words, a form of the manifestation of the existence of entities. Scientific statements merely attest to these states of things, attest that such states of things occur. It would therefore seem more justified to say that scientific statements speak about objects denoted by the terms which occur in such statements, and they do so by attesting to events or states of things. Events are, therefore, in the present writer's opinion, a form of concurrent existence of certain entities. We would, accordingly, treat the opinion that events are the subject matter of the various disciplines as an abbreviation of the opinion described above, which states that they are forms of existence of objects.

We now have to reflect on the terms introduced above: 'event', 'state of things', 'form of existence' (of objects in an event).

Scientific statements usually say that something is happening to the objects referred to in a given discipline, that they bear some relation to one another. This contact between objects will be called their existence in an event or in a state of things. Events are therefore interpreted — as contacts between objects, as the co-existence of configurations of objects — in a word, as a form of their existence.

Not all types of objects can co-exist. Existence is an attribute of certain types of objects only. For instance, we shall speak about the co-existence of a two-place relation and two elementary objects. Likewise, an event or a state of things may consist in the co-existence of a one-place relation (a property) and an elementary object. Examples of states of things include the co-existence of the two-place relation $<$ and the elementary objects 2 and 3, the co-existence of gravitational force and the Earth and the Moon as elementary objects, and the co-existence of the property N, namely that of being a natural number, and the object 2. These three events are described by the statements:

$2 < 3$

$N(2)$,

the Earth attracts the Moon.

On the other hand, two such entities as the Earth and the Moon, which are both elementary objects, and hence objects of the same kind, cannot co-exist (in the specifically technical sense of the word as used in this paper). The same applies to the relation of being less than and the property of being a natural number, the former being a two-place relation, and the latter, a one-place relation — this conclusion being in harmony with our intuitions.

II. EXPRESSING SCIENTIFIC STATEMENTS AND ITS CONSEQUENCES

Consider now how we usually record the occurrence of a state of things, the co-existence of a relation and some objects. Usually, when we want to express the idea that a state of things has occurred which consists in the co-existence of a two-place relation and two individuals, we write the symbol denoting that relation between the names of the two objects. If we want to indicate that a state of things which consists in the co-existence of a property and an individual has occurred, we do so by writing the symbol which denotes that property, and to the right of it we write the symbol which denotes that object. Examples: Let / be the symbol for denoting the two-place relation of divisibility; let 2 and 6 stand for specified natural numbers; let P stand for the one-place relation, namely the property of being a prime number. By making use of these symbols we can describe the state of things wherein 2 is a divisor of 6 and the state of things wherein 2 is a prime number by writing:

$$2/6,$$

$$P(2).$$

There is one consequence of this way of recording states of things which deserves to be mentioned here. When we describe states of things, which consist in the co-existence of certain objects, we do not treat all the participant entities alike; rather we differentiate them. This differentiation is manifested in classing the corresponding symbols in different syntactic categories. In doing this we class those symbols which denote relations and properties as sentence-forming functors (each with corresponding numbers of arguments), and those which denote elementary objects, as terms.

Thus, the occurrence of an event, i.e., the co-existence of certain entities, is usually described by writing the symbols which denote these entities one after another, usually in a specified order, and we interpret them syntactically as follows: the symbols of relations and properties are treated as sentence-forming functors, and the symbols of elementary objects as terms. No special technical term for indicating the co-existence of these entities is used.

This method of formulating statements and differentiating entities by classing their symbols in syntactic categories of one kind or another is connected with a certain other issue and results in a specified manner of interpreting and understanding scientific statements.

Let us look at two tendencies in interpreting statements. In the first, when we are dealing with a statement in which both symbols denoting relations or properties and symbols denoting elementary objects occur, we are inclined to conclude that such a statement informs us about individuals, and hence, that it is these individuals which are the subject matter of the statement. Our approach leads us to interpret a statement in such a way that the objects symbolized by the terms occurring in that statement are the objects to which that statement refers. We are not inclined to give the same status to all the entities denoted by the symbols which occur in a statement and to conclude that the statement informs us both about certain elementary objects and about certain properties and/or relations, i.e., to conclude that the statement refers to all the entities whose symbols occur in it.

The other tendency is closely linked with the former. When we are dealing with a statement in which both symbols referring to elementary objects and symbols referring to properties or relations occur, we are inclined to interpret it in such a way that it informs us about elementary objects and does this by referring to properties and/or relations, so that properties and relations are used to describe elementary objects. Our approach is such that properties and relations do not participate in an event or a state of things on an equal status with elementary objects, while their function is descriptive. We are inclined to treat certain entities as the subject matter of a description, and to treat others as something auxiliary, as the means by which information is conveyed to us about the former. We would, accordingly, have two kinds of entities: those predicated about, i.e., elementary objects, and those which predicate about others, i.e., properties and relations. The symbols of the former are called terms, and those of the latter, sentence-forming functors.

There is, as a rule, no opposition to such an approach. It does not usually happen that we interpret a statement in such a way that it informs us about a property or a relation by referring to elementary objects, by predicating those elementary objects about relations and properties.

Practically everyone, when asked about the subject matter of the sentence

'Romeo loved Juliet',

would reply that it is about Romeo and Juliet, and would not be inclined to say that it is about Romeo, Juliet, and love, even though the corresponding word occurs in that sentence together with the other two.

The advocates of the ontology outlined at the beginning of this paper, in favour of which the present author here declares himself, are far from enthusiastic about a language of science which leads to the consequences mentioned above. They are far from enthusiastic about the non-uniform, *undemocratic*, manner of treating entities, about singling out some of them as those to which statements refer, and treating the others merely as those which are used in speaking about the former.

According to the ontology outlined here, all entities have the same status and all may participate in the process of co-existence (in the specific sense of the term as adopted here), and on an equal footing. There is therefore no reason why, in addition to the natural, intuitive classification of entities into elementary objects, properties, and relations, we should ascribe them the functions of predicating or being predicated about. The assumptions underlying a language which would be adequate to such an ontology must be such that they do not result in the aforementioned tendencies in the interpretation of statements. These assumptions must be such as to bar the interpretation that a simple statement is only about those entities which are denoted by terms, and also such as to bar the interpretation that some entities are predicated about, while the others are used to predicate something about the former.

A language which is to be adequate to the ontology accepted here must be based on the assumptions that all kinds of entities are treated in the same way, i.e., that when we encounter a statement we are inclined to think that it is about all the entities whose symbols occur in it; and also, if we want to express it like this, that we can equally conclude that a statement refers to elementary objects through the intermediary of properties and relations, and that it refers to properties and/or relations through the intermediary of elementary objects. This means that every entity whose symbol occurs in a statement is referred to through the intermediary of the remaining entities whose symbols occur in that statement.

The advocates of the ontology outlined here may also be dissatisfied with the fact that the occurrence of an event — the co-existence of more than one entity — is indicated only by syntactic means, i.e., by writing the symbols which denote those entities next to one another. One is conscious of a lack of a special term that would indicate the co-existence of entities.

What then must be the assumptions underlying a language which is to meet these requirements?

As we have seen, the interpretation of statements which we have found unsatisfactory derives from the classification of those symbols which denote or represent entities in terms of syntactic categories. The syntactic stigma on the symbols which denote entities is unnatural, unjustified, and unnecessary. It is also the cause of that interpretation of statements which we find unsatisfactory, and of an inadequate reflection of ontological assumptions. A language which is to be adequate to the ontology that we have adopted cannot differentiate syntactically the entities referred to in statements formulated in that language, nor can it treat the symbols of certain entities as terms, and those of certain other entities as sentence-forming functors. Furthermore, in such a language the co-existence of more than one entity would not be expressed by the fact that the symbols which denote such entities are written next to one another; a special symbol must be used for that purpose. Its introduction will both guarantee a uniform, *democratic*, treatment of all entities and help to prevent the above-mentioned undesirable tendency in the interpretation of statements; it will also enable us to classify the terms occurring in scientific language in a way which we find interesting, and which will be described later.

III. THE BASIC CONCEPTS AND THEOREMS OF THEORIES ADEQUATE TO THE ONTOLOGY OF THE LANGUAGE OF SCIENCE

We shall now expound the principle of a language adequate to the ontology described above, a language in which all symbols that denote entities are of the same syntactic category.

Every state of things or events consists in the co-existence of some properties and elementary objects, or some relations and elementary objects. A property co-exists with only one elementary object, a two-place relation, with two elementary objects, a three-place relation, with three elementary objects, etc. The co-existence of entities is one of the fundamental concepts of the ontology outlined in this paper. To express this concept we shall use a special term R_i ($i = 1, 2, \ldots$), which takes on the form of a predicate with a corresponding number of arguments, as required in a given case: $R_1(x)$, $R_2(x, y)$, $R_3(x, y, z)$, ...

The fact that a property W co-exists with an elementary object a is expressed by the statement $R_2(W, a)$, where R_2 serves to indicate the co-existence of any property and any elementary object, whenever such co-existence takes place. Let N stand for the property of being a natural number, and *Str*, for the property of being a straight line. States of things which consist in the co-existence of the property N and an elementary object a, i.e., the fact that a is a natural number, are expressed in our language by the statement $R_2(N, a)$, whereas any state of things which consists in the co-existence of the property *Str* and an elementary object b, i.e., the fact that b is a straight line, is expressed by the statement $R_2(Str, b)$.

The co-existence of a two-place relation and two elementary objects is expressed by the predicate R_3 of three arguments. To express the co-existence of a three-place relation and three elementary objects we use the predicate R_4 of four arguments. Let S stand for a two-place relation, T for a three-place relation, and a, b, c, for elementary objects. The co-existence of the two-place relation S and two elementary objects a and b (in other words, the fact the S holds between a and b) is expressed by the statement $R_3(S, a, b)$, and the co-existence of the three-place relation T and three elementary objects a, b, c is expressed by the statement $R_4(T, a, b, c)$.

We shall use an analogical procedure to describe states of things which consist in relations of four, five and more places with the corresponding number of elementary objects in each case.

The predicate $R_1(\ldots)$ requires a brief explanation. It is intended to express the generalized concept of the co-existence of entities. Its introduction may seem pointless or even non-intuitive in view of the fact that no less than two corresponding entities can co-exist. Yet the presence of this predicate in the system is justified for two reasons. The first is ontological in nature: the predicate $R_1(\ldots)$ can be used to express various ontological theses, e.g., that no entity co-exists alone. The other is rather formal in character: the predicate under consideration makes it possible to formulate certain statements in a general form. In science, we often encounter such artificial, but useful, generalizations of concepts; in this connection, mention may be made, for instance, of the zero power, a^0, and the first power, a^1, of the number a.

Although we use one and the same letter R to describe the states of things referred to in the examples given above, nevertheless in each case we are dealing with a different predicate. These predicates differ from one another in their syntactic category, which in this case is indicated by the number of arguments which accompany a given predicate. To state that two entities co-exist we shall use predicates of two arguments each, to state that three entities co-exist we shall use predicates of three arguments each, etc. In our language we must assume, according to our needs, a corresponding number of predicates:

$$R_1 (...),\ R_2 (...,\ ...),\ R_3 (...,\ ...,\ ...),\ ...,$$

which will be the terms used to indicate that states of things or events occur. These terms will be called *co-existence predicates.*

Not all the objects referred to in scientific statements are of the same kind. We single out such types as: elementary objects (e.g., points, numbers, physical bodies); properties (e.g., being a prime number, being a physical body in a liquid state); relations (e.g., "less than" between numbers, attraction with such and such a force between physical bodies), etc. Hence the ontology which underlies the language being discussed here must include concepts that will classify the various entities. These are the concepts of: elementary object; property, or one-place relation; two-place relation; three-place relation; etc.

To express these concepts in the language of the ontology we have adopted here we shall use predicates of one argument each:

$$P_1,\ P_2,\ P_3,\ ...,$$

to be interpreted thus:

$P_1(x)$ — x is an elementary object,

$P_2(x)$ — x is a property (one-place relation),

$P_3(x)$ — x is a two-place relation,

. . . .

By way of example we have:

> $P_3(<)$ is read as: $<$ is a two-place relation;
> $P_2(N)$ is read as: N is a property (in this case N denotes the property of being a natural number).

The number of such predicates to be adopted in a given case must be adjusted to the needs of the theory under consideration, i.e., to the variety of the entities investigated in that theory — the theory itself being based on the ontology described here. The terms P_1, P_2, P_3, ..., which are used to classify the various entities, will be called classifying predicates. The classifying predicates $P_1(...)$, $P_2(...)$, $P_3(...)$, ..., and the co-existence predicates $R_1(...)$, $R_2(..., ...)$, $R_3(..., ..., ...)$, ... are the basic concepts of the ontology discussed in this paper.

It is obvious that not all entities can co-exist. For a given state of things, i.e., a given event, to occur, certain conditions must be satisfied concerning the categorization of the elements of that event as entities of specified kinds, and concerning the specified number of entities. For instance, an event consisting of two properties cannot occur; on the other hand, an event one element of which is a property and the other an elementary object, can occur. Neither three elementary objects nor two two-place relations can co-exist, but a system of one two-place relation and two elementary objects can occur.

To put this in general terms, entities can co-exist in accordance with the following ontological law, which we shall call the *principle of the possible co-existence of entities*:

> Co-existence is possible only in the case of systems of $n+1$ entities, of which exactly one is an n-place relation and the remaining n entities are elementary objects. (The variable n ranges over the set of natural numbers 1, 2, 3, ...).

The above principle, is of course, a necessary condition of the co-existence of entities, but it is not by itself a sufficient basis for their co-existence.

The principle states in which entities can co-exist: only those which form a system consisting of one n-place relation and n elementary objects. In the language of the theory which we are adopting here this is expressed in the form of statements about the combinations of entities that cannot co-exist. To formulate such statements we use classifying predicates and co-existence predicates. The statements which express the principle of the possible co-existence of entities are treated as axioms of our theory. The number and kind of such statements depend on the number and variety of the entities in a given case. Let us suppose that a theory with which we are concerned provides only for two-place relations, properties, and elementary objects. In the language of such a theory we must have the following two co-existence predicates: the co-existence predicate $R_2(\ldots, \ldots)$ of two arguments, which is necessary for describing the fact that a property is an attribute of an elementary object, and the co-existence predicate $R_3(\ldots, \ldots, \ldots)$ of three arguments, which is necessary for describing the fact that a two-place relation holds between two elementary objects. We also need three classifying predicates, $P_1(\ldots)$, $P_2(\ldots)$, $P_3(\ldots)$, which state that something is an elementary object, a property, and a two-place relation, respectively.

The principle of the possible co-existence of entities is expressed by the following statements:

1. Three two-place relations cannot co-exist under the co-existence predicate $R_3(\ldots, \ldots, \ldots)$ of three arguments. In symbols:

$$P_3(x) \cdot P_3(y) \cdot P_3(z) \rightarrow \sim R_3(x, y, z) .$$

2. Two two-place relations and a property cannot co-exist under that predicate:

$$P_3(x) \cdot P_3(y) \cdot P_2(z) \rightarrow \sim R_3(x, y, z) .$$

3. Two two-place relations and an elementary object cannot co-exist:

$$P_3(x) \cdot P_3(y) \cdot P_1(z) \rightarrow \sim R_3(x, y, z) .$$

4. A two-place relation and two properties cannot co-exist:

$$P_3(x) \cdot P_2(y) \cdot P_2(z) \rightarrow \sim R_3(x, y, z) .$$

5. Three properties cannot co-exist under that predicate:

$$P_2(x) \cdot P_2(y) \cdot P_2(z) \rightarrow \sim R_3(x, y, z) .$$

6. A property and two elementary objects cannot co-exist:

$$P_2(x) \cdot P_1(y) \cdot P_1(z) \rightarrow \sim R_3(x, y, z) .$$

7. Three elementary objects cannot co-exist:

$$P_1(x) \cdot P_1(y) \cdot P_1(z) \rightarrow \sim R_3(x, y, z) .$$

8. Two two-place relations cannot co-exist under the co-existence predicate $R_2(\ldots, \ldots)$ of two arguments:

$$P_3(x) \cdot P_3(y) \rightarrow \sim R_2(x, v) .$$

9. A two-place relation and a property cannot co-exist under that predicate:

$$P_3(x) \cdot P_2(y) \rightarrow \sim R_2(x, y) .$$

10. A two-place relation and an elementary object cannot co-exist:

$$P_3(x) \cdot P_1(y) \rightarrow \sim R_2(x, y) .$$

11. Two properties cannot co-exist:

$$P_2(x) \cdot P_2(y) \rightarrow \sim R_2(x, y) .$$

12. Two elementary objects cannot co-exist:

$$P_1(x) \cdot P_1(y) \rightarrow \sim R_2(x, y) .$$

13. A two-place relation alone cannot co-exist under the co-existence predicate $R_1(\ldots)$ of one argument:

$$P_3(x) \rightarrow \sim R_1(x) .$$

14. A property alone cannot co-exist:

$$P_2(x) \rightarrow \sim R_1(x) .$$

15. An elementary object alone cannot co-exist:

$$P_1(x) \rightarrow \sim R_1(x) .$$

If a given theory were to consider more kinds of entities than have been assumed here by way of example, we should accordingly have to increase the number of statements expressing the principle of the possible co-existence of entities, or even to assume an infinite number of such statements in the case of a theory which considers all entities.

The ontological statements listed above (which express the principle of the possible co-existence of entities) allow us to reject certain statements about the co-existence of entities. It must be emphasized that we reject them as false statements, but not as meaningless formulas. Let us suppose that in a given theory we consider, inter alia: a two-place relation, namely the arithmetical relation "smaller than", symbolized by $<$ the property of being a natural number, denoted by N; and elementary objects, namely natural numbers, denoted by, e.g., 2, 3. Once we have established axiomatically the ontological nature of these entities: $<$ is a two-place relation, N is a property, 2 and 3 are individuals (elementary objects):

$$P_3(<),$$
$$P_2(N),$$
$$P_1(2),\ P_1(3),$$

we can refer to statements 1—15. above, which are concerned with the principle of the possible co-existence of entities, in order to confirm that certain statements about the co-existence of $<$, N, 2, 3 are false (numbers in parentheses indicate the statement, on the strength of which a given statement on co-existence is rejected):

(1) $R_3(<, <, <)$,
(2) $R_3(<, <, N)$,
(3) $R_3(<, <, 2)$,
(4) $R_3(<, N, N)$,
(5) $R_3(N, N, N)$,
(6) $R_3(N, 2, 3)$,
(7) $R_3(3, 2, 2)$,
(8) $R_2(<, <)$,
(9) $R_2(<, N)$,
(10) $R_2(<, 3)$
(11) $R_2(N, N)$,
(12) $R_2(3, 2)$
(13) $R_1(<)$,
(14) $R_1(N)$,
(15) $R_1(3)$.

Of course, the statements listed above are only some of those statements about entities $<$, N, 2, 3, which we could reject.

The principle, described above, of the co-existence of entities has to be supplemented by an additional condition, which, however, is not due to ontological considerations and in no way modifies the principle of the co-existence of entities. It results rather from our mode of formulating statements, or from the structure of language; specifically, it is due to the need of unifying the manner in which statements are to be recorded. Let us suppose that we wish to formulate in our language the arithmetical statement which says that $2 < 3$. We can do this in a number of ways, all of which are in agreement with the principle of the co-existence of entities:

$$R_3(<, 2, 3),$$
$$R_3(2, <, 3),$$
$$R_3(2, 3 <).$$

In order to avoid this lack of uniformity we complete the aforementioned principle with the following necessary condition:

> Of those combinations of entities which satisfy the principle of the possible co-existence of entities, only those combinations can occur in which a relation or a property appears first and is followed by elementary objects.

If we assume that in the theory under consideration we have only the following kinds of entities: two-place relations, properties, and elementary objects, we can write the above condition in the form of the following statements:

1) The combination of entities: elementary object, two-place relation, elementary object, in that order, does not comply with the co-existence predicate of three arguments $R_3(..., ..., ...)$:

$$P_3(x) \cdot P_1(y) \cdot P_1(z) \rightarrow \sim R_3(y, x, z).$$

2) The combination of entities: elementary object, elementary object, two-place relation, in that order, does not comply with the same co-existence predicate:

$$P_3(x) \cdot P_1(y) \cdot P_1(z) \rightarrow \sim R_3(y, z, x).$$

3) The combination of entities: elementary object, property, in that order, does not comply with the co-existence predicate of two arguments $R_2(..., ...)$:

$$P_2(x) \cdot P_1(y) \rightarrow \sim R_2(y, x).$$

We shall adopt these statements as axioms. Their insertion into the theory appears indispensable, because an arbitrary arrangement of multiplace expressions denoting entities in statements which describe the co-existence of entities might lead to ambiguities in descriptions of facts. Even though these axioms are not strictly ontological in nature, we can consider them to be ontological because, on account of a certain uniformity and naturalness of description of the theory, they are in harmony with those axioms which refer to the co-existence of entities.

To return to the example discussed above, which we have used to illustrate the principle of the co-existence of entities, we can treat the following statements as false on the strength of the foregoing axioms (the numbers on the right indicate the axiom on the basis of which a given statement is rejected):

(1) $R_3(2, <, 3)$,

(2) $R_3(2, 3, <)$,

(3) $R_2(2, N)$.

IV. STRUCTURAL PRINCIPLES OF THE LANGUAGE OF A THEORY ADEQUATE TO THE ONTOLOGY OF THE LANGUAGE OF SCIENCE

In the preceding section we have described the two fundamental terms of the language under consideration as well as theorems which apply to them. We shall now discuss the formulation of theories in that language. We shall do this by comparing the description of an ordinary theory with that of an adequate theory (i.e., such as now concerns us), and by formulating hints at how an ordinary theory can be formulated as one which is adequate.

For comparison we shall consider the description of a theory T, which is formalized in the standard way[3] and also satisfies the condition that its specific constants include no function terms. It is evident that function symbols can be eliminated from any theory without thereby modifying its semantic aspects.

When describing the language of a theory we usually list: (1) the symbols used in that language, (2) the rules which classify those symbols from a syntactic of view, (3) the syntactic rules which indicate which combinations of symbols are accepted as meaningful statements.

The symbols used in the language of T include the following kinds of expressions:

variable symbols:

(a$_1$) $x_1, x_2, x_3, \ldots,$ (denumerable number of)

constant symbols:

(a$_2$) $a_1, a_2, \ldots,$

(a$_3$) $b_1, b_2, \ldots,$ (finite or denumerable number of) (a₄) $=$,

(a₅) $\rightarrow, \equiv, \vee, \cdot, \sim$,

(a₆) $\bigvee, \bigwedge$.

These symbols have specified syntactic categories assigned to them: those in (a_1) and (a_2) are classed as terms; those in (a_3) are classed as predicates, each with a specified number of arguments; the symbol in (a_4) is treated as a predicate of two arguments; the symbols in (a_5) are classed as sentential connectives, and those in (a_6), as quantifiers.

Familiar syntactic rules are formulated next; they state which combinations of symbols are accepted as meaningful expressions (statements) in *T*. These rules will not be described here.

When proceeding to describe the language of an adequate theory T^* (i.e., a theory adequate to the ontology of the language of science) as the counterpart of the language of *T* (as described above), we must first mention that these two languages differ, in principle, only in their syntactic and classifying rules, whereas their sets of symbols (vocabularies) are essentially the same, that of T^* being richer only in classifying terms and co-existence terms.

The expressions of the language of T^* will be classified in a somewhat different way. We shall single out

variable symbols:

(b_1) $x_1, x_2, \ldots$;

symbols which denote entities:

(b_2) $=$,

$a_1, a_2, \ldots, a_k$,

$b_1, b_2, \ldots, b_f$,

classifying predicates:

(b_3) $P_1, P_2, \ldots, P_k$;

co-existence predicates:

(b_4) $R_1, R_2, \ldots, R_k$;

sentential connectives:

(b_5) $\equiv, \rightarrow, \vee, \cdot, \sim$;

quantifiers:

(b_6) $\bigwedge, \bigvee$.

The rules which assign syntactic categories to these expressions are different, too. The expressions in (b_1) and (b_2) are in one syntactic category, namely that of terms; in the language of *T* they are syntactically differentiated into terms and predicates, the latter in turn being classed in various syntactic categories. The expressions in (b_3) are classed as predicates of one argument each. Those in (b_4) are predicates of one, two, three, and more arguments, respectively. Those in (b_5) and (b_6) are assigned the same syntactic categories as in *T*.

The syntactic rules of *T** differ from the corresponding rules of *T*. However, the difference is limited to the construction rules of atomic statements, and hence the construction rules for compound statements are the same as in *T*. The syntactic rule for the construction of atomic statements in the adequate language of *T** is as follows: an atomic statement is an expression which consists of (the correct linear order being: a predicate followed by its argument in parentheses) either (i) an expression in (b_3) and an expression in (b_1) or (b_2), i.e., a classifying predicate and a variable or an expression which denotes an entity, or (ii) an expression in (b_4) with a corresponding number of expressions in (b_1) or (b_2), i.e., a co-existence predicate and variables or expressions which denote or represent entities, the number of such expressions corresponding to that of the arguments of the co-existence predicate in question. The syntactic rules which apply to compound statements are, as has already been said, analogical to those generally adopted in ordinary languages.

Constant terms in the languages of theories formalized in the standard way are usually classified into logical constants and specific constants. This distinction usually refers to the difference in the nature of the axioms which characterize the concepts corresponding to those terms, or is due to the facility of singling out the objects to which a given theory refers, or else is accounted for by other aspects of a metalogical description of a given theory. In the language of *T* the terms in (a_4)–(a_6) are logical constants, and those in (a_2) and (a_3) are specific constants. For the same reasons, the terms of the language of an adequate theory *T** can be conveniently classified into three categories: logical constants, specific constants, and ontological constants. The expressions in (b_5) and (b_6) and the sign $=$ in (b_2) are logical constants; the expressions in (b_2), except for the sign $=$, are specific constants; the expressions in (b_3) and (b_4) are ontological constants (these are classifying predicates and co-existence predicates).

The expressions in the language of an adequate theory T^* can also be conveniently classified into those which directly refer to the entities to which that theory pertains (expressions, constants or variables, which denote or represent entities) and which cover those in (b_1) and (b_2), and descriptive expressions, which include classifying predicates, co-existence predicates, sentential connectives, and quantifiers, i.e., the expressions in (b_3)–(b_6). This classification will be discussed in greater detail in another place.

V. THE CONSTRUCTION RULES OF THE AXIOM SYSTEM OF AN ADEQUATE THEORY

In the preceding section we have described the main rules of construction of the language of an adequate theory, and now we shall discuss the construction of an axiom system of an adequate theory. In doing this we shall at the same time show how to transform an axiom system of a theory formalized in the standard way into an axiom system of the corresponding adequate theory.

The axioms of a standard theory can be classified into three groups:

A) the specific axioms of the theory in question,

B) the logical axioms (those of sentential calculus and functional calculus),

C) the axioms of the identity theory.

In an adequate theory we adopt the following groups of axioms:

A) the specific axioms of the theory,

B) the logical axioms (those of sentential calculus and functional calculus),

C) the axioms of the identity theory,

D) the ontological axioms (classed into three subgroups, a, b, c).

The specific axioms of T^* are obtained by a translation, in accordance with the rules given above, of the axioms of T. When this is done, the atomic statements of T are translated into the corresponding statements in T^*, the logical structure being left unchanged.

The logical axioms of T^*, which include those of sentential calculus and functional calculus, are the same as in T in the ordinary language.

The axioms of the identity theory for T^* are obtained in a similar way, by translation of the axioms of the identity theory of T.

The list of axioms of T^* must also be enlarged by the three groups of axioms which characterize the ontological terms.

First, the classifying axioms state that the various entities denoted by the constants in T^* are in certain categories and not in others. We adopt as many of these axioms as there are specific constants, including the sign of identity, in T^*, If t_0 is the sign of identity, and t_1, t_2, ..., are specific constants in T^*, then those axioms are in the form

$$P_a(t_0), \qquad P_b(t_1), \ldots,$$

where a, b, ..., are the corresponding natural numbers.

The second group of ontological axioms consists of those which state, loosely speaking, that the kinds of entities to which T^* refers, are disjunctive. The number of these axioms is determined by the number of the kinds of entities in T^*. If T^* is concerned with a finite number of kinds of entities, and hence with a finite number of predicates P_{i_1}, P_{i_2}, ..., P_{i_n}, then the number of those axioms is[4]

$$\sum_{j=1}^{n-1} j$$

and their form is:

$$P_{i_k}(x) \rightarrow \sim P_{i_l}(x),$$

where $k = 1, 2, \ldots, n-1$, and $l = k+1, \ldots, n$. If T^* is concerned with infinitely many kinds of entities, we adopt infinitely many axioms of the above form, and the subscripts of the classifying predicates, namely i_k and i_l, must satisfy the following conditions: $k = 1, 2, \ldots,$ and $l = k+1, \ldots$

The third group of ontological axioms in T^* consists of those which state which entities cannot co-exist. These axioms express the principle of the possible co-existence of entities, analysed earlier in this paper, and refer to the co-existence predicates.

The number of co-existence predicates occurring in the language of T^* depends on the variety of the kinds of entities investigated in that theory. If we intend to investigate an infinite number of kinds of entities, then we have to introduce into the language infinitely many co-existence predicates $R_1(\ldots)$, $R_2(\ldots, \ldots)$, $R_3(\ldots, \ldots, \ldots)$, ... If we consider a finite number of kinds of entities, then these include entities with the greatest number of components. Let this number be k; we then introduce $k+1$ co-existence predicates: $R_1(...)$, $R_2(..., ...)$, ..., $R_{k+1}(..., ...)$.

We then use those co-existence predicates and the classifying predicates to formulate the axioms which determine the principle of the possible co-existence of entities. The number of such axioms is, obviously, determined by the number of co-existence and classifying predicates, and for a finite number of such predicates it equals $\sum_{k=2}^{k+1}(n^k-1)+n$, where n stands for the number of kinds of entities, and k, for the greatest number of components in the entity (entities) investigated in that theory.

If we investigate n (i.e., a finite number) of kinds of entities such that no entity has more than k components, then these axioms will have the following forms (the subscripts a, b, c, ..., n range over the natural numbers 1, 2, ..., n):

$$P_{i_a}(x_1) \rightarrow \sim R_1(x_1)$$

for any i_a.

$$P_{i_a}(x_1) \cdot P_{1_b}(x_2) \rightarrow \sim R_2(x_1, x_2)$$

for $i_a \neq 2$ or $i_b \neq 1$.

$$P_{i_a}(x_1) \cdot P_{i_b}(x_3) \cdot P_{i_c}(x_3) \rightarrow \sim R_3(x_1, x_2, x_3)$$

for $i_a \neq 3$ or $i_b \neq 1$ or $i_c \neq 1$.

. . . .

$$P_{i_a}(x_1) \cdot P_{i_c}(x_2) \ldots P_{i_a}(x_{k+1}) \rightarrow \sim R_{k+1}(x_1, x_2, \ldots, x_n)$$

for $i_a \neq k$ or $i_b \neq 1$ or ... or $i_a \neq 1$, the antecedent of this implication being a conjunction of exactly $k+1$ components.

If an infinite number of kinds of entities is investigated in a given theory, then the number of axioms is infinite, too, and the subscripts k, a, b, ... are arbitrary natural numbers.

CONCLUDING REMARKS[5]

In concluding this paper we shall briefly compare the languages and theories constructed in the standard way with the languages and theories constructed as suggested here. Our comparison will chiefly cover two aspects, the formal and the philosophical, which are not only not incompatible with one another, but, on the contrary, complement one another.

When we compare the standard and the adequate theories from a formal point of view we notice that while in the former the terms which denote or represent the objects described in that theory (elementary objects, individuals, properties, relations, etc.) are differentiated syntactically, in the latter there is no such differentiation; on the contrary, all expressions which directly refer to the objects described in a theory have one and the same syntactic category. A syntactic differentiation of the expressions which denote objects was considered necessary because of the possibility of arriving at a contradiction within a theory. The present paper shows that this is not the case, and that appropriate modification of the language of a given theory permits a contradiction-free description of objects without the need of syntactically differentiating the expressions which refer to them. This possibility has, as we have already mentioned, certain advantages when it comes to the intuitive nature of theorems and their philosophical interpretation.

It is also to be emphasized that adequate theories are essentially richer than their standard analogues. We can formulate a number of facts which cannot be described in languages of the standard theories. For instance, in languages of the standard theories we cannot express the arithmetical fact that the relation 'smaller than' is not less than itself: the formula $\sim (<(<, <))$ is simply meaningless. On the other hand, in an adequate theory this can be formulated as: $\sim R_3(<, <, <)$, which says that three relations 'smaller than' cannot co-exist, in other words, that the relation 'smaller than' cannot hold between two objects each of which is the relation 'smaller than'. Many similar examples could be given.

When discussing formal aspects we should also mention the fact that adequate languages include variables which range over all the objects discussed in a given theory, whereas in standard languages a variable can range only over a certain class of investigated objects.

As far as the philosophical aspects of the concepts of language and theory presented here are concerned, we should emphasize above all the fact that these languages and theories better reflect certain philosophical assumptions underlying science. Adequate theories harmonize particularly well with the opinion that in science we investigate objects, as well as their properties and the relations between them. Standard languages are not in harmony with this position, for the syntactic differentiation of expressions which denote or represent the objects under investigation obscures the ontological interpretation of theorems and the fundamental ontological assumptions, and may even account for their false interpretation.

As has been shown above, the syntactic differentiation of terms which denote and represent the objects under investigation greatly restricts our possibility of describing a number of facts connected with those objects.

Furthermore, unlike standard languages, adequate ones include a number of special terms, namely classifying predicates and co-existence predicates, which enable us to describe and analyse the process of an event taking place as a result of the co-existence of certain objects.

The applicability of the adequate theories

It is obvious that adequate theories can be used to describe objects with a most logically complex structure (sets and relations of arbitrarily high orders). In particular, they can be used in set-theoretical investigations. Generally speaking, every elementarily formalized — and hence practically every theory can be expressed in the form of an adequate theory.

NOTES

[1] By a scientific term we mean here a simple expression in the language of a given discipline, which is neither a variable, nor a punctuation mark, nor a bracket. By a scientific term in the narrower sense of the word we mean an extra-logical constant in the language of that discipline.

[2] In the terminology adopted here 'symbolizes' means the same as 'denotes'. Only constant expressions 'symbolize' (denote); variable expressions (variables) 'represent'.

[3] Our own interpretation of this term agrees with that of Tarski, as formulated in his book *Undecidable Theories*, Amsterdam 1953.

[4] To avoid misunderstandings let us remind the reader that the symbol Σ used in the formula above is the operator which denotes the arithmetical sum of a finite sequence of numbers

[5] Chap. VI (a description of Peano's adequate arithmetic) has been omitted. (Ed.)

A FUNCTIONAL APPROACH TO THE LOGICAL SEMIOTICS OF NATURAL LANGUAGE

by

Jerzy Pelc

Introduction

The term 'functional' and the related 'functionalism' have in the last several decades become associated with the history of linguistic research on natural language. In the course of that time, the meaning of the term has undergone certain changes: from Jean Baudouin de Courtenay who ninety years ago wrote about functional and non-functional elements in language and who constructed functional morphemics, to de Saussure's *Cours de linguistique générale* of half a century ago and to the rise of modern functional linguistics in Roman Jakobson's Prague School (1928–1930). Even today, after a lapse of thirty years, 'functionalism' and 'functional' may have not as yet become fully univocal and strictly distinct terms in linguistics. Lacking, however, the required professional competence, I do not feel qualified to enter the dispute on this subject. Therefore, in order to avoid misunderstandings, I accept the terms suggested by an authority in linguistics.[1] He characterizes modern linguistics as functional if the following properties can be jointly found in it: synchronism in the description of the language, treatment of the language as a system or as a structure in the broad sense of the term (de Saussure), the text as the main object of linguistic study (Bühler) and the resulting anti-psychologizing attitude (Bloomfield), stress on the conventional nature of the language (de Saussure) and recognition of its representative and communication functions as most important (Bühler) and finally treatment of language as social and not as individual. It may be that after this explanation the terms 'functional' and 'functionalism', though they are not yet entirely precise, may have attained a degree of clarity that will prevent serious misunderstandings in what I shall say below.

The term 'functional', as used in the title of the present paper, is neither synonymous with, nor equivalent to, the term 'functional' as used in linguistics. However, the two denotations overlap and common properties of connotations may be pointed out. These remarks anticipate the reflections which are to follow. I intend to confine myself to examples illustrating the functional approach to *the analysis within the scope of the logical theory of natural language.* I do not intend to give either a univocal or exhaustive characteristic of functionalism in logical semiotics of natural language, or a nominal definition of the corresponding term.

The terms *logical semiotics* and *natural language,* occurring in the title, refer to descriptive semiotics as employed by logicians or to the *logical theory of natural language.* Without going into the question of whether there ought to be one or more kinds of semiotics, the well known fact is asserted here that in past research, semiotic reflections and publications of the following logicians: Mill, Peirce, Frege, Carnap, Ajdukiewicz, Kotarbiński, Tarski, Quine and many others — existed side by side with the reflections and publications of linguists, such as those mentioned at the beginning and of many others. Although somewhat awkward and veiling a number of underlying puzzling problems, the term 'natural language', used currently in this type of research, is employed here as the opposite of the 'hybrid artificial language' of certain specific branches of science, such as psychology or sociology, and also as the opposite of 'symbolic language', e.g. of formal logic and algebra. The terms 'colloquial language', 'ordinary language' and so on are, occasionally, though not in every context, synonyms of 'natural language'.

Both linguists and logicians are concerned with this kind of language. But they are so in different ways, and they often concentrate on different problems. In ancient times, the same scholar appeared in two and even three capacities: as grammarian, logician and philosopher. Later the separate fields became more distinct and as specialization progressed, the history of linguistic studies became to an ever increasing degree a history of anachronisms and delays in benefiting from achievements in related fields. Today it happens quite frequently that when a linguist looks to logicians for help, he goes back to Mill and particularly to Peirce, hence to the classics who, though undeniably valuable, are not quite new. Rarely does he rely on the results of recent research. Similarly, a logician rarely refers to linguistic studies written more recently than fifty years ago, that is the period of de Saussure. Consequently, only exceptionally can we find an author who quotes linguistic works written a 'mere' twenty-five years ago, as did Hans Reichenbach in *Elements of Symbolic Logic* (1947) when he referred to Jespersen's *The Philosophy of Grammar* (1924).

The result of this state of affairs is that we do not mutually possess the latest information from the neighbouring field. An unnecessary wall is being raised between the specialized groups, and it is becoming increasingly difficult to find a common language in a group of people who ought to work together.

It is not difficult to foresee that under these circumstances a given information or suggestion may be a revealing discovery on one side while on the other it is like forcing an open door. The same applies to the present case. A logician who has never shown an inclination to make diachronic descriptions of language, may shrug his shoulder at the linguist who states that synchronic treatment of subject matter is a revolutionary discovery in the science of language. In studying the language, further, the logician by the nature of his interests has always concentrated on problems of semantics and the representative function of a sign, neglecting the expressive and emotive functions. That is why the appeals to recognize the representative and communicative functions as the main functions of language (Bühler's *Darstellungsfunktion*), appeals which in the inter-war period were so timely as far as linguistics is concerned, could be justifiably treated by logicians as exhortations to rediscover America. A linguist, on the other hand, would accept as a truism the stress on the necessity of giving consideration to linguistic context. Yet in practice it would be worthwhile to remind logicians, and repeatedly at that, of the significance of context as a modifier of the semiotic function of the sign under consideration.

As can be seen from the above, the logical theory of natural language is in certain respects, by the very nature of things, so to speak, functional in its nature in the sense already mentioned here. I have in mind its synchronic approach: concentration on the signs that appear in a text and not, at least for many years, on the mental experiences of the sign-producer and sign-perceiver; further, recognition of the conventional nature of the expressions of a language as distinct from iconic signs; and especially concentration on the representative functions of a sign: designation, denotation and, with some reservations, connotation. In logical semiotics there has been a marked growth of its branch called semantics, which is concerned with certain functions of the sign, specifically its relation to extra-linguistic reality; this fact testifies that the logical theory of natural language is — in a sense — functional.

It would be unnecessary to repeat these facts as both logicians and linguists are familiar with them. Besides they are applied in practice — in logic anyway. On the other hand, I shall not try to avoid repetition of matters which, though they are known in logical semiotics, yet have made themselves neither completely nor prevalently at home in this field. If these matters are indeed important because of the actual state of research in studies on the functional logical theory of natural language, then no doubt it is worthwhile to mention or to recall them even if in the eyes of the linguists these matters should seem commonplace. Let us not hesitate to force an open door, nor to rediscover America as long as there still are people to whom this door may be shut and to whom America is not yet discovered.

1. *Use of an expression*

What is *functional approach to the logical semiotics of natural language* concerned with? If I were asked to give a most concise reply to the question, confining myself to the enumeration of the most important points, I would say that it is concerned with *how the expression analysed is used.*

Thus, such-and-such *use* of a given word, phrase, sentence or group of sentences comes to the forefront. It must be made clear at the very outset that the word 'use' is ambiguous, vague and perhaps obscure even in this context. It would be advisable therefore to pinpoint its meaning in the manner a word is used is to play a central role in the functional concept of language analysis. We shall try to do so. It may, perhaps, suffice to realize the dangerous misunderstandings that lie in wait.[2]

The first is not especially dangerous, particularly in the Polish language. It may loom larger in English where the term 'use' also appears in the meaning of 'usefulness' and 'utility'. I am concerned with such-and-such an employment of a word, phrase or sentence and such-and-such an application or *use*, and not with its *usefulness*, adaptability and advantages.

Another misunderstanding may result from the fact that the phrase "I use such-and-such a thing in such-and-such manner", "I employ such-and-such a thing in such-and-such manner" can describe three different situations: when the thing is an *organ*; when the thing is an *instrument*; or when the thing used is an *ingredient*. For example, I write with my hand, but also with a pen; and to prepare a cocktail I use the ingredient vermouth. Expressions of natural language may appear in all three roles. When I refer to a given thing by a word, I use that word as an instrument. (One could argue whether it is then always an instrument. Perhaps it is such only when I have used it in a manner approximating the general usage in a given discipline, that is when the word is a precise term and when the thing it refers to is univocally determined by the context and extra-inguistic situation. In cases of spontaneous usage, when the word is not used with precision and when it is not determined by context and situation, it perhaps functions as an organ.) When I place a given word in a syntactic construction, I use it also as an ingredient. But the situation becomes complicated in this case, too. For example, in a metalinguistic sentence: " 'Dog' is a one-syllable word" — 'dog' occurs not only as an instrument and as an ingredient: its part within the sub-quotes occurs as an object, a designatum. Once we are aware of the various possibilities mentioned here, they will perhaps cease to cause trouble. There is no reason to eliminate any of them. For the use of a word as a designated *object* is useful in *semantic* considerations; the use of a word as an *ingredient*, in *syntactic* considerations; the use of a word as an *instrument* (or organ), in *semantic*, *syntactic* as well as *pragmatic* considerations. In my opinion, however, from the point of view of the functional approach to the logical semiotics of natural language, the most important is the use of the word as an instrument.

The third misunderstanding is the most dangerous. For instance, when speaking of such-and-such a use of a given term, I may bear the following in mind. The term, the word 'dog', for instance, has its defined meaning in a given language and in accordance with this meaning

I can use the word 'dog' in such-and-such a way to refer to, let us say, my dog Trot. This use will be determined, among other things, by context or extra-linguistic situation. I may make frequent use of the word 'dog' in the same manner. The class of the tokens of the word 'dog', each used in reference to Trot, determines this-and-this use of the word.[3] At other times, however, I may use the word 'dog' in such a way that it would stand for Arthur Conan Doyle's *The Hound of the Baskervilles*, and I could do this more than once. Here again context and extra-linguistic situation determine the given use of the word. Each of the uses of the word 'dog' mentioned here agrees with the *prevailing usage* of that word. Thus, as we see, speaking of such-and-such a usage of an expression, I have in mind a *linguistic practice*, an established *linguistic custom*. The nominal analytical definition of the word 'dog' refers to this usage. The usage now discussed is a generalization of the uses enumerated before. There are many uses for a given word and comparatively few usages. The word 'orchid', for example, is used in one way only, in the sense of the term 'usage'. The English word 'bay' has more usages. It means: (a) a laurel, (b) a creek, (c) a part of a bridge, (d) to bark, (e) a reddish-brown color. It is therefore a polysemic word. 'Orchid', on the other hand, is a monosemic word. In the case of the word 'bay' the context and extra-linguistic situation perform a dual role. First, they jointly determine the usage of the word, hence, whether reference is made, e.g., to the laurel tree, or to the creek. Then they determine use of the word, that is what laurel tree or what creek is meant here. In the case of the word 'orchid' the role of context and extra-linguistic situation reduces to the use of the term.

The term *usage* may, therefore, be accepted as a synonym of the term *meaning*, in one of its definitions. It is sometimes said that the meaning of an expression is clear to one who knows how to use the expression in a given language, that is, to what objects outside of that language the expression can refer and in what respect and in what grammatical construction it can correctly be used. However, the following danger lies in wait. Owing to a slight shift of meaning, that is instead of treating meaning of an expression as usage, it is asserted that "the meaning in which a person understands an expression is rather the way, defined in some respects, in which he *understands* that expression".[4] It is considered here that "the understanding of an expression consists in a thought of the person who heard the expression, a thought which in his mind became intertwined into a single whole with the hearing of that expression", as well as that "such a thought is the *process of understanding* an expression".[5] Consequently, the way of understanding, i.e. the manner of thinking, becomes the meaning of an expression. The meaning of an expression as understood by a person is not identified here with the thought on which the process of understanding is based, but is identified only with some of its properties. However, this too may raise the objection that this is a psychological interpretation.[6] Although the term 'usage' is pragmatic and not psychological in character, yet, owing to its kinship with the term 'way of understanding', it may become a source of certain undesirable consequences.

'Usage' as a synonym of 'meaning' is troublesome for other reasons. At times the word 'to mean' was understood as if it referred to a certain 'relation' between a given expression and an *entity* which is disparate from it, namely the meaning of the expression, and such that that expression constitutes the name of the entity.[7] Consequently, a meaning is either hypostasized or confused with the designata of the word. The same applies to 'usage'.

Difficulties of a similar nature may arise from the fact that *meaning* is frequently identified with *concept*. It is asserted that "The meaning of a term will be called a concept in the logical sense of the word, and, in particular, a nominal concept in the logical sense of the word".[8] At the same time, it is said that "It is also legitimate to speak about the *designata and the extension of* a concept, in a way similar to that in which we speak about the designata and the extension of a term. A designatum of a term of which a concept is the meaning will be called a designatum of that concept. The extension of a term of which a concept is a meaning will correspondingly be called the extension of that concept".[9] Further, the same is said of the content of a concept which is said of the content, or the intension of terms: "The intension of a *concept* is that which is the intension of the term of which that concept is the meaning".[10] Consequently, the concept, defined initially as a meaning of a term, or as certain properties of thought on which this meaning rests, comes to resemble the term itself, since, like the term, it has a designatum, denotation and content. And in consequence one is tempted to resort to a hypostasis, for since the terms, or at least the concrete tokens of a given term, are physical individuals, why then should we deny the existence of concepts — so similar to the terms. One step further would lead us to speculate, in the spirit of Plato or Locke, on the subject of status and derivation of concepts, a speculation which is like the metaphysical disputes on the subject of the status and derivation of ideas.[11]

For this reason the term 'usage' may become a source of one trouble or another. I think it is important clearly to distinguish between use and usage. And when I said that the functional approach to analysis in logical semiotics of natural language is concerned with, to put it most concisely, consideration of how an analysed expression is used, I had in mind *use* rather than *usage*.

2. *Use of an expression and the classification of terms with regard to the number of designata*

From now on whenever we speak of the *manner in which an expression is used*, we shall mean the *use* of an expression.

Taking a few examples, I shall try to explain why the study of *how* a given expression is *used* plays such an important role in the *functional* treatment of the logical analysis of natural language.

The first example will concern problems of classification of appellations, especially of *terms*. The distinctions customarily made in logic are, among others, the following: a term may be *general*, *singular* or *empty*, depending on whether it has, respectively: two or more *designata*, one designatum only, or no designatum at all. The designatum of a term is the object *designated* by the term, that is each and only that object about which that term *can be truthfully predicated* when it has such-and-such a meaning. One-word *appellatives* as well as *adjectives* are accepted as terms, hence '(a) dog', 'red'; so are one-word *proper names*, hence 'Socrates', 'Apollo'; proper names of more words than one, hence 'Mont Blanc'; *descriptions* of more words than one, hence 'the largest Polish city', '(a) member of parliament', and so on. There is a tendency to expand and to modify the meaning of 'to designate' in a manner which enables one to say that functors, especially sentence-forming ones, have their designata. Thus, if the sentence-forming functor 'runs' applies to John and not to Warsaw, in that case John is one of the designata of that functor, while Warsaw is not.

This concept of designating calls for certain ontological decisions. A diagnosis establishing whether a given term may be truthfully predicated about a certain object depends, among other things, on its existence. Hence, when I classified the word 'Socrates' as a singular term, I had first to accept that the long dead Socrates exists. And when I assert that one of the designata of the term 'dog' is Fido which will be born in ten years, I automatically assert that Fido, who is not yet alive, does exist. If, finally, I accept the term 'Centaur' as empty, I assert that Centaurs do not exist. This opens a whole field for discussion on existence and on designata.

The *meaning* in which a given term appears is essential to the above concept of *designation*. E.g., 'Siren' as a name of a mythological character is an empty term, and as a name of an instrument which produces sound it is a general term. Thus the *usage* of the word intervenes here.

It may be noted on the margin of the problem of designating that the sentence 'two persons understand certain expressions in the same meaning' demands precision. It is said that one of the necessary conditions of this kind of understanding — when it depends on thinking of a certain object — is that the persons in question have the same object in mind.[12] And yet everyone will agree that John and Peter can understand the word 'dog' in the same way although each one has a different dog in mind. There is a further problem. The person who classifies the term 'siren' as a name of more than one object, does so because he understands it in such-and-such a meaning. But on the other hand, one of the necessary conditions of understanding the term in that particular sense is that the thoughts on which that understanding hinges refer to more than one such object. Here again greater precision is necessary, for one may easily be caught in a vicious circle.

But for the time being let us set these problems aside. We are at present interested in the fact that meaning, or usage, plays an essential role in the explanation of designating and of the classification of terms with regard to the number of the designata. I do not deny that meaning is important in considerations of this kind, but I should like to note that as far as analysis of natural language is concerned relativization with regard to meaning is *inadequate*. For it would then seem that empty terms, and besides that singular, and besides that general terms, appear in language in the same way as people, and besides that dogs, and besides that cars, appear in extra-linguistic reality. This classification does not adequately sum up the actual situation in natural language. It classifies terms *out of context* and gives consideration only to the *lexical meaning* of each of them, without taking into consideration their *functioning* in a language. As a result, we receive a static and non-functional representation of the state of affairs.

A *functional* description ought to take account of the following circumstances, made familiar by experience. The word 'dog' may be used in a variety of ways, depending on extra-linguistic situations and contexts. Let us assume that I found my dog Trot sleeping in the wardrobe on top of my suit which he had pulled down from the hanger. Shaking my finger at him, I say, 'Bad dog; what has he done? The dog slept in the closet'. In this context and situation the word 'dog' functions *with regard to its being used* — as a singular term. When I inform someone, 'Trot is a dog', or when I say, 'You should talk to a dog', then the expression 'dog' functions with regard to its being used — as a general term.[13] Then again when Arthur Conan Doyle refers to the dog hero of *The Hound of the Baskervilles*, then the term used in this manner functions as an empty term. Finally, all the sentences given above were used by me in the same way in a certain respect, that is, they were used as examples. In this situation, in the text of the present article, I did not use a single sentence in order to *talk* about a concrete dog, or about every dog, or even about a fictitious dog. The word 'dog' did not once *refer* to a dog; it could not really be predicated about anything. Therefore, the word functioned *with regard to the fact that it was used* as an example, as an expression which is neither empty, nor singular, nor general, because the question simply does not arise.

The term 'dog', taken out of context, isolated from the situation in which it is spoken or written, that is, devoid of relativization with regard to such-and-such use, is also neither singular, nor general, nor empty, as long as it appears as an item in a dictionary. Or, if one prefers, it is at the same time both the first, and the second, and the third. That is, it may be used in such a manner as to form a potential predicate of a true proposition about every dog, or some dogs, or about one given dog; and finally, in such a manner, where there would be no object about which that term could be truthfully predicated.

I would contest the view which holds that since the term 'dog', treated as an isolated lexical item, is suitable, owing to its usage, that is, owing to its meaning, for the role of a predicate of a true assertion about more than one animal, then this fact determines its general character. I feel that we are dealing here with a misunderstanding similar, though not so glaring, as the one below:

"The word '*Pustelnik*' is a Polish place name, but it is also an appellative which means 'hermit'. In the second case it maybe correctly predicated about many persons. In that case the word will always be a general term, hence also in first case, when it is a proper name."

An equivocality arises owing to the impermissible change in the meaning of the word 'Pustelnik'. In our examples, the word 'dog' did not change meaning. The very meaning, or usage, identical in all the given examples, hence also when the word was an isolated lexical item, determines its various uses: as a general term, or as a singular term, at that referring at one time to this and at another time to that concrete dog, or finally, as an empty term. The meaning is characteristic of a given *word-type*, or of the class understood in the distributive sense, of all *tokens of the word*. On the other hand, emptiness, singularity or generality are characteristic correspondingly of this, that or another use of a given word, or of a certain sub-class composed of only some of its tokens. The word 'dog' appeared in the examples given above in a different use each time. Because in one of its uses the word 'dog' was proved a general term ('You should talk to a dog'), one cannot correctly infer that in every use it will be a general term. Specifically, therefore, in the example 'Bad dog; what has he done? ...' the word 'dog' refers to my dog Trot only, and owing to its use it cannot be made a predicate of the sentence 'Trot is a dog', without first changing its given use. In the last sentence whose equivalent is 'Trot is an element of the class of dogs', the term 'dog' was used in a general manner which differs from the first.

It is evident, therefore, that it is not so that there are general terms, and also singular, and also empty terms. But that the same term is once used as a general, once as a singular, once as an empty term, and once in such a way that the question of its generality, singularity or emptiness does not arise at all. The generality, singularity or emptiness is a *relative* trait, made relative to a given use of a term. There is a difference between terms classified from this point of view and the classification of things as people, dogs, cars and others. More like the classification of women according to the degree of kinship: we do not distinguish mothers, and besides that sisters, and besides that aunts, since one woman may be a mother, a sister and an aunt at the same time.

3. *Use of an expression and the classification of terms according to their singular or general intention*

Here is another example testifying to how important it is that the use of an expression be given consideration in a functional approach to the semiotic analysis of natural language.

It is known that names or terms or nominal expressions may also be classified according to *meaning intention*, as general, singular and at times empty expressions, but in a different sense than previously. It is said that "'Zeus' is a *genuine name with a singular intention*, and 'centaur' is a *genuine name with a general intention*, though both are empty names if their denotations are considered."[14] 'Denotation' is understood as the class of all designata of a term, in the meaning defined previously. Thus, the singular or general intention of a term does not depend upon its denotation but on *usage*: it is — so to speak — built into the meaning of the term. Qualifying the term 'centaur' as general, we refer to denotation in another meaning, which Lewis[15] calls *comprehension*, or a classification of all consistently thinkable things to which the term would correctly apply. And classifying the term 'centaur' as empty, we refer to its denotation. According to the present subdivision, 'red object on my desk' will be a general term. 'The red object on my desk' will be singular regardless of whether one red object or more than one was, is or will be on my desk or finally whether it never was, it is not and none will ever be.[16] In the present understanding, the 'square circle' may be recognized as an empty term, for its meaning determines zero-comprehension, since nothing can be consistently thinkable as a square circle. It will be an empty term in the previous understanding too, since it also has zero-denotation because square circles do not exist.

As can be readily seen, here too the problem is formulated as if generality, singularity or emptiness in the present understanding depend on the meaning of a term, on its usage. This case, however, is analogous with the former. *Use* is decisive. In the sentence about the Cyclops Polyphemus, 'The Cyclops was blinded by Odysseus', the word 'Cyclops' is used with singular intention, while in the sentence, 'Polyphemus was a Cyclops', the intention of the word is general. The expression 'my brother', analysed as an isolated lexical item, has general intention. But context or situation, hence such-and-such use, may make it singular as in the sentence, 'My brother gave me this book.'

Thus, *mutatis mutandis*, in this case too considerations of the *use* of a given expression blazes the path for *functional* analysis of natural language. As long as the classification of terms as discussed here is based on the meaning of expressions, on usage, the expressions are shown to us as dead, static and isolated lexical items. Application of analysis with regard to use — not only assures that the general or singular character of expressions is a relative property, but, and this is more important, puts them into a *context* and a situation which accompanies the expression, and demonstrates how that expression actually *functions*.

4. *Use of an expression and classification of terms according to their subjective or predicative character*

Since the days of Aristotle a distinction has been made between *individual* and *general* terms in still another, fairly puzzling, sense. The former can be used only as *subjects of elementary individual propositions*, that is propositions that speak of individuals; the latter can be used as *predicates*.[17] A number of interpretations is possible. The word 'term' may be taken as an equivalent of the phrase 'an expression which appears in an atomic categorical sentence as a subject or a predicate' or as referring to extra-linguistic entities, namely — those elements of a proposition, and not of a sentence in the grammatical sense, which correspond to a subjective or predicative expression. In this connection the words 'subject' and 'predicate' must be understood, respectively, either as a *grammatical subject* and a *grammatical predicate* or as a *logical subject* and a *logical predicate*. Moreover, as each of these is ambiguous and vague, other doubts arise regarding interpretation. This is neither the time nor place for seeking a solution or even discussing this problem at length. It will suffice to assert that according to the fairly prevalent exemplification of the above distinction, an example of the individual term is the so-called pure proper name, 'Socrates', and of the general term — the predicates: 'man', or 'an immortal creature', or 'the longest river in Poland', or 'the youngest Olympic god'. The latter category covers, according to the first classification, both the names as well as the descriptions, both the general and the singular as well as empty terms, and, according to the second classification mentioned here, terms distinguished by *singular intention* and by *general intention*. The matter is posed as if the classes of individual terms and general terms were *mutually exclusive*, as if, therefore, individual terms did exist and besides them also general terms.

This opinion could be formed owing to a failure to give consideration to the use of these expressions, not to speak of the fact that the relativization of these terms to an elementary individual sentence, or a sentence which speaks of a certain individual, must evoke protest. In Chapter 7 I shall try to explain that such sentences do not exist, for talking about something is not a property of a sentence but of such-and-such a use of a sentence. It may be sufficient to call to mind at present the examples given by Quine:[18]

(1) *Lamb* is scarce,
(2) Agnes is a *lamb*,
(3) The brown part is *lamb*,

where the word 'lamb', a general term in each of the understandings discussed so far, changes its role depending on its use. First, it appears as a *cumulative*, or *collective* term, used as an *individual term*; secondly, as a *general* term which refers to every individual of the *Ovis aries* species; thirdly, as a *cumulative* term used in a *general* way so that it refers to every serving of lamb.

Incidentally, it becomes clear that the distinction between cumulative and non-cumulative is, if the analysis takes account of use, relative. The same expression may be used in a collective or non-collective manner.

As regards distinguishing *individual* (that is *subjective*) and *general* terms (that is *predicative*), the above examples would indicate that the above traits are connected with use of expressions. The same word may be used in each of these roles. One or another potential syntactic character — suitability for the role of a subject, suitability for the role of a predicate — is not here a characteristic trait of usage, or of the meaning of a word, but, simply, of its use. Whenever a given word is analysed as an isolated and static lexical item, account being taken of usage only, it always appears to be ambivalent in character from the point of view of the analysis: if it is suitable for the role of a subject, it is also suitable for the role of the predicate of a sentence, and vice versa. Only a study of use, hence functional analysis of a word in a certain type of context and a certain type of situation, discloses whether the word appears as an individual or as a general term in the meaning discussed at present.

5. *Use of an expression and the distinction: proper name — name — description*

Since the origin of Russell's[19] theory of descriptions and even earlier, beginning with Frege's "Über Sinn und Bedeutung",[20] a distinction has been made, especially in Anglo-Saxon logic, between *proper names, names, definite descriptions* and *indefinite descriptions.* This distinction is not popular among Polish logicians; it has been used neither by Ajdukiewicz nor by Kotarbiński.

I cannot at present devote myself to the defects of this classification or to the differences of standpoint represented by various authors. But it may suffice if I explain my position. I shall take examples from the English language which has definite and indefinite articles and is therefore more suitable for this purpose than a language that has no articles.

A 'pure' *proper name* is usually considered a one-word name of a real individual, e.g. the word 'Socrates'; however, there is a difference of opinion as to whether a one-word name of a fictitious individual, or a name, consisting of two or more words, of a concrete or fictitious individual is a proper name. A *definite description* is a substantival phrase having 'the so-and-so' form, where 'so-and-so' appears in the grammatical singular and where the whole is singular in character in the first or in the second of the two meanings of the adjective 'singular' discussed above, there being no full agreement as to in which of the two. Nor is there full agreement with regard to the following matters: (1) Is 'the river' as well as 'he', 'she', 'it' a definite description, as Quine would have it,[21] or must it be a phrase composed of independent parts, as 'the king of England', as Russell would have it.[22] (2) Can singular phrases beginning with demonstrative pronouns 'this', 'that', e.g. 'this queen of England', or relative phrases composed of a predicate and a proper name, e.g. 'Napoleon's mother', also be definite descriptions, as Reichenbach proposes,[23] or not, as Quine says? Russell is prepared to accept as an *indefinite description* a phrase which has 'a so-and-so' form where this 'so-and-so' must be composed of independent symbols. Reichenbach, however, also accepts as an indefinite description an expression in which the article 'a' is followed by one-word predicate, i.e. 'a man'. Both agree that a definite description must be strengthened by the requirement of singularity which does not apply to indefinite descriptions. They also note that indefinite descriptions are connected with pure assertions concerning existence. The stand on *names* depends on what is considered a proper name, a definite description and an indefinite description. Most often the following are accepted as such: 'the lion', 'a lion', 'man', 'water', and, with reservations about its syncategorematic character, 'red'. It is a matter for discussion whether the pronouns 'this', 'that', 'he', 'she', 'it' and others should be included among the names. The popular distinction 'names and descriptions'[24] usually pertains, on the one hand, to *pure* proper names ('Socrates'), and on the other hand, to definite descriptions of the Russell type ('the author of *Waverley*').

The differences enumerated here are a result not only of the subjective divergencies of opinion but also of the fact that the distinction is not satisfactory. One of the reasons for this is the notorious failure to give consideration to the use of the expressions here discussed.

Yet the singular character of proper names and definite descriptions, each of which is understood to refer to one object only or to have singular intention, is bound with the use of a given word. Hence, from that point of view, one cannot correctly distinguish a class of proper names or a class of definite descriptions, but only at best a class of expressions *used as proper names with regard to singular intention*, or as a class of expressions *used — owing to their singular character — as definite descriptions*. Further, singularity cannot be used to distinguish definite descriptions from indefinite descriptions or names, for expressions are not singular once and for all. They can only be *used in the singular manner*. This refers, e.g., to the definite description 'the king of England'.

I can use it in a way in which it would not mention any individual, for instance, when the description occurs in the sentence given as an example. Then again, I may employ an indefinite description or a name as singular, as in the sentence, 'I met a man', where 'a man' refers to some person. Definite and indefinite descriptions as well as names, construed with regard to the manner in which they are used as singular in character, do not constitute mutually excluding classes. It would be correct to state the following: a given expression, owing to its singular character, is used here as a definite description, or as an indefinite description, or as a name.

It may be that a certain kind of expression which, owing solely to its use, appears in a given case in the role of a definite description, may not, from another point of view, be qualified as an expression used in a descriptive sense. Here is another example:

(1) The author of *The Tempest* was a glove-maker's son.
(2) The author of *The Tempest* is also the author of *Romeo and Juliet.*

n each of the above sentences 'The author of *The Tempest*' was used as a singular expression, for it refers to Shakespeare and to him only. But in the first case I used the periphrase solely for the purpose of *identifying* a given person and to avoid monotony I tried not to repeat the proper name 'Shakespeare' — for instance, in a long article. In the second sentence, taking the content of the whole sentence into consideration, I am interested not only in pointing to the person of Shakespeare, but also in *informing* that he wrote *The Tempest.* In the first case, I could have equally well written 'Shakespeare' or 'the author of *Hamlet*' instead of 'the author of *The Tempest*'. I could not do this in the second case because I was interested in giving the following information about the same person: that he wrote *The Tempest* and that he wrote *Romeo and Juliet.* Hence, in the first sentence, I employed the phrase 'The author of *The Tempest*' as a *proper name.* In the second instance, however, I used it to a greater extent *descriptively.* In turn, in the second sentence the predicative expression 'the author of *Romeo and Juliet*' was used more descriptively than the subject 'the author of *The Tempest*'. For the subject also performs a *demonstrative* function, that is, it performs the role of an indicative gesture, it performs a *referential, identifying* function, and that is considered characteristic of proper names and not of descriptions. The phrase 'the author of *The Tempest*' was used in a mixed manner: *as a proper name and as a description.* And the predicative expression 'the author of *Romeo and Juliet*' is used exclusively in an *ascriptive* role; it does not serve to indicate the person of Shakespeare because that has already been done by the subject. It may at best be correctly or wrongly predicated, and that depends on the empirical state of affairs. All these properties are ascribed to descriptions. We have here then a purely descriptive use.

It appears again that consideration of use of an expression, hence of its context and situation, leads to a break with absolute distinctions, in this case with the antithesis: proper names versus descriptions. Instead one ought to speak of the use of an expression as a proper name or of its use as a description. Moreover, one may here indicate the gradations, whereby a given phrase may to a greater or smaller degree appear in the role of a definite description or in the role of a proper name.

Descriptions are at times distinguished from names on the grounds that the first are made up of two or more words while names should be made up of one word. According to this view, the expression 'therapy by injecting the patient's own blood' would be an indefinite description, while the word 'autohaemotherapy' would be a name (of an abstract entity). We shall analyse the example taking into account the use of the second expression. We compare the following two sentences:

(1) A doctor in a clinic explains to the students: 'Autohaemotherapy is effective in cases of *furunculosis* (furuncles or boils).'

(2) Another doctor corrects a student by saying: 'This is not an ordinary blood transfusion; it is autohaemotherapy.'

In the second sentence, although the word 'autohaemotherapy' is a single word, it is used as a description. This fact is indicated by the situation, context and the syntactic position of the word. In order to grasp the contradistinction which the doctor wishes to make it is necessary to go into the etymological structure of the word 'autohaemotherapy' and to know the meaning of its separate parts, that is, the word must be treated as a three-word description or at any rate more closely analysed than in the first example where it performed a referential function, characteristic of demonstratives.

The same applies to proper names. They are used at times as descriptions. This pertains to what are called *nomina-omina*, that is proper names that have a meaning, as for example the names of comedy characters ('Mrs Malaprop' from '*mal à propos*'), and to predicates in sentences like 'this man is Socrates', 'This city is Warsaw'.

As can readily be seen, the same expression may be used as a proper name, as a description or as a name. Thus, there is no situation in natural language in which there are proper names, and besides that names (terms), and besides that definite descriptions, and besides that indefinite descriptions. There are not *mutually exclusive* classes of expressions, but *various uses*. Moreover, there is a *gradation* within the area of each of these uses, that is a greater or lesser intensity of descriptiveness, nominal character or proper-nominal character. Beyond the *pure* uses there are also *mixed* uses, with some descriptiveness and some proper-nominal character. It is necessary to state each time in what respect a given expression was used in a given manner. From the point of view of *functional* treatment of the logical semiotics of natural language, it is necessary to put an end to the rigid division that has so far prevailed in the subject under discussion and instead to classify the given expression *as used in a given respect, in such-and-such a manner and in such-and-such a degree.*

6. *Use of an expression and the sentence*

Before I discuss the *use of sentences*, I should like to emphasize that something which I shall call *denominalization* — is an essential point of the program which sets out to *functionalize* semiotic analysis of natural language. Logical tradition in this type of study, deriving from Mill, Frege and the young Russell, treats language as if it were composed exclusively of *terms*, or principally of terms, or at least as if the terms constituted the "salt of the earth". Natural language repudiates this experience. One of its basic functions, the *communicative* function, is realized primarily by means of *sentences*. The same applies to *expressive* and *emotive* functions. Traditional logical theory of language, focused on certain *representative* functions — *designating*, *denoting*, *connoting* — and absorbed in the contemplation of *ideas, concepts* and *meanings*, noticed this empirical fact in too small a degree. Analysis concerned with the usage of terms, and consequently concentrated on isolated expressions and static lexical items, overestimated the role of terms in language. Although the disproportion was less harmful in the study of artificial languages within the area of pure semantics, it was nevertheless painfully noticeable in logical descriptive semiotics of natural language.

I claim that by taking account of the use of an expression in semiotic analysis, and in this connection by fully appreciating the importance of contexts and extra-linguistic situations, a contribution may be made toward the denominalization of semiotics: ti will ascribe to sentences the place they deserve in language and thus also will restore the upset balance.

The time has come to give some examples. Analysed from the point of view of use, the distinction between *sentences and non-sentences* loses its absolute character. Ajdukiewicz has remarked that expressions, which have the structure of nouns, and hence terms, are sometimes used to make certain statements; for example the noun 'fire' is used in the same meaning as the sentence 'Something is burning'. From our point of view these expressions are declarative sentences for in that meaning they assert a certain fact."[25] I would like to modify this remark and replace the relativization with regard to meaning by a relativization with regard to use. This follows from our earlier considerations. Therefore I would say that in certain circumstances (context and situation) I use the word 'fire' so that it functions as a proposition. Similarly, when a teacher calls, 'Silence!' in a class, he uses the term as an imperative sentence, hence, as a sentence in the grammatical sense, which, however, is not a proposition.

The opposite of this is possible. In the sentence, 'Ceasar said: *Alea iacta est*', the expression '*alea iacta est*', being a *declarative sentence* from the grammatical point of view, is used as a *name*, namely as a name of what Caesar said.

There are instances where a grammatical interrogative sentence, a rhetorical question, occurs as a declarative sentence and at the same time as a proposition, owing to such-and-such a use, and where a grammatical declarative sentence, e.g. 'All the children will write', is used as an imperative sentence.

I shall now dwell for a moment on the subject of the distinctions between *declarative sentences* and *propositions*, and on the subject of the *truth-value* of the latter. It is usually asserted that a proposition is a *true or false* statement, and an equation mark is placed between grammatical declarative sentences and sentences understood as propositions.[26] It is added that a non-declarative sentence, for example an interrogative or an imperative sentence, is not a proposition. In a word, the class of propositions is treated as a sub-class of the class of grammatical sentences.

It was evident from the foregoing remarks on the subject of declarative sentences used imperatively or in the role of a name, that the identification of the declarative sentence with the sentence in the logical sense cannot be sustained. This fact is borne out by further observations in which the use of a sentence is taken into account.[27]

For example, 'In 1966 Warsaw is the capital of Poland', can be used so as to say something about Warsaw; the expression will then be used to make a true assertion. I may, however, use the same sentence only as an example, as I did a moment ago. In that case, the sentence is not used to speak about Warsaw or about any other thing. Nor am I making a true or false assertion. Consequently, it is not the sentence that has a truth-value, that is, it is either true or false (from the point of view of the classical two-valued logic), but only the sentence used in such-and-such a manner.

More controversial is the case of the sentence 'The emperor is naked'. In Andersen's tale it is used as an expression which speaks about a fictitious character, or, in other words, speaks about no one; for if the model of language is modified, the meaning of the word 'about' undergoes an essential change. Consequently, used in the above manner, the sentence does not make an assertion, whether a true or a false one. When, however, the sentence is used, and not in the metaphorical sense, to say something about Franz Josef, the emperor of Austria, as he reviewed the military parade, then the above expression is used to make a false assertion.

And when a person uses it to speak about the present emperor of Ethiopia as he is in his bath, then the sentence is used to make a true assertion.

Finally, in all these cases, I employed the sentence exclusively as an example: with its aid I did not speak about anyone, nor did I make an assertion. Consequently, it is not the sentence that has such-and-such truth-value, but the sentence used in such-and-such a manner.

It is not hard to predict what the objections to the above analysis will be. First, it is possible to point out that we are dealing here with the error of an incomplete proposition. Removing this error, we have the following sentences: 'Franz Josef is naked at t_1 time', 'Haile Selassie is naked at t_2 time', and so on, and these, as it is asserted, simply *are* true or false and are not *used* as true or false. This, however, does not refute the argument that each of the sentences, even when they are completed and are precise, may be used not for the purpose of speaking about Franz Josef at t_1 or about Haile Selassie at t_2, but as examples. Then the sentences will be neither true nor false. Moreover, the interpretation whereby a sentence such as 'The emperor is naked' is considered incorrect because it is incomplete, seems to ignore the general linguistic practice which would then be consistently commiting such errors, hence also logical errors in expressing thoughts. Closer to the needs and to the spirit of natural language is an analysis in the style of Strawson[28] or Bar-Hillel,[29] presented here by me; according to that analysis, there is no need to disqualify the majority of natural language sentences as erroneous.

A second kind of argument against the stand represented here by me would be that the word "emperor" in the sentence "The emperor is naked" — is an *occasional* term. It would be claimed that in that case we have sentences which, though homomorphous, are not synonymous, a fact which may lead into the error of equivocality. When completed, they become heteromorphous, as they were formerly. In reply to this argument, I should like to state that I have no objections to the view that the word 'emperor' is occasional. However, I do object to the opinion that the above sentence was polysemic. I shall endeavor to enlarge upon this remark.

I assume that *occasionality is a property of a vast majority of expressions of natural language*. Together with *ellipticity* of expressions it constitutes its characteristic property. Contrary to accepted opinion, not only words like 'I', 'here', 'now' are occasional, allegedly in contrast to other non-occasional words. Also occasional are words, like, 'table' or the phrase 'eligible for voting' and what is called a *pure* proper name

'John', as well as the verb 'stand' and others. One may at best note the difference in the *degree of occasionality* and the difference in the degree of the dependence of referential function upon context and situation. Being a general property of expressions of natural language, occasionality should not be classified as a defect in the communicative function of the language. From a certain point of view, it is even an asset, for it enables one to employ a given expression in a wide range of circumstances. Occasionality is not a property of word meaning, it is not governed by changes in the usage of the term, by changes which depend upon new circumstances. It does, however, concern *use* and depends upon changes which occur within the *referential function*. Thus, the word 'I' may refer to different persons, according to its use, whereas its meaning, that is its usage, is — within certain limits of time and space — constant in a given language. The rule by which the word ordinarily refers to the person who employs it, applies to various uses.

The same applies to the sentence 'The emperor is naked'. The sentence is occasional because it can speak about different persons, and at times it speaks about no one. The word 'emperor', within the sentence, may refer to various persons, it may mention, point out, identify them or it may indicate and mention no one. But the meaning of the sentence 'The emperor is naked', or the meaning of the word 'emperor', the usage of the expressions, remains unaltered under the changing circumstances of their use.

Hence, truth-value is not connected with a given sentence solely because of such-and-such meaning, such-and-such usage of the sentence, but is connected with a given sentence also because of such-and-such use. The declarative sentence must not be identified with the proposition. When the statement 'The emperor is naked' is used so that it speaks about no one and does not make an assertion, it does not, nevertheless, cease being a declarative sentence. It is not, however, a proposition, for truth-value does not apply to it. With this relativization, we need not, however, renounce the *law of excluded middle*. True, in a given meaning a sentence may be true or false or it may be neither the one nor the other; but with an additional relativization with regard to a given use, it turns out to be either only true or only false, which fact depends on the actual state of affairs, or else the question of its truth-value does not arise.

Thus, analysis of the use of expressions leads to the following conclusions. It should not be said that a given expression is, e.g. a sentence, or even that it is a sentence in a given language and in a given meaning, or usage. This kind of relativization is not adequate. It must be added that is it so because of a given use. Briefly: expressions *are not* sentences, but are sometimes *used* as sentences. The same applies to the classifications of types of sentences. The same expression may be used as a declarative sentence, or, e.g., as an imperative sentence; also, at one time as a term and at other times as a sentence. Declarative sentences must not be identified with propositions: I can employ a declarative sentence as a proposition, but I do not have to. In the same manner, I can at times employ a term or an interrogative sentence. And finally comes the truth-value problem. It applies to an expression with regard to its use, such use namely which in the given case turns a sentence into an assertion.

7. *Use of expressions and functional logical semiotics of natural language*

By means of examples I tried to demonstrate wherein lies the essential aspect of a *functional* approach to analysis within the scope of logical semiotics of natural language. The examples pertained only to some problems, that is to selected questions in terms of classification; further they concerned the problem of distinguishing the latter from proper names on the one hand and from descriptions on the other and, finally, of distinguishing expressions which are not sentences from sentences, and the classification of the latter, especially with regard to purported identification of declarative sentences with propositions. In each of these instances it was shown that absolute division or distinction ought to give way to relative distinction. For the characteristic situation of natural language is that a given expression may appear in various roles, depending on its use. Language is not, therefore, composed of mutually exclusive classes of expressions, such as terms and sentences, and among terms — empty, singular and general terms, or proper names and descriptions, and among sentences — declarative, interrogative and imperative sentences. A given expression is not so much such-and-such, for example, definite description, as in certain cases it had been used as such-and-such in such-and-such respect and in such-and-such degree. It seems that this phenomenon is more general in character. It may be extended beyond the scope of the examples analysed here. Hence, it may be discovered in the distinction between genuine *names* and *onomatoids*, i.e. names of abstract entities, in the distinction among grammatical *parts of speech* and among *parts of sentences*, hence within the scope of *syntactic* relations, etc. (Such relations are also noted within the structure of a single word as occurring between its etymological elements.)

In the analysis made here, I devoted my attention chiefly to the *use of an expression*. I think that it is the most essential, if not the central, aspect of the *functional* approach to the *logical theory of natural language*. This by no means indicates a retreat from the meaning of expressions, from usage. On the contrary, it must be borne in mind that there are certain essential relations between use and usage. For when I encounter the Polish expression '*dwukilowy jacht*', I can decide only by analysing its use, whether it means 'a yacht with two keels' or 'a model (of a yacht) weighing two kilograms' (the confusion arises only in Polish where the adjective '*dwukilowy*' can equally well mean one and the other). Thus, *use* serves to establish *usage*. It must also be borne in mind that relativization with regard to use is complementary in relation to relativization with regard to the meaning of the word; it occurs within the scope of the second. That is, a given expression may be used in such-and-such a use, only when it appears in such-and-such a meaning (usage). Thus, there is no underestimation of meaning. The point I wish to make is that relativization only with regard to meaning, which may be sometimes adequate in pure semantics and also in descriptive semiotics of artificial language, gives a static image when we confine ourselves to it in the analysis of natural language: it discloses the expressions of that language as isolated lexical items and fails to disclose the actual functioning of words, phrases and sentences in language context and in extra-linguistic situation.

Consideration of the use of expressions does not shift the direction of language analysis but is a step forward in enriching it. Consequently, I would not wish to have the things I said understood in the following manner: "The classification of terms into empty, singular and general ones is a mistake; distinction between proper names and descriptions is a mistake; distinction between sentences and non-sentences is a mistake." I do not assert this. However, I do think that the classifications or distinctions mentioned here are, as far as natural language is concerned, more superficial because they are less functional; the characteristics pointed out in the examples probe deeper into the nature of that language since they are more functional. The traditional ones have a reason for existence because they construct certain ideal types of kinds of expressions. This idealization is useful not only for didactic reasons, since it makes typology possible, typology understood as an arrangement of concrete acts of speech with regard to the degree of intensity of properties occurring in them. In this sense, the traditional classifications and characteristics were the point of reference of our analysis, a sort of springboard. For example, since Russell has established the meaning of the term 'definite description', I was able to say that a given expression may be used as more or less descriptive in character. Hence, in studies on natural language one cannot do without those traditional principles. But I think they are not sufficient. On the other hand, in the construction of artificial languages they are not only necessary but sometimes sufficient.

Logical semiotics of natural language makes other demands on the researcher than does logical semiotics of artificial language. In the former, as I have tried to make clear so many times, it is necessary to take the use of expressions into consideration. The importance of that requirement is closely related to a certain characteristic property of natural language, to which I drew attention in passing: that by the very nature of things this language is occasional. That explains why in semiotic analysis relativization is becoming so important with regard to language contexts of the expression analysed and to extra-linguistic situations in which it was used, hence relativization with regard to use. Without it we would have to classify the majority of the expressions of that language as incomplete and hence faulty in the logical sense. That verdict will be passed by those analysts who rest satisfied with relating an expression to its meaning. I am afraid that such a verdict would be unjust. It would be based on the failure to understand the manner in which a natural language functions. Its expressions are not supposed to play their representative, communicative, expressive and emotive role alone. They can do so only in a group, together with their linguistic environment, and together with extra-linguistic situations. That is the normal state of affairs. A normal state of affairs should not be treated as a fault. It would be wiser to accept the fact that it is so and to draw appropriate conclusions.

I think that the acceptance of the use of expressions, which constitutes the foundation of the functional analysis of natural language, is precisely this kind of conclusion, and a very important one. The examples I have discussed here do not exhaust all the possibilities of the functional approach to the logical theory of that class of languages. They do not even exhaust the possibilities which are connected with the analysis of use; and, after all, functional analysis does not stop at the study of use. Thus, for instance, I had no time to mention the fact that the consideration of the use of an expression leads to further promising distinctions and characteristics. One may distinguish, among others, such aspects of the expression as the primary and secondary conceptual content symbolized, i.e. present and evoked; the propositional attitudes (with regard to these), expressed and evoked; emotions and conative attitudes expressed and evoked; the emotional tone; the emotions and attitudes revealed; other kinds of effects; the purpose. One may further distinguish the cognitive and non-cognitive elements, etc.[30] I shall only make here the general remark that the study of the use of expressions may help fill the gap which is evident in analysis in the field of the logical theory of natural language as usually limited to the study of its representative functions. It may enrich them by giving proper consideration to *expressive*, *emotive*, *performative* and *other functions.*[31] Based on the consideration of the use of an expression, functional logical analysis of natural language is, by its nature, a *semiotic* analysis and not only a *semantic* analysis, while the traditional logical theory of language was confined mainly to semantics.

This is a separate and important problem which ought to be developed. Although this cannot be done here, it can neither be completely ignored. When I analysed the semantic function of an expression with regard to its use then by the same token I introduced pragmatic problems into those considerations. The same occurs within the scope of the study of syntactic relations. I think that the *addition of the pragmatic factor is a necessary condition of a functional analysis of natural language.* Logical theories of artificial languages may at times do without pragmatic relations. On the other hand, a logical theory of natural language would be crippled without pragmatics. Metaphors and personifications of the kind: 'the term designates something', 'the sentence speaks about something', which are so common in the description of natural language, always imply the man who employs that term or sentence. Thus, if an analysis concerned with the use of expressions helps to decode that metaphor or personification, if the true relation between semantics or syntax and pragmatics is discovered, and if a way is paved for reducing semantic or syntactic problems to pragmatics, then this fact is worthy of emphasis and of a penetrating study. I think that this is the field where a solution is to be sought to the problem of the specific character of natural language and accordingly to the specific traits of its semiotics.

It now remains to me to apologize to the logicians who "hope to replace philosophizing by reckoning",[32] for not having fulfilled in these remarks what Gilbert Ryle called "the formalizer's dream",[33] a thing Ajdukiewicz cautioned against when he prophesied, and was partly right, that after he is dead, participants in logical conferences, instead of talking like normal people, would cover blackboards with formulae and symbols. This was not necessary here. Besides I incline toward the opinion that the logic of a natural language, and even the logic of the statements of scientists, cannot be adequately represented by means of formal logic.[34] It does not mean, however, that I think that reflections in the style of Church,[35] who tries to make an artificial language a model for natural languages, are fruitless. But studies within the scope of functional logical semiotics of natural languages neither start nor end here.

In conclusion, to banish the spectre of banality which, it may be, haunts my reasonings, I shall repeat that we should not fear to force open doors or to rediscover America as long as there is someone to whom the door is tightly shut and to whom America is not yet discovered.

8. *Summary*

Introduction. — The term 'functional', which occurs in the title of this paper, is neither synonymous nor equivalent with the same term as used by linguists. But the extensions of these two terms overlap.

1. *Use of an Expression.* — What is the functional approach to the logical semiotics of natural language concerned with? It is concerned with how a given expression is used. I mean thereby a certain use of an expression, and not its usefulness. Moreover, we are mainly concerned with the use of an expression as an instrument, but we shall also consider its use as an organ, ingredient, or designated object. The distinction between use and usage is important. When I speak of use I mean the following situation. A given expression, be it 'dog', has its established meaning in a given language, and in accordance with that meaning I may use that word so that it should refer to my dog Trot. Such a use is determined by a context and by an extra-linguistic situation. I may, of course, use the word 'dog' many a time. Now the class of the tokens of the expression 'dog', each used with reference to Trot, determines a certain use of that expression. Such a use of this word is in agreement with its usage, with the use of that word common in a given language, with the linguistic practice established for the word 'dog'. The term 'usage' may be considered a synonym of the term 'meaning' in one of the senses of the latter. When saying that the functional approach to the logical semiotics of natural language is concerned with how a given expression is used I mean above all the use of such an expression.

2. *Use of an Expression and the Classification of Terms with Regard to the Number of Designata.* — Names may be general, singular, and empty, according to the number of designata they have. But this classification does not adequately reflect the state of things in natural language. The names thus classified are separated from their context, and it is only their dictionary meaning that is taken into account, whereas a functional analysis ought to take into account the fact that the same word may be used in different ways, according to the extra-linguistic situation and the context. The word 'dog' is being used sometimes as a general, sometimes as a singular, and sometimes as an empty name, and sometimes so that the problem of its generality, singularity, and emptiness does not arise at all. The general, singular, and empty nature of a name is a relative property, connected with a given use of a given word.

3. *Use of an Expression and the Classification of Terms According to their Singular or General Intention.* — Nominal expressions are also classified according to their meaning intention by making a distinction between those to which we ascribe a singular intention ('Zeus'), those to which we ascribe a general intention ('centaur'), and those to which we ascribe an empty intention. But here too the use, which depends on the context and the extra-linguistic situation imparts to the same expression different intentions on different occasions. For instance, 'my brother' is an expression with a general intention when examined in isolation as a dictionary item, but in the sentence "This book was given to me by my brother" the expression acquires a singular intention.

4. *Use of an Expression and Classifications of Terms According to their Subjective or Predicative Character.* — Since Aristotle a distinction has been made between individual and general names. The former (e.g., 'Socrates') may occur only as the subject of an elementary singular proposition, while the latter (e.g., 'philosopher') may occur as predicate. But the functional analysis of words occurring in definite contexts and definite extra-linguistic situations reveals that these properties are associated with the use of expressions and that the same word may be used in either of these two functions.

5. *Use of an Expression and the Distinction: Proper Name — Name— Description.* — Since Frege and Russell a distinction has been made between proper names, names, definite descriptions, and indefinite descriptions. But from the point of view of the functional analysis of words we do not have here to do with mutually exclusive classes of expressions, but with different uses of expressions. Moreover, each of these uses is subject to gradation, so that a given expression can be used to a greater or a lesser extent as a description, a name, or a proper name, and also in a mixed function. This is why a given expression ought to be classified as used in a given respect in a given way in a given degree.

6. *Use of an Expression and the Sentence.* — The logical tradition dating back to Mill, Frege, and Russell treats language so as if it consisted solely or mainly of names. This is contrary to facts as far as natural

language is concerned, since the basic functions of that language are performed by sentences. This is why something I would term 'denominalization' is an essential item in the programme to make the semiotic analysis of natural language more functional. The distinctions into sentences and non-sentences, and also that into declarative, interrogative, and imperative sentences, lose their absolute character when examined from the point of view of the use of expression (i.e., in conformity with that programme). Declarative sentences ought not to be identified with propositions: I may use a declarative sentence to express a proposition, but I need not do so. When I use such a sentence so that it does not refer to anyone and does not serve to construct an assertion, it does not cease to be declarative sentence, but is not a proposition and the problem whether it is true or false does not arise at all.

7. *Use of Expressions and Functional Logical Semiotics of Natural Language.* — A given expression, when examined from the functional point of view, is not such-and-such, but happens to be used so-and-so in such-and-such respect, and in such-and-such degree. This is more general and can cover other examples than those discussed in the present paper. Relativization to use is something complementary to relativization to meaning, that is, to usage. Hence any disparagement of meaning is here quite out of the question. The requirement that use be taken into consideration is closely associated with this in that natural language by its very nature is occasional. This is why the relativization of the expressions examined as to their linguistic context and the extra-linguistic situation in which they occur is so important in semiotic analysis. Otherwise we would have to consider most expressions belonging to natural language as logically defective. Such a verdict would do injustice to natural language.

Notes

[1] Cf. Zawadowski, Leon, 'Główne cechy językoznawstwa funkcjonalnego' (The Principal Properties of Functional Linguistics). Introductory article in *Podstawy języka* (Fundamentals of Language) by Roman Jakobson and Morris Halle. Authorized Polish edition, revised and expanded, edited with footnotes and introductory article by Leon Zawadowski, Wrocław 1964, pp. 7–31.

[2] Cf. Ryle, Gilbert, 'Ordinary Language', *The Philosophical Review*, LXII (1953), 167–186; and in the anthology *Philosophy and Ordinary Language*, ed.

by Charles E. Caton (Urbana Ill., University of Illinois Press, 1963), pp. 108–127. Only the distinction between 'use' and 'utility' is according to Ryle. In distinguishing the use of a word in the role of an instrument, organ, ingredient and designatum I have gone beyond Ryle. In the distinction 'use' – 'usage' only the point of departure is common. Our views part company after that.

[3] Cf. Strawson, P.F., 'On Referring', *Mind*, LIX (1950), 320–344; and in the anthologies: *Essays in Conceptual Analysis*, ed. by Antony Flew, (London, 1956), pp. 21–52; and in *Philosophy and Ordinary Language*, pp. 162–193 (see footnote [2]).

[4] Ajdukiewicz, Kazimierz, *Pragmatic Logic*, (Warsaw–Dordrecht, 1974), p. 12. Cf. also pp. 7–56.

[5] See footnote [4].

[6] Cf. Ajdukiewicz, *Pragmatic Logic*, pp. 13–14 "To understand an expression *E* in its meaning *M* is the same as to understand it by means of a thought which, in respects that are essential to meaning, has certain properties. It will be said about those properties that they are expressed by the expression *E* in its meaning *M*. In other words, an expression *E expresses*, in a meaning *M*, a property of a thought *P* is the same as: if a person at a given moment understands the expression *E* in its meaning *M*, then he understands it by means of a thought which has the property *P*."

[7] See footnote [2].

[8] Ajdukiewicz, *Pragmatic Logic*, p. 32.

[9] *Op. cit.*, p. 34.

[10] *Op. cit.*, p. 47.

[11] See footnote [2].

[12] Cf. Ajdukiewicz, *Pragmatic Logic*, pp. 7–12.

[13] It might be objected that the example fits well only into those languages which, like Polish, have no articles; while in English and in other languages which make use of articles it is not the word 'dog' alone which occurs in sentences given above as instances, but 'the dog' or 'a dog', and it is the change of an article which accounts for the singular or general character of the use of the expression. I admit that languages without articles provide best examples of the change in use, discussed in this chapter. On the other hand, however, it seems that those examples could be expanded so that they would fit into languages which do have articles, e.g., English. In order to do that it would be necessary to consider articles 'the' and 'a', which occur in our examples, as parts of the context in which the analysed word 'dog' is embedded. Then it would be one and the same word 'dog', and not 'the dog' or 'a dog', which is used in different ways – as a singular, or general, or empty term, or as none of these.

[14] Kotarbiński, Tadeusz, 'Z zagadnień klasyfikacji nazw' (Selected Problems of the Classification of Terms), Łódź 1954, *Rozprawy Komisji Językowej Łódzkiego Towarzystwa Naukowego* (Proceedings of the Language Commission of the Łódź Scientific Society). Vol. I; reprinted in Kotarbiński, *Elementy teorii poznania logiki formalnej i metodologii nauk* (Elements of Gnosiology, Formal Logic, and Methodology of Sciences), Warsaw, 1961, pp. 461–462, and in *Gnosiology* –

the Scientific Approach to the Theory of Knowledge. Transl. by O. Wojtasiewicz; transl. ed. by G. Bidwell and C. Pinder, Oxford–Wrocław, 1966, Pergamon Press – Zakład Narodowy im. Ossolińskich, p. 392.

[15] Cf. Lewis, C.I., 'The Modes of Meaning', *Philosophy and Phenomenological Research*, IV, No. 2 (Buffalo, New York, 1943, pp. 236–249.

[16] See footnote [15]

[17] See footnote [14].

[18] Cf. Quine, Willard van Orman, *Word and Object* (New York, The Technology Press of M.I.T. and John Wiley, 1960), pp. 90–185).

[19] Cf. Russel, Bertrand, *Introduction to Mathematical Philosophy* (London, Allen and Unwin, 1919), Chapter XVI 'Descriptions'; and 'On Denoting' by the same author, *Mind*, XIV (1905).

[20] Cf. Frege, Gottlob, 'Über Sinn und Bedeutung', *Zeitschrift für Philosophie und Philosophische Kritik*, 100 (1892).

[21] See footnote[18].

[22] See footnote[19].

[23] Cf. Reichenbach, Hans, *Elements of Symbolic Logic* (New York, The Macmillan Company, 1948), pp. 256 ff.

[24] Cf. Ayer, Alfred J., 'Imiona własne a deskrypcje' (Proper Names and Descriptions), *Studia Filozoficzne* [Philosophical Studies], 5(20) Warszawa 1960, 135–155; and 'Names and Descriptions' by the same author, in the collection: *Thinking and Meaning: Entretiens d'Oxford 1962 Organisés par l'Institut International de Philosophie*, 5,20 (Paris, Ed. Neuvelaerts, 1963), pp. 199–202.

[25] Ajdukiewicz, *Pragmatic Logic*, p. 16,

[26] See footnote [4].

[27] See footnote [3].

[28] See footnote [3].

[29] Cf. Bar-Hillel, Yehoshua, 'Indexical Expressions', *Mind*, LXIII, No. 251 (1954), pp. 359–379.

[30] Cf. Frankena, William K., 'Some Aspects of Language' and: 'Cognitive and Noncognitive', in the volume *Language, Thought and Culture*, ed. by Paul Henle (Ann Arbor, The University of Michigan Press, 1958), Chapter V and VI, pp. 121–172.

[31] Cf. Austin, L. J., 'Performative – Constative', in the volume *Philosophy and Ordinary Language* (see footnote [2]), pp. 22–54.

[32] Ryle, 'Ordinary Language', p. 125: "The appeal to what we do and do not say, or can and cannot say, is often stoutly resisted by the protagonists of one special doctrine, and stoutly pressed by its antagonists. This doctrine is the doctrine that philosophical disputes can and should be settled by formalizing the warring theses. A theory is formalized when it is translated out of the natural language (untechnical, technical or semi-technical), in which it was originally excogitated, into a deliberately constructed notation, the notation, perhaps of *Principia Mathematica*. The logic of a theoretical position can, it is claimed, be regularized by stretching its non-formal concepts between the topic-neutral logical constants whose

conduct in inferences is regulated by set drills. Formalization will replace logical perplexities by logical problems amenable to known and teachable procedures of calculation (...). Of those to whom this, the formalizer's dream, appears a mere dream (I am one of them), some maintain that the logic of the statements of scientists, lawyers, historians and bridge-players cannot in principle be adequately represented by the formulae of formal logic. The so-called logical constants do indeed have, partly by deliberate prescription, their scheduled logical powers; but the non-formal expressions both of everyday discourse and of technical discourse have their own unscheduled logical powers, and these are not reducible without recourse to those of the carefully wired marionettes of formal logic. The title of a novel by A.E.W. Mason *They Wouldn't be Chessmen* applies well to both the technical and the untechnical expressions of professional and daily life. This is not to say that the examination of the logical behaviour of the terms of non-notational discourse is not assisted by studies in formal logic. Of course it is. So may chess-playing assist generals, though waging campaigns cannot be replaced by playing games of chess. I do not want here to thrash out this important issue. I want only to show that resistance to one sort of appeal to ordinary language ought to involve championing the programme of formalization. 'Back to ordinary language' can be (but often is not) the slogan of those who have awoken from the formalizer's dream. This slogan, so used, should be repudiated only by those who hope to replace philosophizing by reckoning".

[33] See footnote [32].

[34] See footnote [32].

[35] Cf. Church, Alonzo, 'The Need for Abstract Entities in Semantic Analysis', *Proceedings of the American Academy of Arts and Sciences*, No. 1 (1951). Also cf. Copi, Irving M., 'Artificial Languages', in the anthology *Language, Thought and Culture* (see footnote [30]), pp. 96–120.

THE PRINCIPLE OF TRANSPARENCY AND SEMANTIC ANTINOMIES

by

Leon Koj

Introduction

The task I have set myself in this paper is to investigate semantic antinomies and the names of expressions which occur in them in the light of what is termed the principle of transparency. This principle states that when we use words as signs of any objects, our attention is focussed on those objects, and not on the words we are using.

The principle of transparency has pride of place in discussions on signs. On the other hand, semantic antinomies are the main problem of semantics. This is why investigating the relationship between the two should prove interesting. The issue has not until now been taken up systematically, which makes such an investigation even more justified.

Our analysis of semantic antinomies and of the names of expressions that occur in them will be carried out as follows: I shall construct a miniature axiomatic system whose axioms are analogues of the various components of the principle of transparency; the theorems (obtained in that system) concerning names of expressions will be used in our investigation of semantic antinomies.

The system I suggest here is pragmatic and psychological in nature. The issues with which it is concerned have often been raised by philosophers and psychologists. The analysis I shall carry out makes use of certain methods developed in logic.

1. Axioms

The principle of transparency has been formulated in various ways. Three of these formulations will be selected for closer analysis, on the basis of which we shall establish our first axioms.

Here is Husserl's version of the principle of transparency:

"...wenn wir unser Interesse zunächst dem Zeichen für sich zuwenden, etwa dem gedruckten Wert als solchem (...) so haben wir eine aüssere Wahrnehmung (...) wie irgendeine andere, und ihr Gegenstand verliert den Charakter des Wortes. Fungiert es dann wieder als Wort, so ist der Charakter seiner Verstellung total geändert. Das Wort (...) ist uns zwar noch anschaulich gegenwärtig, es erscheint noch, aber wir haben es darauf nicht abgesehen, im eigentlichen Sinne ist es jetzt nicht mehr der Gegenstand unserer 'psychischen Betätigung'. Unser Interesse, unsere Intention, unser Vermeinen (...) geht ausschliesslich auf die im sinngebenden Akt gemeinte Sache."[1]

Ossowski writes as follows:

"We can adopt in turn two attitudes toward them (i.e., semantic entities — L.K.): we can use them as semantic entities, and a moment later treat them as ordinary material objects, without any semantic function whatever. Semantic entities have a peculiar property: once we want to predicate something about an object which is a semantic entity we immediately deprive it of its semantic function and adopt with respect to that object another attitude, i.e., we treat it as an object which is not a semantic entity. To be more picturesque we may say that a semantic entity is transparent to such operations as predicating, indicating, and naming, in a manner similar to that in which glass is transparent to sun-rays: through the intermediary of such an object predication may refer to another object, but we cannot predicate anything about it unless it ceases to perform the function which turns it into a semantic entity."[2]

It can be seen clearly that Ossowski's formulation is modelled on that of Husserl's, and does not contribute anything essentially new. In his paper Ossowski gives a kind of definition of a semantic entity, where we find a somewhat different description of transparency, but this latter description would appear to be incorrect.[3]

Finally, Schaff writes:

"'Transparency' to meaning, so characteristic of verbal signs, appears precisely when we completely cease to perceive the material shape of a sign (except for cases of disturbance in the normal process of communication) and are conscious only of its semantic aspect."[4]

Schaff's formulation turns out to be the most radical of the three:

he claims that when we use words as signs we practically do not perceive the words at all, but only their meanings. His statement differs from those made by Husserl and Ossowski in other respects too. While the other two are of the opinion that when we use a sign our attention is focussed on the object, Schaff says that our attention is concentrated on meaning. Furthermore, in Schaff's interpretation only spoken signs, i.e., certain sounds, are transparent;[5] according to Husserl and Ossowski, all language signs (semantic entities) — both sounds and inscriptions — are transparent.

Despite these differences the three authors agree on the essence of the phenomenon: (i) when something is a sign, our attention is directed to the fact indicated by that sign (or to the meaning of that sign, and through its intermediary to the fact); (ii) when something is a sign, we ignore it (in Ossowski's formulation: we neither predicate anything about it nor indicate it). By slightly modifying the terminology and making it resemble that of Ossowski rather than that of Husserl, we obtain the following formulation of the principle of transparency:

(1) If an object x is a sign of an object y, then we ascribe to y a property f.

(2) If an object x is a sign of an object y, then there is no property f such that we ascribe it to x.

Theorems (1) and (2) will be analysed and expanded until they yield the first axioms of our system. They are more suitable as an object of analysis than the original formulations of the principle of transparency, and they are not at variance with the basic intentions common to all three authors.[6]

On reflection we have to modify (1) and (2): the verb forms which occur in them indicate that if, for a person v, an object x is a sign of an object y, then v ascribes to x a property f; and if, for a person v, an object x is a sign of an object y, then v does not ascribe to x any property f.

Furthermore, Husserl's text makes it clear that the time factor should be taken into consideration in both (1) and (2). For we do not focus our attention on x only when x functions as a sign; when it is not a sign, i.e., when it does not function as such, it may be ascribed various properties.

Consequently, (1) and (2) become:

(3) If, at a moment t, an object x is, for a person v, a sign of

an object y, then there is a property f such that v ascribes f to y at t.

(4) If, at a moment t, an object x is, for a person v, a sign of an object y, then there is no property f such that v ascribes f to x at t.

Formulations (3) and (4) are rather lengthy, and hence the following abbreviations are to be recommended:

(5) '$Z(t, v, x, y)$' will be used for 'at a moment t, an object x is, for a person v, a sign of an object y'.

(6) '$P(t, v, fy)$' will be used for 'v ascribes f to y at t'.

We shall also adopt '$\rightarrow$' as the implication symbol and 'Σ' as the symbol of the existential quantifier. We can now write (3) and (4) as

(7) $Z(t, v, x, y) \rightarrow (\Sigma f)\, P(t, v, fy)$,

(8) $Z(t, v, x, y) \rightarrow \sim (\Sigma f)\, P(t, v, fx)$.

Now let us examine these formulas. The verb 'ascribes' occurs in both of them, and it is to be observed that it can be interpreted in two ways: we can ascribe something to something with, or without, conviction. In (7) and (8) we have in mind an ascription of properties where conviction is not necessary. It is our intention that ascribing a property to an object amounts to thinking that that object has that property. In order to avoid possible misunderstandings we may read '$P(t, v, fy)$' as follows: 'a person v ascribes (not necessarily with conviction) a property f to an object y at a moment t'. Since we are using Quine's logic as our basis (see below), the formulation 'y has a property f' is synonymous with the formulation 'y is in a class f'. Hence 'fy' means the same as '$y \in f$,' where 'f' ranges over the set of any classes one chooses. To simplify matters we retain 'fy' in '$P(t, v, fy)$' and in '$W(t, v, x, fy)$'. The latter formula will be explained later. To indicate that 'fy' is a sentential function we shall not insert a comma between 'f' and 'y'.

As well as specific constants, (7) and (8) include logical constants. These are used in the senses given to them by Quine in his *Mathematical Logic*. His system forms the logical basis of the theory under construction here. Quine's system, while not introducing types, makes it possible to avoid expressions that would be typically ambiguous. Variables can range over very comprehensive sets, which greatly simplifies substitutions. We shall avail ourselves of this property of Quine's system; for instance, the variable 'y' will range over a set that consists of classes and individuals (strictly speaking: of any classes, including singular classes of individuals such that each class is identical with its (only) element, i.e., an individual).

We must now form a clear understanding as to which sets the variables occurring in (7) and (8) range over. Variable 'v' is the easiest to describe: it ranges over the set of human beings.

Variable 't' ranges over the set of some moments. Are these time points? The texts quoted above are not explicit on this issue, but it seems legitimate to surmise that the principle of transparency does not refer to such moments. The time during which a sound functions as a sign is certainly not over long, but there are certainly moments which are shorter. For instance, the time during which the word 'Constantinople' functions as a sign in 'Constantinople was taken by the Turks in 1453' is presumably longer than the time during which the word 'dog' functions as a sign in 'the dog is an animal'. It would seem that the time during which we use a word as a sign depends on the time required to pronounce the sentence in which that word occurs. The variable 't' should therefore range over the set of time segments. The question naturally arises whether these can be time segments of arbitrary length, for instance, as long as a week or even a year. Let us reflect once more on the principle of transparency. One part of it states that during the time when a word functions as a sign it is not ascribed any property. If this were to be a whole week, we should not be in a position, during a whole week, to ascribe any properties to that word. It seems obvious that during a week a word can function as a sign and that nevertheless during that week we can ascribe certain properties to it. In the case of such long periods the principle of transparency would be false, since its antecedent would be true (for v, x would be a sign of y for a week), and its consequent would be false (v could ascribe some properties to x during that week). Since the authors of the principle of transparency accepted it as true, they must have interpreted it differently than we have just done.

Note that the formulation 'at a moment t, for a person v, an object x is a sign of an object y' is ambiguous. The first meaning is as follows. If, for instance, the word **'Peter'**[7] is a sign of Peter at 10:32 a.m. on

December 31, 1960, then according to this first meaning we may say that 'Peter' is a sign on December 31, 1960, that it is a sign in December 1960, that it is a sign in 1960, that it is a sign in the 20th century, etc. According to the first meaning, x functions as a sign in every time segment which contains that time segment in which x is used.

According to the second meaning of the formulation 'at a moment t, for a person v, an object x is a sign of an object y', what is meant is only that brief time segment in which x is actually used as a sign, i.e., that time segment which is as it were filled by the action of using x as a sign, in other words, that short time segment which is contained in all those time segments in which x functions as a sign in the first meaning of the formulation 'at a moment t'. In the second meaning, the formulation 'at a moment t, for a person v, an object x is a sign of an object y' is synonymous with 'for a person v, an object x is a sign of an object y during a whole period t'. For an object x functions as a sign during a whole period t if and only if t is a sufficiently short time segment, contained in all those time segments during which x functions as a sign in the first meaning of the formulation 'x is a sign at a moment t'. And if x is a sign during such a short period, then according to the second meaning it is a sign during the whole of that period.

We have concluded that the principle of transparency is false in the case of sufficiently long time segments if we give the first meaning to the formulation 'at a moment t, for a person v, an object x is a sign of an object y'. If this formulation is given the second interpretation, the principle of transparency proves not to be false even if names of arbitrary time segments are substituted for 't'. Hence the formulation under consideration is to be given the second meaning (interpretation), and 't' can then range over the set of all time segments. Accordingly, new conventions become valid, and the conventions (5) and (6), which might have led to difficulties, are annulled.

(9) '$Z(t, v, x, y)$' will stand for 'an object x is a sign of an object y for a person v during a whole period t'.

(10) '$P(t, v, fy)$' will stand for 'a person v ascribes a property f to an object y during a whole period t'.

A further analysis of the formula '$Z(t, v, x, y)$' shows that semantically it amounts to "for a person v, an object x indicates an object v during a whole period t", and yet, to avoid frequent changes in symbolism, we shall retain our notation. But, beginning with the paragraph preceding formula (12) and beginning with footnote [12], where the set over which 'y' ranges is fixed and where certain terminological conventions are introduced, 'Z' in '$Z(t, v, x, y)$' is to be interpreted as 'indicates', and not as 'is a sign'.

Note that, in their present interpretation, '$Z(t, v, x, y)$' and '$P(t, v, fy)$' can be defined by reference to '$Z(t, v, x, y)$' and '$P(t, v, fy)$', respectively, as interpreted in accordance with (5) and (6). To prevent misunderstandings, let us give these two formulas, interpreted in accordance with (5) and (6), a slightly different form, namely

$$Z_1(t, v, x, y)\,,$$
$$P_1(t, v, fy)\,.$$

We shall also let the symbol 'c' stand for the inclusion of segments. In the light of what has been said above, we can formulate the following definitions:

$$Z(t_1, v, x, y) \equiv Z_1(t_1, v, x, y)\cdot \sim (\Sigma t_2)\,(t_2\, c\, t_1 \cdot t_2 \neq t_1 \cdot Z_1(t_2, v, x, y)),$$
$$P(t_1, v, fy) \equiv P_1(t_1, v, fy)\cdot \sim (\Sigma t_2)\,(t_2 c\, t_1 \cdot\, t_2 \neq t_1 \cdot P_1(t_2, v, fy)).$$

'$Z_1(t, v, x, y)$' and '$P_1(t,v, fy)$' can be defined in a similar manner by reference to '$Z(t, v, x, y)$' and '$P(t, v\; fy)$', respectively.

In accordance with the interpretation adopted for '$Z(t, v, x, y)$', the statement 'during the whole of 1960, for a person A, the word "**Peter**" was a sign of Peter' is false. But a statement which says that during the whole of 1960 the person A had a disposition to treat the word "**Peter**" as a sign of Peter may be true. By adopting the formulas '$Z(t, v, x, y)$' and '$P(t, v, fy)$' as translations of the corresponding parts of the formulation of the principle of transparency we declare ourselves in favour of the actual, and not potential, interpretation of that principle.[8]

Next we have to consider the set over which the variable 'x' should range. The quotation from Husserl indicates that he means any objects that can function as signs — apparently inscriptions, sounds, and other kinds of objects as well. The situation is different with Schaff, for whom the validity of the principle of transparency is restricted to sounds.[9] Regardless of whether the principle of transparency is true only for sounds, or for other kinds of signs as well, let 'x' range over the set consisting of sounds and inscriptions.

It is to be emphasized that 'x' is meant to range over the set of singular, specified sounds and inscriptions, that is, *sign-events*, and not *sign-designs*.[10] We may substitute for 'x' only names of specified sounds or inscriptions, located in time and space. We cannot substitute for it singular names of shapes, i.e., singular names of certain abstract entities, singular names of classes of sounds or classes of inscriptions (if we define a shape as a class of equiform inscriptions or a class of equisonant sounds).

As we know, both single words and linguistic entities consisting of several words each can function as signs. Linguistic entities consisting of several words include, for example, compound names such as 'Isaac Newton', 'the winner in the fifth international chess tournament', etc. If we agree that sentences are signs, then we have to conclude that most sentences, and signs in general, are compound entities. Since 'x' is meant to range over the set of sounds and inscriptions, the point is that this set should include all signs. We must therefore define a sound or an inscription so that not only single vowels and consonants and not only single non-compound words, but also compound names and sentences, will be sounds or inscriptions, as the case may be.

One more consideration should be taken into account when a sound or inscription is being defined. As we shall see later, the concept of part will prove necessary. This concept can be introduced if we base the system we are constructing here on mereology. This, would, however, greatly complicate the formal aspect of our analyses. But in certain situations the concept of part can be identified with that of segment. For instance, a segment of a straight line is part of that straight line. In order to make such an identification in the case of a sound (an inscription) we have to assume that (i) a sound (an inscription) is an ordered set, so that, for instance, a sound can be interpreted as a set of vibrations of the air, ordered according to the relation of precedence; (ii) that any vibration of the air which is later than any other vibration which is a component of that sound and earlier than any other vibration which is a component of that sound is also a component of that sound. Finally, we shall assume that a sound must have a first and a last element. The definition of an inscription must satisfy the same conditions.

Formulating the definitions of a sound and an inscription that will meet these requirements is a very intricate, but nevertheless realistic task. Definitions of a sound and an inscription will not be required in the further course of our considerations. It is enough to assume that (a) compound sounds and compound inscriptions are sounds and inscriptions, respectively; (b) every element which lies between any components of a sound (an inscription) is also a component of that sound (inscription);[11] (c) every sound (inscription) has a first and a last element. These assumptions suffice to demonstrate, by reference to the appropriate definition of a segment, that if x_1 and x_2 are sounds, and if x_1 is contained in x_2, then x_1 is a segment of x_2. Hence, if 'x' ranges over the set of sounds and inscriptions, then the formula '$x_1 \subset x_2$', which usually indicates containment, may be interpreted as 'x_1 is a segment (part) of x_2'.

Do (7) and (8) exhaust the meaning of the principle of transparency? Apparently they do not. Schaff claims that we fail to perceive physical aspects of a sign when we make use of it. The statements by Husserl and Ossowski reveal the same intention, even though their formulations are not so radical. But if we fail to perceive the physical aspects of a sign, then we should not perceive its parts either. The *sui generis* "prohibition" of perceiving the parts of a sign is certainly implicit in the principle of transparency, but it follows from neither (7) nor (8). Furthermore, we should also fail to perceive those expressions of which a given sign is part. Whenever we use a sign, it becomes more or less invisible as we know. It is then impossible to perceive the whole in which a sign might be embedded, since otherwise we should perceive the sign, too, or at least we should be in a position to perceive it. We thus have the second prohibition implied by the principle of transparency. We may look for other prohibitions as well in this principle. Note, for instance, that if a sign x denotes a class y, then a certain property is ascribed to y. Suppose now that the property ascribed to y is: x is an element of y. That property is identical with a property of x, namely the property that x is an element of y. In that case, however, contrary to (8), x would be a sign and would be ascribed a property, namely the property of being an element of y. In short, the class y denoted by x cannot have x as its element. Finally we have the last prohibition: the object y denoted by x cannot be in x. This is obvious: since x is a class of vibrations of the air or a class of particles of chalk, if y

were in x, then it would be a vibration of the air or a particle of chalk, and thus would be part of x. We do know, however, that we are not in a position to perceive any part of a sign while it is being used by us as a sign. By making use of the symbolism introduced earlier in this paper we can write these four prohibitions as a single theorem:

$$(11) \qquad Z(t, v, x, y_1) \rightarrow (y_2\, cx \vee xcy_2 \vee y_2 \in x \vee x \in y_2) \rightarrow \sim (\Sigma f)\, P(t, v, fy_2)).$$

We thus have three theorems which, taken together, express the meaning of the principle of transparency, namely (7), (8), and (11). They may be reduced to two, since (8) follows from (11). Hence (7) and (11) are required to express the intuitions inherent in the principle of transparency.

None of the three authors quoted above says precisely what kinds of facts can be indicated by signs.[12] We can only guess that the facts indicated by signs can be classes, individuals, or states of things. In that case 'y' would be at the same time a term variable and a sentential variable. Such an interpretation of 'y' could result in various difficulties.[13] It is better to disregard in our axioms the sweeping generalization implied by the selected formulations of the principle of transparency. Therefore, let, 'y' range over the set of any classes. In this way (7) and (11) are deprived of the generalization implied by the principle of transparency. In order to grasp the whole sense of this principle we shall introduce a new primitive concept, namely that of referring, in the formulation 'an object x refers to a state of things fy for a person v during the whole period t' — in symbolic, notation, '$W(t, v, x, fy)$'. The introduction of this new concept enables us to complete (7) and (11). So far we have been speaking about names or terms, and now we can also characterize sentences as signs. The meaning of the principle of transparency, as applied to sentences, is wholly analogous to (7) and (11). The former has its analogue in

$$(12) \qquad W(t, v, x, fy) \rightarrow P(t, v, fy),$$

the sense of which is more or less as follows: during the time when a sentence x refers to fy one thinks that y has the property f. The latter has its analogue in

$$(13) \qquad W(t, v, x, fy_1) \rightarrow (y_2 cx \vee xcy_2 \vee y_2 \in x \vee x \in y_2 \rightarrow \sim (\Sigma f)\, P(t, v, fy_2)).$$

Our interpretation of (13) resembles that of (11): we do not think about parts and elements of the sentence which we are using at a given moment; nor do we think about the wholes (sets) of which the sentence which at a given moment functions as a sign, is a part (element).

It seems that (7), (11), (12) and (13) exhaust the meaning of the principle of transparency. They will be adopted here as axioms.

The constants 'Z', 'W' and 'P' are not sufficiently characterized by the axioms adopted above; but the intuitions inherent in the principle of transparency do not tell us anything about any possible strengthening of the system of axioms. This can be done by reference to intuitions quite different from those mentioned so far. For instance, it would appear that the equivalence 'Shakespeare is the author of *Macbeth* ≡ ≡ Shakespeare is the author of *Hamlet*' does not yield the equivalence 'I think that Shakespeare is the author of *Macbeth* ≡ I think that Shakespeare is the author of *Hamlet*'. Hence it would appear that the constant 'P' is intensional in nature and that the following theorem holds:

(14) $$\sim ((f_1 y_1 \equiv f_2 y_2) \rightarrow (P(t, v, f_1 y_1) \equiv P(t, v, f_2 y_2))).$$

It would also appear that during a sufficiently short period we are able to have only one thought; if this is true, then the following theorem holds:

(15) $$P(t, v, f_1 y) \cdot P(t, v, f_2 y) \rightarrow f_1 = f_2.$$

It would be advantageous to formulate a theorem concerning a connection between the relation Z and the relation W. This could perhaps be formulated as follows: if x is a sentence which refers to fy, then there is part of that sentence (which need not be its proper part and need not be singled out graphically as a separate word) which indicates y. The corresponding theorem would be:

(16) $$W(t, v, x_1, fy) \rightarrow (\Sigma x_2)(x_2 c x_1 \cdot Z(t, v, x_2, y)).$$

The intention underlying this statement is very simple: every sentence speaks about something — about an object — and hence it must contain a term, even if it is not marked explicitly in the grammatical sense of the word.[14]

As we are interested here mainly in the principle of transparency, we shall not include theorems (14)–(16) in our axiom system, but leave them as informal suggestions as to how the terms '*Z*', '*W*', and (especially) '*P*' might be given more precision.

Our proposed axiom system is consistent. Let the variables range over non-empty sets of objects described in any consistent theory T, and let the constants "Z", "W" and "P" stand for empty relations in that consistent theory. Formulas (7), (11), (12), (13) will then have false antecedents for all substitutions for variables, and being implications with false antecedents, they will be true in T. Our axioms thus have an interpretation in a consistent system, and therefore they themselves form a consistent system.

2. *Theorems and definitions*

The principle of transparency is one of the few statements which are almost universally accepted. In Sec. 1 we strove to offer an accurate and precise formulation of this principle. As the principle itself does not raise great doubts, the axioms, as its exact reflections, should prove equally intuitive. In fact it is otherwise: even the axioms may give rise to objections, and theorems may be still more controversial. Is it the principle of transparency which is responsible for the doubts, or is it the interpretation of it suggested in this paper? This question will be left unanswered here. In any case it would be well to reflect on these theorems, which seem to be at variance with our intuitions.

Let us first recall the axioms:

A1 $$Z(t, v, x, y) \rightarrow (\Sigma f)\, P(t, v, fy),$$

A2 $$Z(t, v, x, y) \rightarrow (y_2 c x \vee x c y_2 \vee x \in y_2 \vee y_2 \in x \rightarrow \sim (\Sigma f)\, P(t, v, f y_2)),$$

A3 $$W(t, v, x, fy) \rightarrow P(t, v, fy),$$

A4 $$W(t, v, x, fy) \rightarrow (y_2 c x \vee x c y_2 \vee x \in y_2 \vee y_2 \in x \rightarrow \sim (\Sigma f)\, P(t, v, f y_2)).$$

Below are some of the theorems; their proofs, being very simple, are omitted.

T1 $$Z(t, v, x_1, y_1) \cdot x_2 c y_1 \rightarrow \sim Z(t, v, x_2, y_2),$$

T2 $$Z(t, v, x_1, y_1) \cdot y_1 c x_2 \rightarrow \sim Z(t, v, x_2, y_2),$$

T3 $$Z(t, v, x_1, y_1) \cdot x_2 \in y_1 \rightarrow \sim Z(t, v, x_2, y_2),$$

T4 $$Z(t, v, x_1, y_1) \cdot y_1 \in x_2 \rightarrow \sim Z(t, v, x_2, y_2),$$

T5 $Z(t, v, x, y_1) \rightarrow \sim xcy_1$,

T6 $Z(t, v, x, y_1) \rightarrow \sim y_1cx$,

T7 $Z(t, v, x, y_1) \rightarrow \sim x \in y_1$,

T8 $Z(t, v, x, y_1) \rightarrow \sim y_1 \in x$,

T9 $\sim (Z(t, v, x_1, x_2) \cdot Z(t, v, x_2, y))$,

T10 $\sim (Z(t, v, x_1, x_2) \cdot Z(t, v, x_2, x_1))$,

T11 $Z(t_1, v, x_1, x_2) \cdot Z(t_2, v, x_2, y) \rightarrow t_1 \neq t_2$,

T12 $\sim Z(t, v, x, x)$,

T13 $\sim W(t, v, x, fx)$,

T14 $W(t, v, x, fy) \rightarrow \sim ycx$,

T15 $W(t, v, x, fy) \rightarrow \sim xcy$,

T16 $W(t, v, x, fy) \rightarrow \sim x \in y$,

T17 $W(t, v, x, fy) \rightarrow \sim y \in x$.

Theorems analogous to T5–T17 are obtained on the strength of the definitions of denoting, designating, and naming. The definitions, and some of the theorems, will be discussed later.

D1 $Den(t, v, x, y_1) \equiv Z(t, v, x, y_1) \cdot (\Sigma y_2)(y_2 \neq y_1 \cdot y_2 \in y_1 \cdot$
$\cdot Z(t, v, x, y_2)) \cdot (y_3)(Z(t, v, x, y_3) \cdot$
$\cdot \sim (y_3 \in y_1) \rightarrow y_3 = y_1)$,

D2 $Des(t, v, x, y_1) \equiv Z(t, v, x, y_1) \cdot (\Sigma y_2)(y_1 \in y_2) \cdot$
$\cdot Den(t, v, x, y_2))$,

D3 $Nam(t, v, x, y_1) \equiv Z(t, v, x, y_1) \cdot (y_2)(Z(t, v, x, y_2)$
$\rightarrow y_1 = y_2)$,

T18 $\sim Den(t, v, x, x)$,

T19 $\sim (Den(t, v, x_1, x_2) \cdot Den(t, v, x_2, y))$,

T20 $Den(t, v, x, y) \rightarrow \sim xcy$,

T21 $Den(t, v, x, y) \rightarrow \sim ycx$,

T22 $Den(t, v, x, y) \rightarrow \sim x \in y$,

T23 $Den(t, v, x, y) \rightarrow \sim y \in x$,

T24 $Den(t, v, x_1, y_1) \cdot x_2 \in y_1 \rightarrow \sim Den(t, v, x_2, y_2)$,

T25 $\sim Des(t, v, x, x)$,

T26 $\sim ((Des(t, v, x_1, x_2) \cdot Des(t, v, x_2, y))$,

T27 $Des(t, v, x, y) \rightarrow \sim xcy$,

T28 $Des(t, v, x, y) \rightarrow \sim ycx$,

T29 $Des(t, v, x, y) \rightarrow \sim x \in y$,

T30 $Des(t, v, x, y) \rightarrow \sim y \in x$.

Similar theorems are obtained when we make use of the definition of naming; they are not worth listing here, since they can easily be reconstructed.

Further definitions introduce semantic concepts of descriptive semantics. The nature of the theorems changes correspondingly.

D4 $x \; Den \; y \equiv (\Sigma t, v) \; Den \, (t, v, x, y)$,
D5 $x \; Des \; y \equiv (\Sigma t, v) \; Des \, (t, v, x, y)$,
D6 $x \; W fy \equiv (\Sigma t, v) \; W \, (t, v, x, fy)$,
T31 $x \; Den \; y \rightarrow \sim xcy$,
T32 $x \; Den \; y \rightarrow \sim ycx$,
T33 $x \; Den \; y \rightarrow \sim x \in y$,
T34 $x \; Den \; y \rightarrow \sim y \in x$,
T35 $\sim x \; Den \; x$,
T36 $x \; Des \; y \rightarrow \sim xcy$,
T37 $x \; Des \; y \rightarrow \sim ycx$,
T38 $x \; Des \; y \rightarrow \sim x \in y$,
T39 $x \; Des \; y \rightarrow \sim y \in x$,
T40 $\sim x \; Des \; x$,
T41 $x \; W fy \rightarrow \sim xcy$,
T42 $x \; W fy \rightarrow \sim ycx$,
T43 $x \; W fy \rightarrow \sim x \in y$,
T44 $x \; W fy \rightarrow \sim y \in x$,
T45 $\sim x \; W fx$.

Let us now analyse the intuitive sense of some of these theorems, and let us also consider examples which would appear to contradict them.

We begin with T5–T8. Their essential meaning is as follows: if, at some time, x is a sign of y for someone, then x is not part of y; nor is y part of x; nor is x an element of y; nor is y an element of x.

We now focus our attention on examples which seem to refute T5–T8.[15] Let y be a speech, and x, a name of (reference to) that speech used in the speech itself, for instance:

(1) My present speech outlines the basic goals of our organization.

Now (1) includes a name x of the speech y:

(2) My present speech.

By virtue of T5, if (2) is in fact a name of the speech in question, then (2) cannot be part of that speech. And if we assume that x is part of the speech y, then x is not a name of the speech y. Do not (1) and (2) refute T5? They do, on condition that (2) denotes the whole speech, including (1) and (2). But if we assume that (2) denotes the whole speech except for (1), and hence except for (2), then T5 is not refuted, for in that case (2) would denote a whole of which (2) is not part.

Let us consider now an example which illustrates the difficulties connected with T6:

(3) The only capital letter in line 9.

It is the letter 'T' which should be the designatum of (3), and that letter is, as we can see, part of (3). We thus have an example of a name which supposedly denotes its own part. But such a possibility is excluded by T6. If (3) is a name of 'T', then, by virtue of T6, 'T' is not part of (3). And if 'T' is part of (3), then (3) is not a name of 'T'. Since 'T' certainly is part of (3), then, by virtue of T6, (3) is not a name of 'T'. On the other hand, it is a fact that, having written (3), we know that it is 'T' which should be the designatum of (3). Thus (3) seems to be a name of 'T', despite T6. An argumentation in favour of the validity of T6 will not be easy in view of this state of things.

The present writer personally has no doubt that T6 reflects the intentions underlying the principle of transparency, namely the claim that we ignore x (and hence also the parts of x) when x functions as a sign. It would not seem that the difficulties we encounter here can be due to a misinterpretation of the principle of transparency. If this principle accurately describes properties of signs, then the difficulty can only be due to a superficial analysis of (3). Let us therefore return to (3).

When we read (3) we of course do not know what letter is meant, because we do not know where line 9 is. We therefore look for that line and repeat the name 'The only capital letter in line 9', i.e., we use the following sign-events:

(4) The only capital letter in line 9.

(5) The only capital letter in line 9.

Etc.

Every new copy of that name is used later than (3), and if it were to be written down, it would occur in a line that comes after line 9. These other copies of the name 'The only capital letter in line 9' do not occur in line 9 and can, therefore, unreservedly denote the only capital in (3). What, however, does (3) denote? Let us consider how we look for line 9. In a natural way, we look for it outside line 9, since in accordance with the principle of transparency the semantic intention is directed outside the word(s) we are reading at a given moment. Obviously, there is no line 9 that could be situated in any other place than (3). There is, accordingly, no only capital letter that could occur in such a non-existent line. It follows from this that (3) denotes an empty class. Now, (3) and (5) being equiform, we do not distinguish one from the other and are therefore inclined to think that (3) denotes its own part, namely the letter 'T'. The identification of (3) and (5) is made even easier by the fact that (5) is repeated only in our minds, and the lapse of time between our using (3) and (5) is usually very small.

A similar process presumably takes place in the case of examples which more or less differ from (3).

Note that the analysis we have carried out does not prove the validity of T6, since it is based on the principle of transparency, from which T6 follows.

It has not been proved that (3) is in fact an empty term. It has only been shown that the supposition that (3) might be an empty term is not groundless. Hence the objections against T6 have been deprived of some of their strength. Further arguments in favour of T5 and T6 will be given later.

In connection with T7 let us consider a new example:

(6) Expression

Now (6) should denote a set y of all expressions. Obviously, (6) is itself an expression, and hence it should be in the set y denoted by (6). In other words, the following theorem should hold:

(7) $(\Sigma t, v)\, Z\,(t, v, (6), y) \cdot (6) \in y.$

Now (7) is in conflict with T7. Resolving this obstacle depends in this case on our assuming that (6) denotes the class of all expressions except (6). In that case it is not true that (6) is an element of y, and T7 remains valid. The consequences of the solution we have suggested are far-reaching, for consider

(8) Expression

Now (8) denotes the class of all expressions except (8). The class denoted by (8) includes (6), and the class denoted by (6) includes (8) but does not include (6). The classes denoted by (6) and (8) respectively are different, even though (6) and (8) are equiform. If we were to quote other words equiform with (6), then each of them would denote a slightly different class. None of those equiform words would have the same extension as any other, and, to make matters worse, it would not even be possible for the extensions of two such words to be identical. The difference of extensions, and accordingly of meanings, of equiform words is extremely embarrassing, since it is at variance with the fundamental principle on which, among other things, the rule of detachment (the *modus ponens*) is based: equiform expressions are synonymous, and thus equiform statements are equivalent. The foundation on which formal logic to a certain extent rests — namely that operations on equiform expressions are operations on the same meanings — is incompatible with the solution that the various copies of the word 'Expression' should denote different classes and thereby have different meanings.

We shall try to remove this obstacle. First of all, we shall assume unreservedly that equiform expressions are synonymous. Note that (6) denotes a class y_1, which does not include (6); furthermore, (8) denotes a class y_2, which does not include (8). This has already been established. We also know that y_1 and y_2 should be equal in view of the equiformity of (6) and (8). Hence y_1 should not include (8), and y_2 should not include (6), for then y_1 and y_2 would not differ by their elements (6) and (8). Moreover, if we were to take into consideration more words equiform with (6), and hence also equiform with (8), y_1 should not include any expression equiform with (6) and (8). Likewise, the denotata of y_3, y_4, ..., y_n, the latter being equiform with (6), should include neither (6), nor (8), nor any expression equiform with (6). In such a way the denotata of terms equiform with (6) would not differ as to their elements: the denotata of expressions equiform with (6) would be identical. This shows a way in which T7 can be brought into line with the principle that equiform expressions are synonymous. The removal of the obstacles encountered earlier is possible, even though it can take place only at the cost of an additional restriction of the extensions of some expressions.[16]

Having commented on T7 and having demonstrated that it need not necessarily be false we shall proceed to discuss other theorems.

T8 creates the same problems that T6 created, and it can be defended in a like manner. Furthermore,

T12 $\sim Z(t, v, x, x)$

seems to be at variance with our linguistic intuitions. We shall discuss it now in connection with the principle of the synonymity of equiform expressions. According to T12, no expression can be its own sign. We can immediately adduce examples which suggest that some names indicate themselves, for example:

(9) That which is written in line 9 from the top.

We want to demonstrate first that (9) resembles (3) and that the comments made about (3) also apply to (9). As in the case of (3), in (9) the semantic intention is directed outside (9). This is why (9) is an empty term, for there is no line 9 (on this page) outside the line in which (9) is written. Written and spoken expressions which are equiform with (9) but have been formulated later than (9) can, of course, indicate (9). Non distinction between (9) and those expressions which are equiform with (9) accounts for the impression that (9) indicates itself. Here we are again dealing with an infringement of the principle that equiform expressions are synonymous. The solution which we shall suggest here will apply to both (9) and (3).

How can we make (9) synonymous with the expressions which are equiform with it? In view of T12 we cannot assume that (9) indicates that which is indicated by those expressions which are equiform with it, i.e., that (9) indicates (9). The other possibility left is to assume that the expressions equiform with (9) indicate that which is indicated by (9), that is, an empty class. This second solution is, however, artificial. Why should the name

(10) That which is written in line 9 from the top

be an empty class? Is it only because in line 9 of that page there is an empty term which incidentally has the same shape? And if something else were written in that line, would (10) not be an empty class? Both solutions are unacceptable. We can only assume that (9) is incomplete as a formulation, and that it should look like this:

(11) That which is written in line 9 from the top, somewhere before the colon or after 'W'.

Now, (9) and (11) differ from one another in that (11) includes an explicit formulation of the semantic intention which in (9) was merely implied. Since (9) is written after the colon and before 'W', the additional formulation "somewhere before the colon or after 'W'" simply states that the designatum of (9) is outside the line in which (9) is written. We can expand (10) in a similar manner, thereby yielding:

(12) That which is written in line 9 from the top, somewhere before the colon or after 'O'.

On comparing (11) with (12) we immediately see that they are not equiform, and hence they may indicate different objects. Thus, incomplete formulations, which unfortunately happen to have identical forms, may also indicate different objects.[17]

We have seen that the consequences of the interpretation adopted above of the principle of transparency give rise to very serious doubts. The difficulties which we have observed have not been completely removed, and only the possibility of doing so has been indicated. The present writer intends to treat the objections raised here in greater detail in two separate papers: one dealing with token-reflexive terms, and the other, with quoted terms.

Regardless of the doubts which may arise, the theory constructed here has certain advantages, which are brought out in full relief when this theory is compared with the theory of degrees of language.[18] First of all, it must be said that all the objections raised against T5 – T12 are equally applicable to the theory of degrees of language. In the latter theory, (2), (3), (6), (9) are meaningless. Let us recall the examples which were supposed to refute T5–T12. First we have:

(2) My present speech.

Now, (2) is a name of expressions and according to the theory of degrees of language forms part of a certain metalanguage. The speech indicated by (2) belongs to a corresponding object language, and hence the metalinguistic expression (2) cannot be part of that speech. The expression

(3) The only capital letter in line 9

is a metalinguistic term. Its designatum should belong to object language, and as such it cannot be part of (3). The word

(6) Expression

is likewise a metalinguistic entity, and it cannot indicate itself, since (6) indicates expressions in object language. Likewise, in the case of

(9) That which is written in line 9 from the top

metalanguage is confused with object language.

And yet the sense of T5–T12 does not coincide with the corresponding prohibitions formulated in the theory of degrees of language. One of the differences consists in the fact that T5–T12 have a certain psychological justification in the form of the principle of transparency. The theory of degrees of language is an *ad hoc* structure intended to bar semantic antinomies. Moreover, T5–T12 restrict the operation of constructing names of expressions less than the theory of degrees of language does. In accordance with T5–T12, we may construct names of expressions which denote any classes of expressions, provided that elements of the classes so denoted are not their names or other expressions connected with those names by the relation of inclusion or containment. In other words, x may be a name of any class y provided that neither x nor any expression connected with x by the relation of inclusion or containment is in y. According to the theory of degrees of language, not only are x and the expressions connected with x by the relation of inclusion or containment not elements of the class y denoted by x, but neither are any expressions which are in the same metalanguage to which x belongs.

Let us now consider several other theorems.

The sense of T14–T17 is very simple: an object y, to which a sentence x refers, cannot be contained in, nor be an element of, x. Likewise, x can be neither an element of, nor contained in, y. If y is a sound or an inscription, then y cannot be part of x, nor can x be part of y. T14–T17 do not give rise to any additional doubts over and above those discussed earlier in connection with T5–T12.

We shall refer to T9 later, in connection with the antinomy of heterological words.

Now A1–A4 can be used to prove theorems in which defined terms occur. In accordance with the convention adopted earlier, a sign denotes its extension. The constant 'Z' stands for 'indicates', the meaning of which has also been discussed provisionally. Hence, if x denotes y, then x thereby indicates y. If x denotes y, then it also indicates elements of y. Note, too, that during an appropriately short period of time x

can denote only one class. This yields the definition of denoting given by D1 above. The reservation that we are dealing with elements here, i.e., with classes which are in the full set, is omitted here; this reservation is necessary in Quine's system.[19]

If x designates y_1, then x obviously indicates y_1. Moreover, y_1 is an element of the extension of x. These intuitions suffice to formulate D2, i.e., the definition of designating.

If x names y, then y is the only object indicated by x. This enables us to formulate D3, i.e., the definition of naming.

T5–T8 have their analogues in theorems on denoting, designating, and naming.[20] By way of example, let us return to the theorems on denoting given previously. Now, T27–T30 contribute nothing new as compared with T5–T8, but T27–T30 do offer a good opportunity to move from pragmatics to semantics. To do this we shall make use of D4, which defines denoting as a semantic relation, that is, a two-place relation that holds between an expression and the corresponding object. The concepts thus obtained belong to what Carnap has termed descriptive semantics. The transition from pragmatic concepts to semantic ones is done simply by binding the variables 't' and 'v', so that D1 becomes D4. Furthermore, by reference to T27–T30 we immediately obtain T31–T34.

By introducing other definitions of semantic concepts we can prove the semantic analogues of pragmatic theorems.

So far we have focussed our attention on theorems which are particularly useful in a discussion of semantic antinomies. It is worth while mentioning that the new definitions indicated above enable us to formulate theorems which are of a different nature.

3. *Names of expressions in semantic antinomies*

The role of expressions in semantic antinomies will be investigated by the example of two well-known antinomies, namely that of heterological expressions and that of the liar. The former will be considered in the form given it by I. Copi, who follows Ramsey in this connection;[21] the latter, in the form familiar from Tarski's study of the concept of truth.[22]

The antinomy of heterological terms is based on the definition of those terms, which have the property that none of them belongs to the class it denotes. Here is the definition of those terms, which exactly follows Copi's formulation:

$$x \in het \equiv (\Sigma y)\,(x \text{ denotes } y \cdot (z)\,(x \text{ denotes } z \equiv z = y) \cdot \cdot \sim x \in y).$$

The definiens of this definition states that x denotes something, namely y; that x denotes y only, for whatever is denoted by x is identical with y; and that x is not in y, i.e., is not in the class which it denotes.

Copi obtains a contradiction from this definition. Here is an exact rendering of his reasoning.

Case 1

(a) $\text{`}het\text{'} \in het \rightarrow (\Sigma y)\,(\text{`}het\text{' denotes } y \cdot (z)\,(\text{`}het\text{' denotes } z \equiv z = y) \cdot \sim \text{`}het\text{'} \in y),$

(b) $\text{`}het\text{'} \in het \rightarrow \text{`}het\text{' denotes } y \cdot (z)\,(\text{`}het\text{' denotes } z \equiv z = y) \cdot \cdot \sim \text{`}het\text{'} \in y,$

(c) $\text{`}het\text{'} \in het \rightarrow \text{`}het\text{' denotes } y \cdot (\text{`}het\text{' denotes } het \equiv het = y) \cdot \cdot \sim \text{`}het\text{'} \in y,$

(d) $\text{`}het\text{'} \in het \rightarrow (\text{`}het\text{' denotes } het \equiv het = y \cdot \sim \text{`}het\text{'} \in y),$

(e) $\text{`}het\text{'} \in het \rightarrow het = y \cdot \sim \text{`}het\text{'} \in y,$

(f) $\text{`}het\text{'} \in het \rightarrow \sim \text{`}het\text{'} \in het.$

Case 2

(a) $\sim \text{`}het\text{'} \in het \rightarrow (y) \sim (\text{`}het\text{' denotes } y \cdot (z)\ \text{`}het\text{' denotes } z \equiv z = y) \cdot \sim \text{`}het\text{'} \in y),$

(b) $\sim \text{`}het\text{'} \in het \rightarrow (\text{`}het\text{' denotes } het \rightarrow \sim ((z)\,(\text{`}het\text{' denotes } z \equiv z = \text{`}het\text{'}) \cdot \sim \text{`}het\text{'} \in het)),$

(c) $\sim \text{`}het\text{'} \in het \rightarrow ((z)\,(\text{`}het\text{' denotes } z \equiv z = het) \rightarrow \text{`}het\text{'} \in het),$

(d) $\sim \text{`}het\text{'} \in het \rightarrow (z)\,(\text{`}het\text{' denotes } z \equiv z = het)$ (on the assumption that 'het' is unambiguous),

(e) $\sim \text{`}het\text{'} \in het \rightarrow \text{`}het\text{'} \in het.$

Now, (f) of Case 1 and (e) of Case 2 yield:

$$\text{`}het\text{'} \in het \equiv \sim \text{`}het\text{'} \in het.$$

It would appear that one of the basic steps in the reasoning shown above is the substitution for 'x' of the expression "heterological".[23] As a result of this operation the definiendum has been transformed into

(1) 'heterological' ∈ heterological.

The definition of heterological terms yields a contradiction on the obvious condition that (1) is a statement. Now, (1) has to be a statement both syntactically and semantically. We shall say that x is a semantically meaningful statement if we can truly predicate about x that it describes a certain state of things. Since the predicate in 'x describes a certain state of things' is true, that statement cannot result in a contradiction to the theorems which we hold to be true. If it did result in a contradiction to those theorems, this would show that the statement 'x describes a certain state of things' is false and that x is not a semantically meaningful statement. To return to (1): if it merely possessed the syntactic structure of a statement, but were not a statement in the semantic sense of the term, then there would, of course, be no contradiction at all. We arrive at a contradiction only when we find that two contradictory states of things are said to take place simultaneously. The inscription $\ulcorner x \equiv x \urcorner$,[24] where x does not describe any state of things, even though it may have the form of a statement, does not point to a contradiction, because x does not describe anything (does not convey anything). Usually, however, an inscription of the type $\ulcorner x \equiv \sim x \urcorner$, where x is a statement only in the syntactic sense, is held to point to a contradiction, for it is assumed that every statement in the syntactic sense of the term is a statement in the semantic sense as well. In any case, (1) must be a statement in the semantic sense if we are to obtain a genuine antinomy.

As we can easily see, syntactically (1) is a statement of the type $\ulcorner A \textit{ is } B \urcorner$. For this reason the expressions "heterological" and 'heterological', which occur in (1), should be terms from a syntactic point of view. The ordinary method of using quotation marks makes us suppose that "heterological", being a singular term, names the word 'heterological'. The word 'heterological' should, by definition, denote the class of heterological expressions, i.e., expressions such that none of them is in the class it denotes. This method of understanding the expressions "heterological" and 'heterological' is in agreement with the intentions underlying the formulation of the antinomy now under consideration. Whatever the word 'heterological' indicates, it is certain that it must be a term, because (1) is a statement of the type $\ulcorner A \textit{ is } B \urcorner$, the word 'heterological' occupies the place of B, and (1) is meant to describe the state of things which consists in 'heterological' being heterological.

Thus, if (1) is a statement in the semantic sense of the term, then "heterological" and 'heterological' are terms in the same sense. Note that the word "heterological" must name the expression 'heterological' at the time when "heterological" denotes the class of heterological expressions, since both expressions must be terms at a time when (1) is a statement (cf. (16) in Sec. 1). In brief, if the definition of heterological terms is to yield a contradiction, then there must be a person for whom at a certain time "heterological" names 'heterological' and 'heterological' indicates the class of heterological expressions. If no such person is found, then we shall not find a person for whom the definition of heterological terms yields a contradiction.

We shall now use the concepts introduced previously to record the condition which is necessary for the antinomy of heterological terms to occur:

(2) $(\Sigma\ t, v)\ (Nam\ (t, v, \text{“}het\text{”}, \text{‘}het\text{’}).\ Den\ (t, v, \text{‘}het\text{’}, het)).$

Now, (2) immediately yields

(3) $(\Sigma\ t, v, x, y_1, y_2)\ (Nam\ (t, v, x, y_1) .\ Den\ (t, v, y_1, y_2)).$

By making use of the definitions of naming and denoting we obtain:

(4) $(\Sigma\ t, v, y, y_1, y_2)\ (Z\ (t, v, x, y_1)\cdot(z)\ (Z\ (t, v, x, z) \rightarrow z = y_1)\cdot$
$\cdot Z\ (t, v, y_1, y_2),$

and (4) yields

(5) $(\Sigma\ t, v, x, y_1, y_2)\ (Z\ (t, v, x, y_1)\cdot Z\ (t, v, y_1, y_2)).$

Now (5) contradicts T9. If we accept T9 we have to reject (5) and, accordingly, (2). The condition necessary for the antinomy of heterological terms is not satisfied, and hence no person exists who could obtain a real contradiction from the definition of heterological terms, because the substitution, in the definition of heterological terms, of the word '*het*' for the variable '*x*' yields a semantic nonsense, while a meaningless inscription in the form $\ulcorner x \equiv \sim x \urcorner$ does not state contradiction. The result arrived at here recalls the theory of degrees of language.

The antinomy of the liar is based on two premisses, both of which are definitions in nature. The first is a partial definition of truth:

(6) '*c* is not a true statement' is a true statement

$\equiv c$ is not a true statement.

The second establishes the meaning of the inscription '*c*': it states that '*c*' is a name of the inscription '*c* is not a true statement'. In view of this convention we have:

(7) '*c* is not a true statement' $= c$.

If, on the strength of (7), we replace, in (6), the expression '*c* is not a true statement' by the letter '*c*', we obtain

(8) *c* is a true statement $\equiv$ *c* is not a true statement.

We shall analyse this antinomy in detail. Let the two sides of the equivalence (8) be denoted by (8a) and (8b), respectively. Note that (8a) and (8b) are meant to be statements describing the states of things which consist in *c* being a true statement and *c* not being a true statement, respectively. If (8a) or (8b) do not describe anything for anyone, then, of course, it would be difficult to speak about a contradiction or an antinomy. If an antinomy is to occur, the following statements must be true:

(9) $(\Sigma\ t, v,)\ W\ (t, v,$ (8a), *c* is a true statement),
$(\Sigma\ t, v,)\ W\ (t, v,$ (8b), *c* is not a true statement).

At the time when (8b) describes the state of things which consists in *c* not being a true statement, the word '*c*', which is part of (8b), should indicate (8b), since only in this case would the meaning of '*c*' agree with (7). In a word, it should be true that

(10) $(\Sigma\ t, v,)\ (W\ (t, v$ (8b), *c* is not a true statement)·
·*Nam* $(t, v,$ '*c*', (8b))·'*c*' c (8b)).

(Note that the Roman-type c in (10) above and in (11) and (12) below stands for the two-place relation 'is part of'.)

By virtue of the functional calculus law $fa \rightarrow (\Sigma\ x) fx$, (10) yields

(11) $(\Sigma\ t, v, x_1, fy, x_2)\ (W\ (t, v, x_1, fy) \cdot Nam\ (t, v, x_2, x_1) \cdot x_2\ \mathrm{c}\ x_1)$.

Furthermore, if we use the definition of naming and omit two factors of the conjunctions, (11) yields

(12) $(\Sigma\ t, v, x_2, x_1)\ (Z\ (t, v, x_2, x_1) \cdot x_2\ \mathrm{c}\ x_1)$.

But (12) contradicts T5. Accordingly we have to reject (12), and consequently (11) and (10), too. The third factor of (10) is an empirical statement and is certainly true. We have, therefore, to conclude that either (8b) does not describe any state of things (cf. (11)) or '*c*' does not denote (8b), i.e., '*c*' has a meaning other than that assigned to it by (7). In both cases no antinomy is obtained. The case is obvious if (8b) does not describe any state of things, i.e., is not a statement in the semantic sense of the term. In such a case (8b) simply states nothing. And if '*c*' does not denote (8b), we cannot carry out the replacement that leads from (6) to (8), because the convention formulated as (7) is not valid.

The foregoing analysis of the two most widely known semantic antinomies shows, in the light of the theorems resulting from the principle of transparency, that no genuine contradictions are obtained. It is common knowledge that antinomies are due to the specific reflexivity of the relations of indicating and marking characteristic of expressions which occur in formulations resulting in antinomies. This reflexivity is eliminated by the principle of transparency, and that is why, in view of the theorems which follow from that principle, we can state that no genuine antinomies are encountered and that there are no semantically meaningful statements implying contradiction.

What are believed to be semantic antinomies are due to the identification of certain syntactic concepts with semantic ones. In such a case the syntactic concept of a sentence coincides with the semantic one. It is usually believed that a well-formed inscription automatically refers to a state of things. Likewise, an inscription which has the syntactic properties of a term is identified with a genuine term which indicates something. Regardless of our statement that we never encounter genuine antinomies the problem remains very difficult to solve, for the point is to formulate a set of syntactic rules such as would guarantee every inscription formed in accordance with those rules to be semantically meaningful. Such a set of syntactic rules would enable us to identify operations on inscriptions with operations on meaningful expressions without any risk of obtaining a contradiction. It would appear that the theory of degrees of language is in fact a syntactic procedure which adjusts the form of expressions to the needs of semantics. Expressions in metalanguage are given forms different from those in object language, and inscriptions in which both metalanguage and object language expressions occur, are treated as meaningless not only semantically, but also syntactically. The stratification of language into object language and a corresponding metalanguage can be carried out in artificial languages. It is different in the case of natural languages, where the forms of signs are fixed and cannot be changed in an arbitrary manner. Since in some cases in natural languages names of expressions cannot be given forms other than the designata of those names, the application of ordinary rules of inference results in apparent contradictions. In this connection it has been said that natural languages result in contradictions, and that they are not even languages at all. They have, in a sense, been doomed to "logical death", since inner contradictions have been considered to be part of their very nature.[25] In the light of the results obtained above to conclude that natural languages are innately defective would appear to be somewhat premature. For in a tentative appraisal of current language we have to take into account not only its syntactic rules, which in fact do admit the construction of expressions which result in inscriptions like $\ulcorner x \equiv \sim x \urcorner$,[26] but also rules which are both pragmatic and semantic in nature. The latter rules, based on the principle of transparency, bar the construction of antinomial statements. The formation rules for statements must therefore become more complex. In order to be a statement an inscription x will have to be in a form prescribed by syntactic rules. Moreover, it will not be possible to give a meaning to x which would allow a statement to be made about x in contradiction to the principle of transparency. An investigation as to whether x can be semantically meaningful would have to resemble the analysis of antinomies carried out above. First of all, we should have to determine what x, and possibly also its parts, would have to point to or refer to. Next, the convention establishing the meanings of expressions on the basis of the rule $\ulcorner fa \rightarrow (\Sigma x) fx \urcorner$ would yield a corresponding existential statement, which would have to be compared with theorems resulting from the principle of transparency. In the case of a contradiction we should have to conclude that x is not a meaningful expression.[27] This would be a complex procedure, not to be compared — as regards simplicity — with the operations that bar antinomies and follow from the theory of degrees of language. In any case, there is a method which enables us to distinguish those current language expressions which are semantically meaningful from those which are meaningless and result in contradictions.

Conclusion

At the outset I set myself the task of investigating the role of names of expressions in semantic antinomies in the light of a certain theory of signs.

It would appear that the goal has been attained. I have constructed a consistent system which can be treated as a theory of signs based on the principle of transparency. This principle has been analysed, and the first step has been taken to give it a clear and comprehensible formulation. At the same time, certain consequences of this principle have become manifest which will induce some people not to accept it as true. Bringing out in relief the doubtful points connected with the principle of transparency is a result of its being given a more precise formulation.

I have referred to the axioms I have accepted to prove a number of theorems characterizing the pragmatic concepts of pointing and referring; these theorems have been used to analyse the names of the expressions which occur in semantic antinomies. It has been shown that these antinomies include no genuine names of expressions whatever, in the semantic sense of the term. The antinomies under consideration include merely names of expressions in the syntactic sense of the term, and in that sense only.

I have also obtained certain other results as by-products, even though I have concentrated on the issue of semantic antinomies.[28] We can also see the prospects, dim as they still are, for a closer co-ordination of pragmatics with semantics. It would appear that semantics can be deduced from pragmatics,[29] and this could provide opportunities for a closer linking of linguistic semantics with logical semantics.

References

[1] Ajdukiewicz, K., *Język i poznanie* (Language and Cognition), Vol. 1, Warszawa 1960.

[2] Carnap, R., *Introduction to Semantics*, Cambridge (Mass.) 1946.

[3] Carnap, R., *Meaning and Necessity. A Study in Semantics and Modal Logic*, Chicago 1947.

[4] Copi, I.M., *Symbolic Logic*, New York 1956.

[5] Czeżowski, T., *Logika* (Logic), Warszawa 1949.

[6] Czeżowski, T., *Odczyty filozoficzne* (Lectures on Philosophy), Toruń 1958.

[7] Fraenkel, A., Bar-Hillel, Y., *Foundations of Set Theory*, Amsterdam 1958.

[8] Husserl, E., *Logische Untersuchungen*, Vol. II, Pt.1, Halle 1922.

[9] Kotarbińska, J., 'Pojęcie znaku' (The Concept of Sign), *Studia Logica*, Vol. VI, 1957.

[10] Łukasiewicz, J., *Elements of Mathematical Logic*, Oxford 1963.

[11] Martin, R.M., *Toward a Systematic Pragmatics*, Amsterdam 1959.

[12] Morris, Ch., *Signs, Language and Behavior*, New York 1946.

[13] Ossowski, S., 'Analiza pojęcia znaku' (An Analysis of the Concept of Sign), *Przegląd Filozoficzny* 1926, No. 1–2.

[14] Popper, K.R., 'Self-Reference and Meaning in Ordinary Language', *Mind* 63 (1954).

[15] Quine, W.V.O., *Mathematical Logic*, Cambridge (Mass.) 1955.

[16] Reichenbach, H., *Elements of Symbolic Logic*, New York 1948.

[17] Schaff, A., *Introduction to Semantics*, Oxford 1962.

[18] Szober, S., *Zarys językoznawstwa ogólnego* (An Outline of General Linguistics), Warszawa 1924.

[19] Tarski, A., 'The Concept of Truth in Formalized Languages', in: *Logic Semantics, Metamathematics*, Oxford 1956.

NOTES

[1] Husserl, [8], Vol. II, Part I, p. 40.

[2] Ossowski [13], pp. 30–1. (See also Ossowski's paper in the present book. – Tr.)

[3] Kotarbińska [9], pp. 67–9.

[4] Schaff [17], p. 294.

[5] Schaff [17], pp. 196–7.

[6] The issue of whether x is merely a sound is therefore disregarded here.

[7] The word 'Peter' is printed in bold type to indicate the fact that it is a name of a sign-event, and not of a sign-design. Concerning the terms 'sign-event' and 'sign-design' see Carnap (2), pp. 5-8.

[8] Kotarbińska [9], p. 65.

[9] Schaff [17], p. 196.

[10] Carnap [2], pp. 5–8.

[11] What is meant here is not a spatial relation of lying between, but a relation defined thus: x_2 lies between x_1 and $x_3 \equiv x_1 \, R \, x_2 \cdot x_2 \, R \, x_3$, where R is any ordering relation.

[12] We shall adopt the following informal terminological conventions: a name denotes one non-singular class (i.e., its extension), a name designates more than one object (designatum), a name names an individual, a name points to, i.e denotes, or designates, or names (its denotatum), a statement refers to a state of things, a state of things here means the same as a proposition; (cf. Carnap [3], pp. 26–30), a sign (a name or a statement) indicates a fact, i.e., either refers to or points to a fact.

For brevity, the formulation "throughout a period *t* for a person *v* a sound (an inscription) *x* denotes *y*" we shall just say that "*x* denotes *y*".

[13] For similar reasons Kotarbińska has abandoned the idea of formulating a single definition of sign to cover both symptoms and signs. Cf. Kotarbińska [9], p. 104. For the same reasons she has abandoned the idea of formulating a single definition of a verbal sign to cover both names (terms) and statements. Cf. Kotarbińska [9], pp. 126–7.

[14] For similar comments see Szober [18], pp. 92–5. (This issue concerns those languages, like Polish, in which there are well-formed single-word sentences, because, for instance, the indication of the grammatical person is incorporated into a verb form. In English, this could perhaps apply to single-word imperative sentences, on the assumption that intonation serves to indicate the imperative, or is replaced in that function in writing by the exclamation mark. – Tr.)

[15] Examples of this type are cited by opponents of the separation of language and metalanguage. Cf. Popper [14].

[16] The problem can be assessed with precision only when rigorous definitions of equiformity and the principle of synonymity of equiform expressions are formulated. This problem, by the way, is relevant only in the case of non-ambiguous languages. The simplest solution would be to assume that (6) and (8), although equiform, have different meanings. This is probably the situation in natural languages, where a great many expressions have numerous meanings each.

[17] The solution suggested here is less than sketchy, and in some respects even incorrect. The present writer will analyse such names as (9) in a separate paper on token-reflexive terms. Concerning these terms, above all see Reichenbach [16], pp. 284–7.

[18] Cf. T. Czeżowski's terminology in Czeżowski [5], pp. 17–9.

[19] Quine [15], Sec. 24 and p. 171, T 230.

[20] Concerning the terms 'speaks about' and 'refers to' see Czeżowski [5], p. 157. The terminology has been somewhat generalized. (Not all the expressions mentioned here have been retained in translation in view of language differences between Polish and English. – Tr.)

[21] Irving Copi's formulation has been chosen only because Fraenkel gives a favourable opinion of it in [17], pp. 16 and 18.

[22] Tarski [19], p. 7.

[23] Concerning the use of quotations see Quine [15], pp. 23–6 and 33–7. Even though his are the most comprehensive of all comments on quotations known to the present writer (see also Tarski [19], pp. 9–12), they do not suffice for an exhaustive analysis of antinomies. This is why this analysis will be rather vague on certain points. In particular, no distinction will be made between quotational names of sign-events and those of sign-designs.

[24] In this case the variable '*x*' ranges over the set of names of statements.

[25] Tarski [19], p. 14.

[26] Cf. footnote[24].

[27] This method is also to be applied to substitutions of the axioms and theorems

adopted here. Then it turns out that, for example, '$W(t, v, c, fc)$' is meaningless if $c =$ '$W(t, v, c, fc)$'. Not all substitutions which are syntactically correct lead from meaningful expressions to meaningful expressions. The rule of substitution must be complemented by a rule requiring the semantic meaningfulness of the results of substitutions.

[28] The wealth of problems is shown by Morris [12].

[29] Pragmatics has been superstructured over semantics by Martin [11].

THE SEMANTIC FUNCTIONS OF OBLIQUE SPEECH

by

Witold Marciszewski

I. INTRODUCTION

Veni, vidi, vici. We know quite well to what these words, addressed by Caesar to the Roman Senate, referred: to the events which occurred when the Roman troops came face to face with the army of the king of Pontus. To what, however, does the statement made by a student during his Latin class refer if that student says that 'Caesar said that he came, saw and conquered'? Does it refer to the same events to which Caesar's words refer, and if it does, what does it say about those events? Does it say that they were the subject matter of the report made by Caesar to the Senate? Or is it a statement not about military events, but about linguistic events, namely the pithy formulation coined by the conqueror? Or is it a statement about Caesar's state of mind at the time when he was making it? When we refer to his words by using their English-language version we do not reproduce the sounds which were then heard in the Senate chamber; we merely reproduce the speaker's thought, and hence his state of mind.

Similar questions arise when the words 'to say', 'to state', or 'to utter' are replaced by the words 'to think', 'to believe', 'to know', 'to doubt', etc., and hence words which are sometimes termed epistemic. The words 'to see', 'to hear', etc., come close to them, but they differ from the former in being liable to a physical interpretation in terms of physical stimuli that act upon receptors, which makes another type of description possible. Sometimes it is not verbs which account for oblique speech, but such formulations as 'it is possible that', 'it is necessary that', etc., which are classed as modal terms. If we join to them the class of formulations represented by the phrases 'it is true that', 'it is not true that', 'it occurs that', 'it is the case that', etc., we have an idea of the wealth of the forms of oblique speech. Many of the phrases listed above occur in the company of words other than 'that', namely 'how', 'whether', 'to', etc., which also form structures called oblique speech. In my choice of examples and in my analysis itself I shall confine myself to phrases followed by 'that', but the results which will be obtained can be applied, *mutatis mutandis*, to phrases followed by other words with an analogous function.

Thus the structure of the phrases analysed here can be represented by the schema '*S* that *p*', where *S* stands for one of the words or phrases adduced above, possibly with a genuine grammatical subject, other than the expletive or dummy *it*, and *p* stands for any sentence.

The present paper does not suggest any single solution of the problems posed, problems which concentrate on the semantic function of the phrase 'that *p*'. Several possible solutions will be offered, and the relationships between them remain the subject of further investigation. Our analysis will follow two different lines: the first will be in the direction of a series of partial complementary solutions, based on the assumption that the various types of oblique speech require different theoretical approaches, each approach being adjusted to the specific characteristics of a given type; the second will be to treat all the varieties of oblique speech in the same way. The section concerned with the former approach will be called, somewhat metaphorically, "A pluralistic theory of oblique speech", and the latter, which reveals the author's universal ambitions, will be called "A monistic theory of oblique speech". In both cases the term theory is used somewhat loosely. These will rather be fragmentary drafts of theories. But what logicians have so far said concerning oblique speech is not only fragmentary, but marginal as well, in the form of casual comments in connection with other issues. This is why concentrating on this problem represents a step forward, even if only a very small one.

II. A PLURALISTIC THEORY OF OBLIQUE SPEECH

2.1. *Oblique objective contexts*

Let us compare the following two sentences:

(1) John hears that Peter is calling for help.

(2) John hears Peter's call for help.

Analogous pairs of sentences could be formed with the verbs 'sees', 'feels', etc., that is, verbs which refer to sense data. The transition from (1) to (2) consists in replacing the dependent phrase, traditionally treated by grammarians as a clause, by what is indisputably a noun phrase. In some languages the latter phrase may have the form of the structure known as *accusativus cum infinitivo*, which can also be given a nominal interpretation since the infinitive is often treated as a noun or, to put it more cautiously, as "a verbal form which has the meaning of a noun".[1]

It would not appear hazardous to claim that the relationship between (1) and (2) is that of logical equivalence in the sense of *L*-equivalence as defined by Carnap.[2] This is so because whenever (1) is true, (2) is true, too, and conversely, so that the two sentences are equivalent. Moreover, this is not an equivalence which has to be verified by reference to extra-linguistic facts: to establish that it holds, it suffices to know the English language, or strictly speaking, the semantic and syntactic rules of that language. The test of asking whether empirical investigations in an extra-linguistic domain are required for the acceptance of an equivalence as true is a satisfactory practical criterion of logical equivalence, even though we are not dealing here with a language whose rules are rigorously defined as in the case of artificial languages, to which Carnap's definitions of semantic concepts refer and apply.

Now if (1) and (2) are logically equivalent, and if inscriptions in the form of 'John' and 'hears', which occur in (1) and (2), are, respectively, not only equiform, but also logically equivalent (which in the case of non-compound expressions is tantamount to synonymy), then we have to conclude that the formulations

(3) that Peter is calling for help,

(4) Peter's call for help,

are logically equivalent, too.

In our case this conclusion is important inasmuch as it enables us to identify the semantic category of formulations like (3) in contexts like (1). This is obviously the category of terms, for if (3) is logically equivalent to (4), then it must have the same semantic category as (4), which is a name of an event or of a state of things, and hence an abstract term.

This conclusion may appear rather unexpected in view of the traditional grammatical terminology, according to which oblique phrases that begin with 'that', or 'that-clauses', form a certain kind of sentence:

hence the grammatical term 'dependent (or: subordinate) clause', which is carefully avoided here and replaced by the more neutral terms 'oblique speech' and 'oblique phrase'. (The treatment of 'that-clauses' as sentences used to be standard in traditional Polish grammars, even if it no longer applies to recent interpretations of English syntax. — Tr.) If such usages are not purely a matter of unfortunate terminology, and if they indicate the underlying theoretical approach which assigns to all dependent clauses, i.e., oblique phrases, the semantic category of sentences, then that theoretical approach is wrong in the light of the assumptions formulated above.

The conclusion we have reached also predetermines the semantic category of oblique phrases and the syntactic structure of the contexts in which such phrases occur. An oblique phrase, being a term, has the semantic function of denoting, its denotatum being an object in that ontological category which can best be termed the category of states of things. The concept of a state of things is far from clear, but we shall not explain it here, because it will be discussed in the next section, where it will be the focal concept.

If an oblique phrase is a term, and if the part of it which is left after 'that' has been cancelled is a sentence, then 'that' is a term-forming functor of one sentential argument, and the words which in (1) are represented by the verb 'hears' are sentence-forming functors of two term arguments each. Such a syntactic analysis of oblique speech is given by Bar-Hillel in one of his papers on algebraic linguistics.[3] This syntactic role of 'hears' and similar words reflects their semantic role of describing a two-place relation. In this case this is a relation between the cognizing subject and the object of his sensory cognition, a relation which can be described in physical terms as the effect of air waves on his organ of hearing or, in the case of seeing, the effect of light rays on his organ of vision, etc.

Finally, we have a terminological proposal to make here. Contexts like (1), in which words relating to perception occur as the main functor and determine the role of the oblique phrase as that of a term, will be called objective oblique contexts, as distinguished from other contexts, which will be discussed later. This term seems appropriate because of the triple association connected with the word 'object' and its derivatives. First of all, the oblique phrase occurs here as the term denoting the object of perception (or, more generally, cognition), the latter as distinct from the cognizing subject, denoted by the grammatical subject of a given sentence. Secondly, the oblique phrase functions grammatically as an object. Thirdly, the term suggested here refers to the distinction between object language and metalanguage, which helps us better to distinguish the contexts analysed here (with the oblique phrase as an object language term) from those to be discussed in the next section, in which the oblique phrase plays a role analogous to that of metalinguistic expressions. And since the latter role is also syntactically different (since the metalinguistic expressions are treated as sentences), the contexts of the latter type, to which we now turn, will be called propositional.

2.2. *Propositional oblique contexts*

Let us now take the sentence

(5) John says that Peter is calling for help

as our standard sentence. We can easily see that the transformation of that sentence in a manner analogous to the transformation of (1) into (2) does not yield an equivalent expression but, to make matters worse, results in syntactic disorder: no one who knows English will use or accept as correct the sequence 'John says Peter's call for help'. This shows an essential difference between (5) and (1), despite the identical external structure of the two. What we might term the inner or deep structure will be disclosed only by a semantic analysis which resorts to paraphrasing.

From a semantic point of view, the closest paraphrase of (5) would be:

(6) John says, 'Peter is calling for help'.

This is so because the same fact can be reported in the form of (5) or (6) and the choice of form depends only on whether we have to give our account in writing or in speech. The spoken language does not have quotation marks at its disposal, and this is why that which we would formulate in writing as (6) might in speech take the form of (5).

The relation between the two is such that whenever one may use (6), with a quoted expression, one may also use (5), i.e., oblique speech. The converse does not hold, as can be seen in the case of quotations of statements made in another language, and also in the case of statements that include token-reflexive expressions. Both cases are illustrated jointly by the sentences:

(7) Caesar said, 'Veni, vidi, vici',

(8) Caesar said that he came, saw and conquered.

It is, not of course, possible to pass from (7) to (8) by merely replacing the comma and the quotation marks with 'that'; we have to make two other essential modifications, namely to translate the quoted expression into English and to change the grammatical person of the verbs in the quoted expression from the first to the third.

Our task will now be to describe the contexts of type (5) in such a way as to bring out both the similarities between (5) and (6) and the differences between (7) and (8), as well as other analogous differences.

This can be done if we view type (5) propositional contexts as reproductions of certain texts with a certain degree of exactitude. Now, those who utter (5) and (6) reproduce a text produced by John and at the same time inform others that it was John who produced it. Such a reproduction may be maximally exact, or less exact; in the latter case, only some elements of what is being reproduced are indicated, while the others are disregarded. A maximally exact reproduction is usually placed in quotation marks. For the sake of simplicity, we shall not discuss here all the functions performed by quotation marks. One of these functions is undoubtedly to indicate a *verbatim* reproduction of a given text. A reproduction the maximal exactitude of which the speaker is unwilling to guarantee is formulated by him as oblique speech. This is what the speakers of (6) and (8) are doing. Since oblique speech does not exclude a *verbatim* reproduction, in some cases it may happen that such a reproduction coincides with reproduction by quoting. This is why, in the case of (5) and (6), the two may be used alternately. This cannot occur, however, if it is necessary to move from one language to another or to change token-reflexive words. In these cases we have to give up reproducing the shape or sound of the text we reproduce and confine ourselves to preserving its meaning; sometimes even part of the meaning has to be abandoned. What is omitted in such a case may include such characteristics of a given text as its specific stylistic features, the emotional content, the phonetic features, etc. A change in the emotional content of a text may consist, for instance, in reporting a person's statement ironically, which does not infringe the purely descriptive aspect of that statement (if these things can be separated), but modifies it from a valuational point of view. This accounts for the various differences between an original, i.e., a text being reproduced, and its reproduction(s).

The concept of propositional oblique speech as a reproduction of a text results in ascribing to it other semantic and syntactic properties which we have ascribed to objective oblique contexts. In order to report correctly on those properties we must first do away with a certain prejudice concerning the function of quotation marks, for this prejudice extends, by analogy, to propositional oblique contexts as well. The correct opinion that quotation marks sometimes transform an expression into a name of that expression becomes a prejudice if we replace 'sometimes' by 'always'. There is no doubt that in the statement

'Hamlet' has six letters

the first word is a name of the word 'Hamlet', for since we are predicating something about that word we have to use its name. We could provide analogous examples of how quotation marks transform expressions in other semantic categories, such as sentences, into names of sentences or other expressions. Suppose, however, that we have a book which is an anthology of philosophical texts, the original texts being separated by the editor's comments. We are thus dealing with two levels of language: that of the quoted texts and that of the commentaries on them. The editor may, of course, mark off the texts commented upon from his commentaries by using quotation marks, but he need not do so if it is known by some other means where a quotation ends and a commentary begins. Would we consider such a quoted text of, say, 40 pages to be a single, very long, name just because the text is, or could have been, placed in quotation marks? The length of the quotation would, of course, be no obstacle, since there are no limits in this respect. However, the editor's intention was not to predicate anything about that supposed 40-page-long name, but to provide reproductions to those who did not have the opportunity to listen to the original lectures or to read the original manuscripts of the philosophers whose works are included in the anthology. Sometimes the reproduction can be even closer to the original (e.g., a tape-recorded text of a lecture or a photograph of the manuscript), yet if precise but expensive techniques are not used on the assumption that the readers are not interested in the timbre of the voices or in the handwriting of the philosophers concerned, this does not alter the fact that both a photographic reproduction and a reproduction by printing techniques perform the same function: they provide a more or less exact reproduction of the original, but in no way become a name of that original. The fact that we are dealing with a reproduction can be indicated in various ways, one of them being to place the quoted text in quotation marks. In this situation they perform one function less than when they produce a quotational name: in the latter case they also indicate a reproduction, but at the same time they impart to that reproduction a new, semantic, function by making the reproduced text a name of the original. In certain cases this second step is not required, and then the function of the quotation marks boils down to indicating a reproduction.

What is the function of the quotation marks in the case of (6)? The quotational expression, that is, the quotation marks with the phrase which they comprise, has not been produced with the intention of being used as a name, and hence of being used as a grammatical subject or subjective complement, but merely with the intention of reproducing the text uttered by John. If the criterion of the intentions of the speaker is accepted as satisfactory (and it would be difficult to find any other criterion to replace it), then the quotational expression now under consideration is not a name, but a sentence — the only difference being that it is a sentence in another language, inserted in the language of John's utterance, just as a photograph of a sentence from the first folio may be included in a book on Shakespeare. Nothing could substantiate treating that photographed sentence as a metalinguistic name; likewise, nothing substantiates treating less exact reproductions, made with an analogous intention, as elements of the class of metalinguistic names.

The semantic function of a sentence which is a reproduction is the same as that of the sentence which is the original (relative to that reproduction), with the proviso that it is as it were pushed into the background, since we are interested in the reproduced sentence mainly as a specimen or a copy of a specimen of a linguistic fact, and much less in the relation it bears to what it describes. A sign does not cease to be a sign, nor does it change its semantic properties, merely by being placed in a museum and because some people are more interested in the sign itself (as an artefact, etc.) than in that of which it is a sign.

If we agree that a type (6) sentence can be replaced by a type (5) sentence, as a result of which some exactitude of reproduction may be lost, but the intention to reproduce a text is preserved, then all that has been said about quotational reproductions — i.e., reproductions which also more or less exactly inform about the external characteristics of the text — must apply to reproductions in the form of oblique speech, i.e., those which are not intended to reproduce the external characteristics. A reproduction of a text with regard to its meaning alone, disregarding its external characteristics, does not become a name of that text since it is less exact — more schematic — than a reproduction marked as quotational. Hence in such contexts with oblique speech we are dealing with a sentence which is reproduced and with an indication, analogous to quotation marks, informing us that a reproduction has been made. This indication takes the form of the word 'that', of course, and what follows 'that' is a reproduction or reconstruction of a text, made in the language of the person reporting on that text. The fact that it is the language of that person can best be proved by the necessity of transforming the personal pronouns involved, for instance the 'I' in the original into the 'he' in the reproduced text, or, possibly, vice versa.

As for the semantic function of a sentence which is in oblique speech and is a reproduction of a certain original we have to assume, as in the case of quotational contexts which are not name-forming, that it is the same as the semantic function of the original sentence. We shall not make an investigation of this function of sentences, since this is a separate issue, and not an easy one to solve.

Thus, the syntactic description of propositional contexts starts from the assumption that what follows 'that' is a sentence, and that 'that' itself, being an indicator of the reproduction performed, cannot be subsumed under any category thus far singled out (in this respect it resembles quotation marks). It is rather to be interpreted as a punctuation mark, something like a colon adapted to special purposes: it is used to indicate that what follows is a reproduction whose degree of exactitude is at most that of a reproduction placed in quotation marks, and at least such that some essential meaning is preserved (however this essential meaning may be defined). The word 'says' and the like would then be a sentence-forming functor of one term argument and one sentential argument. Another interpretation would consist in joining 'that' to that functor; in such a case 'says that' would have to be interpreted as a single word which differs from 'that' in that its meaning must be brought out by the following paraphrase: 'says something which in my language can be reproduced with an exactitude which is no less than the preservation of meaning, for instance with the help of the sentence which follows'. In this case (5) would be broken into the term argument 'John', the sentential argument 'Peter is calling for help', and the sentence-forming functor of those arguments 'says that'. Syntactic analyses like the latter are offered by Y. Bar-Hillel, but in another work than that previously cited; this syntactic interpretation of oblique contexts is also mentioned by K. Ajdukiewicz.[4]

When oblique contexts are treated as reproductions of texts with a specified degree of exactitude, we can outline a new approach to the problems connected with carrying out such logical operations on those contexts as replacement, substitution, etc. (the problem of intensionality). In the case of a language which has no variables these operations — baffling when it comes to oblique speech — boil down to one of replacement.

To illustrate the problem let us consider the following example. Suppose that Quintus, a Roman who lived in the Republican period, made the following statement which has somehow been conveyed to us:

(9) *Roma sita est ad Tiberim.*

This statement can be reproduced in English, in oblique speech, with varying degrees of exactitude. The most exact reproduction would be:

(10) Quintus said that Rome lies on the Tiber.

This would be a reproduction with an exactitude of meaning, in the sense that the syntactic structure of the reproduction is the same as that of the original, and the corresponding elements of those structures may be considered synonymous, for instance, on the basis of a Latin-English dictionary. But we may content ourselves with a less exact reproduction, in which only the identity of the denotation is required. This could take the form of:

(11) Quintus said that the capital of the Republic lies on the Tiber.

This sentence may give rise to misunderstandings if we do not take into account the ambiguity of the word 'that' as an indicator of oblique speech. This word does not inform about the intended degree of exactitude of reproduction, but in fact functions at various levels of exactitude. Hence, if the context does not indicate that clearly, or is even misleading, then someone could consider (11) to be an incorrect reproduction of (9) and argue that Quintus, when thinking about Rome, was not necessarily thinking about it as the capital of the Republic, and that we should not, therefore, suggest that he was speaking about the capital of the Republic. This argument would be valid only if the intention which accompanied (11) was to reproduce the original with an exactitude of meaning, as is the case of (10). But if we use 'that' only as an indicator of reproduction with an exactitude which is no greater than reproduction as to denotation, then (11) is a correct reproduction of Quintus' statement.

Moreover, in view of the above we also have to accept as correct the following reproduction:

(12) Quintus said that the capital of the Papacy lies on the Tiber.

While we accept its correctness, we should not recommend such a manner of speaking, because it may result in misunderstandings in view of the aforementioned ambiguity, or multiple function, of the indicator 'that'. But if we explain that our intention is merely to indicate, by using the English language, what object was meant by Quintus in his statement, then (12) serves that purpose as effectively as (10) and (11) do: in other words, it is a reproduction with an exactitude of denotation.

The difference between (10), on the one hand, and (11) and (12), on the other, can also be illustrated by the fact that the transition from the original statement (9) and its reproduction (10) can be made on the basis of information confined to the sphere of vocabulary, whereas the transition from (9) to (11) or (12) requires encyclopaedic information, and hence information covering extra-linguistic facts. This is an analogue of the difference between the knowledge of the identity or difference of meanings, and the knowledge of the identity or difference of denotations.

Although (11) and (12) are at the same level of exactitude of reproduction relative to (9), there may be still another difference between the two: this manifests itself when we expect oblique speech to perform another function besides that of reproducing a text, namely that of informing as to the state of mind of the person who is using direct speech. This issue will be discussed in the subsection that follows.

2.3. *Indirect oblique contexts*

The class of contexts with which we shall now concern ourselves is represented by the sentence:

(13) John thinks that Peter is calling for help

This sentence can be viewed in two ways, either as resembling objective contexts, represented by (1), or as resembling propositional contexts, represented by (5).

In the former approach, the word 'thinks' is treated as a predicate of two arguments, describing a relation between John and the state of things that consists in Peter's calling for help. The idea that thinking is a relation — the same applies to other mental states, such as desiring, hating, etc. — has both determined advocates and determined opponents among philosophers. Its spokesman and in a sense its originator (since he held all acts of mind to be relational in nature) was F. Brentano, and his followers include E. Husserl, A. Meinong, and K. Twardowski. It would also appear that this idea is in harmony — although for other reasons and perhaps as restricted to acts of cognition — with the Marxist interpretation of cognition as a reflection of the real world: when John thinks that Peter is calling for help, then in John's mind or brain (there are various preferences as to terminology) there is a reflection of the state of things which consists in Peter's calling for help. On the other hand, we have to face here the classical problem of how the objective argument of that relation is to be interpreted if no such object exists and the subject is the victim of an illusion or some other error. The problem will not, of course, be analysed here, but it has to be mentioned in order to show how far philosophical issues may interfere with certain linguistic problems.

As an example of opposition to the theory of thinking as a relation we may mention the views of G. Bergmann, who engages in semantic studies very closely connected with ontology.[5]

But there is no need to resort to philosophical arguments to find grounds for the latter approach, according to which (13) is classed as a propositional context. This does not have to lead to a rejection of the theory of the relational nature of cognitive acts; even while accepting it we need not make use of it when analysing a given issue. We may assume that whoever utters (13) is also reproducing a certain text, as in the case of the person uttering (5), the difference being that in the case of (13) we are dealing with a reproduction of a text which might never have been verbalized, but only conceived. The process is sometimes metaphorically called "reading another person's thoughts."

In order to explain this issue in greater detail we shall now introduce the term 'potential text', which will be defined by reference to the concept of a reductive, or operational, definition, i.e., a definition by indicators.[6] Dispositions, or internal states, which are described by such words as 'thinks', 'believes', 'knows', 'doubts', etc., can in some cases be described by indicators which are in the class of non-verbal behaviour. For example, we may say with reference to a stalking cat that 'he thinks that there is a mouse in the hole', even though the sentence that follows 'that' is not a reproduction of a statement made by the cat, not even of a potential one. But when it comes to more complex mental states such behavioural indicators prove insufficient as a rule and we have to resort to verbal indicators. Let us refer once again to Quintus' opinion concerning the geographical situation of Rome. An indicator of Quintus' conviction that Rome lies on the Tiber could have taken the form of a statement made under circumstances guaranteeing that it was an expression of his opinion, and not just a sentence uttered — as a joke or as a linguistic example — namely a statement such as: '*Roma sita est ad Tiberim*'. This relationship is better shown by the reductive statement which functions as a meaning postulate for the phrase written in italics (to simplify the grammar we shall turn the entire formulation into the present tense):

(14) If Quintus is asked where Rome lies, then Quintus *is convinced that Rome lies on the Tiber* if and only if he replies '*Roma sita est ad Tiberim*'.

This statement can be modified and made more realistic by giving the consequent a weaker form, namely that of an implication which formulates only the sufficient condition of Quintus' conviction. In that case room is left for other statements indicating the conviction that Rome lies on the Tiber. Such statements can be listed with the help of appropriate quotations in successive reductive statements, or described jointly, in a general manner, in a single sentence of which the consequent can now be an equivalence without any major infringement of realism. An example of such a statement would be:

(15) If Quintus is asked where Rome lies, then Quintus *is convinced that Rome lies on the Tiber* if and only if he replies by the sentence '*Roma sita est ad Tiberim*' or by any other sentence which, in Quintus' language, is interchangeable with the former.

The concept of interchangeability, which has been left undefined here, can be defined and explained in various ways, but at the moment there is no need to declare oneself in favour of any particular explanation.

Every statement in the class mentioned in a meaning postulate (in the form of a reductive definition) such as (15), i.e., a statement which refers to ideas, judgements, convictions, etc., will be termed an 'indicator of a judgement' (the latter being understood in the psychological sense of the word). The class of such indicators will be termed a 'potential text', expressing a certain judgement. A 'reproduction of a potential text' is a reproduction of one of the elements of that class of judgement indicators. Reproduction of an indicator in respect of its form is not, of course, a reproduction of the potential text in respect of its form, since that text, if we may put it this way, has no form at all, since it is an abstract object. In this case, however, it is a maximally exact reproduction. A less exact reproduction of a potential text takes place if it consists in the reproduction of one of the indicators of judgement — but a reproduction with an exactitude which is less than one of form, for instance, an exactitude of literal translation from one language into another.

To cope with the paradox of replacement in such psychological contexts it is sufficient to resort to the trick of treating the word 'that' as an ambiguous indicator of the reproduction of a next, a potential text in this case.

For instance, is it legitimate to replace 'Rome' by 'the capital of the Republic' in the statement

(16) Quintus was convinced that Rome lies on the Tiber and thus to obtain the statement

(17) Quintus was convinced that the capital of the Republic lies on the Tiber.

The answer is as follows. It is legitimate to do this if we adopt definition (15) as well as the concessive condition that the original indicators of the potential text, which are referred to in (15), should be reproduced with an exactitude of meaning (this formulation is intended to cover both synonymy within one and the same language and adequately correct translations from one language into another). This condition is as it were inherent in the sense of the formulation 'is convinced', since the meaning postulate connected with it does not require that we should always inform about the content of someone's convictions by quoting the formulations used by the person who has those convictions; it requires, however, something which is best rendered by the term 'synonymy'; the various explanations of the latter term result in splitting both the condition formulated above and its generalization (given below) into a series of more precise formulations.

Replacing the word 'Rome' in (16) by the phrase 'the capital of the Papacy' would be an example of an incorrect replacement, i.e., one which does not guarantee the preservation of truth; this is the case because with regard to the Romans in the Republican period it is to be assumed that the statement 'the capital of the Papacy lies on the Tiber' is not a translation (into English) of any expression that could be an indicator of the same potential text as the statement 'Roma sita est ad Tiberim'.

Generalizing the above condition, which applies to only one example we can now formulate the following rule of replacement for oblique psychological contexts. Let the letter *S* stand for any psychological term such as 'thinks', 'believes', etc. The formula '*a S* that *z* (c)' will then be the schema of a statement in which '*a*' is a name of a person or group of persons, and '*c*' is a component of a statement '*z*'. Here is the rule:

> In a statement in the form '*a S* that *z* (*c*)' the expression "*c*" may be replaced (*salva veritate*) by an expression '*d*' if and only if the statement '*z* (*c*)' and '*z* (*d*)' are indicators of the same potential text or are reproductions of indicators of that text with an exactitude of meaning.

The above formulation is, of course, not the only one possible. The rule can be made more liberal by the admission of a reproduction of the indicators with an exactitude of denotation or with still other kinds of exactitude. To do this it suffices to reformulate that part of the rule which refers to the required exactitude of reproduction of indicators as an appropriately constructed disjunction.

The aim of the present paper is not to offer arguments for the choice of one version or another of the rule of replacement for intensional contexts. Such arguments could not even be formulated without a previous explicit indication of the tasks which such a rule would have to serve — for instance, the kinds of interpretative operations to be performed on the existent or potential texts of the authors in whom we are interested. The point here has merely been to outline a method that might enable us to avoid the paradoxical consequences in which inadvertent replacements in oblique contexts, i.e., replacements which do not take into account the specific nature of such contexts, can result. It would appear that the method suggested here is both effective and in the maximum possible harmony with our intuition. It could be further verified by being formulated so as to cover the formulation of the rule of substitution and possibly of other logical rules as well.

Thus the problems encountered in connection with replacement in oblique contexts can be solved by the adoption of a rule of replacement specially adapted to those contexts. Such a procedure would seem to be quite natural. If any of the trends suggested in the early period of logical positivism may be considered by now to have been completely overcome, then it is surely the trend towards a logical unification of language, which would make all kinds of statements and utterances subject to the same rules. It is another matter that the burden of proof that a given kind of context requires a special treatment rests on the person who makes such a claim, and not on his opponent. We have tried to provide here, if not proofs, then at least circumstantial evidence of the logical peculiarities of oblique speech. This was followed by tentative formulations of prescription for handling those peculiarities.

III. THE MONISTIC THEORY OF OBLIQUE SPEECH

In the previous section we have discussed three kinds of oblique contexts: objective, propositional, and indirect; for each we have suggested a different theory to reflect some semantic and syntactic properties of these contexts and to provide measures for solving certain logical problems characteristic of them. There are two reasons why we are not confining ourselves to the proposals formulated above and proceeding to outline another theory, this time one intended to be universal in nature and hence capable of describing all oblique contexts in a uniform way.

First, there are many oblique contexts which are not covered by any of the categories listed above and, moreover, do not form a homogeneous group for which a tentative fourth theory could be formulated. These include: expressions formed by modal expressions 'it is necessary that', 'it is possible that', etc.; also, expressions which are clearly extensional in nature — which distinguishes them from all other varieties of oblique speech — such expressions as 'it is true that', it is not true that', 'it is the case that', 'it has happened that', etc.; and finally, expressions with evaluating phrases: 'it is important that', 'it is advantageous that', etc. (in some cases, the grammatical rules that govern the surface structure of English may require somewhat different formulations, depending on the overall context — Tr.). Other varieties could, perhaps, be found, too. It would be legitimate, and perhaps even plausible, to work out separate methods of description and separate logical rules for each of those varieties, but such a procedure would certainly be neither economical, nor elegant, nor theoretically interesting.

Secondly, even if there were not so many varieties of oblique speech, and if it were therefore possible to confine ourselves to the theories outlined above, it would nevertheless be tempting to find a uniform method of handling all the facts described above, and precisely for the reason mentioned above. We usually have to pay a certain price for making a theory more comprehensive. The costs of the undertaking will not be concealed; on the contrary, they will rather be emphasized, because the realization of this fact represents a theoretical profit. It is not to be decided here whether the advantages in the form of the values listed above, not excluding the aesthetic values, are great enough to outweigh or even just to counterbalance the costs. In such issues we have no scale of measures, and the choice is ultimately a matter of the personal preferences of authors and readers. But an author's task is to provide data on which individual choices can be based, and this is our present purpose.

3.1. *The ontology of states of things*

We shall now refer to an idea formulated earlier in connection with objective contexts. This is the idea that the oblique phrase, together with 'that', refers to a state of things which — in the case of contexts which have been termed objective — bears a relation to a certain perceiving system. It is natural to speak here about a relation, because we are used to treating physical action as a kind of relation, and the contexts involved can also naturally be treated as statements about the physical actions of certain stimuli (i.e., states of things) upon the receptors of a cognizing system.

If we now decide on a broad extension of the concept of relation or property (relations can be interpreted as properties of pairs, triples, etc.), then we can treat all oblique contexts as statements each of which ascribes a certain property, or a relation with respect to something, to a certain state of things. This extension of the concept of property, which on some points may seem unnatural or at variance with philosophical assumptions which are otherwise recommendable, will be the main cost of the present attempt at a solution.

Our first step ought therefore to consist in a precise formulation of the concept of a state of things, a concept which is too obscure in its current use to serve its purpose here.

We shall therefore adopt the following ontological and semantic assumptions:

(Z.1) An object's having a property, a pair of objects being connected by a two-place relation, etc., is an atomic state of things. Let W stand for any property, and x, for any individual. A state of things which consists in x having the property W will be symbolized $\langle W, x\rangle$; the notation for relations in such a case will be analogous: $\langle R^n, x_1, \ldots, x_n\rangle$, where R^n stands for an n-place relation.

(Z.2) Molecular states of things include negative states of things and conjunctive states of things. A negative state of things $\overline{X}$ consists in the fact that X does not hold. A conjunctive state of things $X \cdot Y$ consists in the fact that both X and Y hold (they are called components of a conjunctive state of things).

(Z.3) Names of states of things form a subcategory in the semantic category of terms, along with such subcategories as names of individuals and names of properties and relations.

(Z.4) Certain properties and relations can be predicated about states of things, e.g., 'It is good for the grain crops that it is raining' — in symbolic notation: $S\langle W, x\rangle$.

(Z.5) The atomic states of things are identical when they consist of the same individuals and the same properties. In symbolic notation: $\langle W, x\rangle = \langle U, y\rangle \equiv (W = U \text{ and } x = y)$.

(Z.6) The negative states of things $\overline{X}$ and $\overline{Y}$ are identical when X and Y are identical. Two conjunctive states of things are identical when all their components are identical pairwise.

(Z.7) The identity of individuals: $x = y \equiv \bigwedge F(Fx \equiv Fy)$.

(Z.8) The identity of properties can be interpreted in two ways:
(a) $A = B \equiv \bigwedge x\,(A\,(x) \equiv B\,(x))$ — extensional interpretation,
(b) $A = B \equiv \bigwedge F(F(A) \equiv F(B))$ — intensional interpretation.

(Z.9) The structure of A is a property in the sense of (Z.8.b) of a compound property A (because it is something which can be predicated about A); the former being understood as a complex system of component properties and the relations between them.

(Z.10) The principle of replacement for oblique contexts: If in a statement $S\langle W, x\rangle$,$\langle W, x\rangle$ is replaced by $\langle U, y\rangle$, then the logical value of the initial statement remains the same if $\langle W, x\rangle = \langle U, y\rangle$.

It is the adequately comprehensive concept of property which makes it possible to unify all oblique contexts. This broadening of that concept will probably be the philosophically most controversial issue, since properties and relations will cover the fact that something is possible or necessary, i.e., the modalities, the fact that something exists, the fact that something is or has been thought, uttered, believed, etc. It is common knowledge that the problem of whether existence is a property has been the subject matter of many philosophical controversies. Treating formulations in the form 'x thinks about y' and 'x speaks about y' as descriptions of certain relations may seem philosophically suspect, too. But an admission of doubts does not imply a privilege for answers in the negative — in our case answers which would refuse to existence, modalities, etc., the status of properties. This is why it is worth while to avail ourselves of a maximally broad concept of property for the benefit of those who would be inclined to negate the answer in the negative.

Other difficulties are connected with the concept of a molecular state of things: there are writers who are reluctant to accept the concept of a negative or a disjunctive state of things (the last-named can be defined in terms of a negative and a conjunctive state of things).

These concepts could perhaps be defended on the basis of common sense: the fact that there are no traffic lights at a crossing is also a state of things, one which even has certain consequences in the form of positive states of things, such as a traffic jam. Likewise, the fact that a red or a green light is burning and that neither both are turned on at the same time nor only one of them exclusively, is also a state of things — this time a disjunctive one. Even those states of things to which quantified statements after 'that' refer could probably be handled in some way, for instance by being treated as generalized conjunctions or disjunctions. A more troublesome point is the fact that current language, in which we are interested here and in which states of things are usually described, lacks the clear atomic structure postulated above. For instance, what state of things corresponds to the formulation 'the fact that the capital of the Republic lies on the Tiber'? If the phrase 'the capital of the Republic' were taken as a proper name, we would be dealing with an atomic state. But what if it is treated differently? We are familiar with the interpretation of categorical statements like 'the capital of the Republic lies on the Tiber' as hypothetical ones, whether with or without an existential assumption (concerning the existence of the objects which satisfy that statement). We would then have: 'for every x, if x is the capital of the Republic, then x lies on the Tiber', with 'and there is an x such that x is the capital of the Republic' in the case of the existential interpretation. Could something like that really be the point of a current language formulation? If our answer is in the negative, then the ontology of the states of things, which here, as in the case of Wittgenstein's *Tractatus*, is modelled on the language of logic, will have to be restructured so that it complies with the 'ontological commitment' of natural languages.

So much for our difficulties. But even if these are left unsolved one may ask about the consequences of the proposal made here. These consequences are associated with the semantic and syntactic characteristics of oblique speech. Thus, the oblique phrase 'that $W(x)$' is a term (a name) which denotes a state of things which consists of the thing x and the property W. The function of denoting is not accompanied here by one of connoting, which — as has been noted by J. St. Mill — is characteristic of abstract terms. In this connection it must be borne in mind that the concept of connotation is not to be identified with that of meaning. Connotation is that kind of meaning which characterizes a certain kind of term, namely those which are both concrete and general, i.e., which are not proper names; other kinds of expressions have others types of meaning as their attributes. Thus abstract terms do mean something, since there are rules governing their use in the process of communication; they also denote something, since they are names of states of things.

The syntactic analysis of oblique contexts, based on the assumptions listed above, results in singling out, in statements of the form 'x S that $W(y)$', the functor (predicate) S of two arguments and two arguments, x and *that* $W(y)$. The oblique phrase is sometimes preceded by a functor of one argument, e.g., 'it is true'. Since $W(y)$ lies within the syntactic category of sentences, the word 'that' must be included in the category of term-forming functors of one sentential argument.

Another syntactic analysis is also possible. If we assume that states of things are also, or exclusively, denotations of declarative sentences (statements), and if we also treat 'that' as a component of the main functor, then that functor must be interpreted as a sentence-forming functor of one sentential and one term argument, or, in other cases, of one sentential argument only. This line of semantic-syntactic analysis will not, however, be pursued here because it would take us far beyond the limits of our considerations by making us add a separate section on the assumptions underlying such an analysis and on its consequences.[7]

3.2. *Replacement in oblique contexts treated as names of states of things*

The rule of replacement is the decisive criterion of the correctness of a theory of oblique speech (such a rule being part of that theory). If it can prevent the paradoxical phenomenon whereby replacement by an equivalent expression sometimes turns a true statement into a false one, then a theory which includes such a rule qualifies for the "finals" if there are any rival theories in the field.

The rule of replacement formulated in (Z.10) above splits into two versions according to the interpretation of the identity of states of things, which in turn depends on the concept of the identity of properties. There are at least two such concepts, expressed by formulations (a) and (b) in (Z.8). By translating them into the well-known language of denotation and connotation we can say that states of things can be identical either in view of the identity of the denotations of the concrete terms which are components of the name of a given state of things, or in view of the identity of the connotations of those terms. This is why it is convenient to call the concept of a state of things which is partially defined by the extensional version (a) a denotational one, and that which is partially defined by the intensional version (b), a connotational one.

In order to formulate two versions of the rule of replacement we must first single out the two kinds of contexts to which these two versions would respectively apply. Let us consider, by way of example, contexts with the word 'thinks'. When a friend of mine, seeing that I am preoccupied with something, asks me 'What are you thinking about?', and if I reply 'I am thinking about my cook', then he can interpret that in one of two ways. Either he will treat the phrase 'my cook' as replacing the proper name of that person, let us say 'Mary', and then he will only learn about the subject matter of my thoughts, but not about their content (I may think of Mary as a careless cook or as a buxom blonde), or else he will treat that phrase as information about the content of my thoughts, i.e., he will realize that I am thinking about that person *qua* cook, and am, for instance, considering her vocational qualifications. Let us consider another example. 'The leader of the Hussites was considered to be a brave man.' In this case, too, there are two possible interpretations: (i) the phrase 'the leader of the Hussites' merely replaces the name of John Hus (such a replacement may be recommended, for instance, for stylistic reasons), and hence is used only with the intention of indicating a certain person, without any intention of providing information about him, (ii) the phrase 'the leader of the Hussites' is used to indicate the aspect in which the speaker is interested in the person of John Hus, and then the reference to his being brave is to be understood in such a way that John Hus was brave as the leader of the Hussites and revealed the kind of braveness (the reference could also have been to prudence, cleverness, popularity, etc.) which is expected in a leader of a religious and social movement. By analogy, one could use the phrase 'the capital of the Papacy' only as a replacement for the proper name 'Rome', or else to indicate that one thinks about a given object as the capital of the Papacy.[8]

As can be seen, the word 'thinks' and a number of related words — let us call them epistemic functors — are marked by a certain ambiguity or bifunctionality since on one occasion they are referred as it were to an "entire" object, and on another occasion, to one of its aspects. These roles will be termed the objective and the aspectual function of epistemic functors, respectively.

By availing ourselves of this terminology we can formulate two further theorems which reflect the intuitions illustrated by the examples given above, and which serve as meaning postulates for the terms 'objective function' and 'aspectual function'.

(Z.11) The use of an epistemic functor in its objective function indicates the denotational interpretation of the state of things to which the phrase that is the argument of that epistemic functor refers.

(Z.12) The use of an epistemic functor in its aspectual function indicates the connotational interpretation of the state of things to which the phrase that is the argument of that epistemic functor refers.

We have thus brought about a meeting of two independent trends in semantic analysis concerned with different components of oblique speech. We have availed ourselves of the two functions of epistemic terms, which we have combined with the fact that the two criteria of identity, when applied to properties, yield two conceptions of the identity of states of things, and hence two conceptions of states of things. Each of these two conceptions of states of things has been assigned a different function of epistemic functors.

The natural conclusion is to adopt two versions of the rule of replacement, one for contexts to which (Z.11) applies, and the other for contexts to which (Z.12) refers. The former states that if an epistemic functor is used in its objective function, then the replacement of terms which are components of the second argument of that functor must be made while preserving the denotation of these terms, but not necessarily while preserving their connotation. Thus, the phrase 'Evening Star', 'Morning Star', and 'Venus' may be used alternately; the same applies to the phrases: 'the capital of the Republic', 'the capital of the Papacy' and 'Rome', and to: 'John Hus' and 'the first leader of the Hussites'.

The second version states that if an epistemic functor is used in its aspectual function, then the replacement of terms which are arguments of the second argument of that functor must be made while preserving the connotation of those terms. Hence one cannot replace the phrase 'King Henry VIII' by 'the father of Queen Elizabeth I', but the latter may be replaced by 'the direct male ancestor of Queen Elizabeth I', because these two have an identical connotation for the interpretation of identity which is defined by (Z.8.b).

There is an obvious analogy between these two rules and the rules governing the reproduction of texts, whether actual or potential, discussed in 2.2. and 2.3. The first of the rules analysed above has its analogue in the reproduction of a text with an exactitude of denotation, the second, in the reproduction of a text with an exactitude of connotation. A closer comparison of these two approaches, which must inevitably lead to the question whether the two rules can be observed simultaneously or whether a choice between them must be made in a given case, must be postponed until we are in a position to shed some light on the concept of the intention which guides the use of words. For in our case we should first have to answer the question whether it is possible to use oblique speech with the double intention of describing both a certain state of things and either a certain text or the speaker's mental disposition, or whether only one type of intention can occur in a given case. Nor would it be easy to explain how such intentions might be discovered.

The choice between different theories in semiotics requires an explanation of one more issue. There is certainly a grain of truth in that approach in the philosophy of science which is termed instrumentalism: theories need not carry absolute truth — they serve as tools used for specified tasks, and at a given moment that theory ought to be chosen which can better serve the purpose of the moment. Thus the burden of choice is made to depend on the nature of one's research tasks. But whereas in the natural sciences one has a fairly good idea as to what those tasks are — it suffices to mention here the prediction of future events and discoveries — in the case of less empirical and more analytic theories, which seem to include logical semiotics, these tasks are rather vague. This is why the criteria of choice between the various theories are very difficult to formulate precisely. It is, therefore, one of the most urgent tasks of semiotics to define better the tasks of that discipline.

NOTES

[1] Cf. S. Szober, *Gramatyka języka polskiego* (A Grammar of Polish), Warsaw 1963, p. 101.

[2] Cf. R. Carnap, *Meaning and Necessity*, Chicago 1956.

[3] Y. Bar-Hillel, 'A Quasi-Arithmetical Notation for Syntactic Description' in: Y. Bar-Hillel, *Language and Information*, Addison-Wesley 1964. It is to be noted, however, that on other occasions the same author advances a different proposal for syntactical analysis. A description coinciding with the one quoted here is also suggested by R.C. Jeffrey in *The Logic of Decision*, New York 1965.

[4] A. Fraenkel and Y. Bar-Hillel, *Foundations of Set Theory*, Amsterdam 1958, p. 120; K. Ajdukiewicz, 'Intensional Expressions', pp. 96-125 in the present book.

[5] Bergmann also objects to an analysis of the concept of seeing if the latter is interpreted as a mental fact, and not merely as a physical one. He says in this connection: I do not believe that an instance of seeing, or of any other awareness, is merely the exemplification of a relation, or any other character between two "things", as indeed the seeing is. (G. Bergmann, 'Intentionality' in: G. Bergmann, *Meaning and Existence*, Madison 1960, p. 6).

[6] The idea of such definitions was formulated by R. Carnap in his *Testability and Meaning*, 1936. For a detailed analysis of reductive definitions see M. Przełęcki, 'Pojęcia teoretyczne a doświadczenie' (Theoretical Concepts and Experience), *Studia Logica*, Vol. XI, 1961.

[7] The thesis that the denotatum of a statement is its logical value is well-grounded in logical tradition. On this point the otherwise rival theories of, e.g., Frege and Church, Carnap, and Ajdukiewicz, are in agreement. The same authors also agree that a proposition is the connotation, meaning, or intension, of a statement (these concepts are closely related, even though they occur in different theoretical contexts). The assumption adopted here (that which is the connotation of a direct statement '*p*' is the denotation of the corresponding oblique formulation 'that *p*') comes closest to the idea of Frege and Church, the difference being that a proposition is treated as a state of things of the real world, and not as an abstract, or ideal, object. For the views of the authors mentioned above concerning the semantic functions of statements see G. Frege, 'Ueber Sinn und Bedeutung', *Zeitschrift für Philosophie und Philosophische Kritik*, 1892, No. 100 (the English-language version is to be found in H. Feigl and W. Sellars, *Readings in Philosophical Lin-*

guistics, New York 1949); A. Church, *Introduction to Mathematical Logic*. Vol. 1, Princeton 1956; R. Carnap, *Meaning and Necessity*, Chicago 1956; K. Ajdukiewicz, 'Proposition as the Connotation of Sentence', pp. 81-95 in the present book.

[8] The various referents of one and the same term which change according to situation or context were discussed by the mediaeval theory of suppositions (cf. I. Bocheński, *Formale Logik*, Munich 1956). The issue has been revived, independently of the theory of suppositions, in certain analyses carried out by contemporary authors; see, for instance, P.F. Strawson, 'On Referring', *Mind*, Vol. 59, 1950; J. Pelc, 'A Functional Approach to the Logical Semiotics of Natural Language', pp. 342-75 in the present book.

THE SEMANTIC CONCEPTION OF TRUTH IN THE METHODOLOGY OF EMPIRICAL SCIENCES

by

Ryszard Wójcicki

Introduction

This paper is concerned with the problems of logical semantics, more precisely, with the truth theory. It brings an attempt to give an extension of this theory. The definition of the phrase '*is a true sentence*' which was given by Tarski seems to supply us with a notion whose usefulness for the methodology of empirical sciences is limited. To put it more exactly let us give a rough outline of Tarski's conceptions.[1]

In a certain language sufficiently rich for this purpose another language is subject to consideration together with its domain, i.e. this fragment of reality (e.g. a fragment of the real world or a system of mathematical concepts) which is spoken about. Let us denote the first of these two languages by the symbol *ML*, the latter by *L*, and finally the domain of *L* by $\mathfrak{M}$. The language *ML* is thus a metalanguage for the language *L*. We shall assume that *ML* satisfies the following two conditions: (i) it contains the names of all the expressions of the language *L*, in particular it contains the names of the sentences of this language, (ii) every sentence *a* of *L* has its meaning equivalent (a translation) in *ML*, thus, all that can be stated about $\mathfrak{M}$ in *L* can as well be stated in *ML*. It follows that for each sentence *a* of the language *L* we are able to form in *ML* a statement of form:

(*) ... *is a true sentence of the language* L *if and only if* - - - ,

where the name of the sentence *a* should be put in place of the dots, and the translation of the sentence should be put in place of the dashes.

Note that the first part of the equivalence (*) pertains to *L* and the second to the domain of the language *L*. If there is a particular kind of relationship between the sentence *a* and the domain to which it pertains, i.e. if the sentence states the facts which take place in the domain, or, to put it in other words, if the sentence conforms with the reality, then it is true. In his definition Tarski aims at reconstructing the intuitions of the classical conception of truth.

Every metalinguistic statement of the form (*) defines the phrase "is a true sentence of the language *L*" relative to a sentence of the language *L*. Thus, such statements are partial definitions of truth. The complete definition must be a sentence which entails all the partial definitions. Let us note that if we wanted to define the notion of a true sentence for a natural language (e.g. a fragment of the Polish language) and using for that purpose some other natural language (e.g. English), then the problem would be reduced to finding a finite number of prescriptions completely determining the rules of translating the statements of the first language into the statements of the other.

Proceeding this way we might hope to succeed in reducing the infinite class of the statements of the form (*) e.g.:

'*Śnieg jest biały*' *is a true sentence in Polish if and only if snow is white.*

to a finite number of general rules which establish principles of adequate translation. These rules would amount to an implicit definition of the phrase *is a true sentence in the Polish language*.

Imprecise and whimsical nature of the syntactic and semantic structures of natural languages makes us expect that for such languages it is impossible to establish completely adequate and at the same time sufficiently general rules of translation. Anyway the investigation of natural languages is the task of the linguists rather than logicians. Dealing with the concept of truth, the logicians focus their attention on the artificial languages. They construe different set-theoretic patterns of domains, forming at the same time the languages in which those domains can be described. Some set theory is always used to provide the formal framework for these considerations. The universality of set theories (regardless of which one is selected) guarantees considerable generality of such considerations; at the same time the mathematical character of these theories guarantees them the sufficient standard of precision. If we use the term 'model' in the most common sense, we can say that the logic investigates the relationship between language and reality by using various models. The languages investigated in logic can inherit in some idealized form the properties of the languages which were "naturally" developed. The models of the reality, i.e. the domains constructed with the help of set-theoretic notions, can resemble the domains investigated in the various branches of science. To the extent to which these similarities take place, logical semantics refers to natural languages (e.g. the languages of particular branches of science), to the domains of these languages, and to the relationships between these languages and their domains. One might think that the semantic truth theory should be most suitable for the description of the formal aspects of the relationships holding between the mathematically grounded empirical sciences and their domains.[2] In fact the languages of the sciences of that kind are particularly close to those investigated in logic. Note, however that the theorems having the form of mathematical equations are, as a rule, of approximative character, i.e. they are true only in certain intervals of exactness. Grasping the discovered regularity in the form of a mathematical equation usually assumes some idealization of the processes ruled by this regularity. This may serve to explain the presence of such concepts as 'material point' or 'ideal gas' in handbooks of physics. If we treat literally, e.g. the concept of material point, then the respective branch of theoretical mechanics will refer to fictitious objects and thus the problem of truth in this branch of physics will not exist at all. However, if we consider some physical objects of finite dimensions or other as the material points, then the laws of mechanics of material points will appear merely approximatively true. Still Tarski's theory offers the rigid opposition 'truth — falsehood'. There is no room for approximate truth in it. Certainly, if one insists, it may be assumed that the majority or even all basic laws of those disciplines which utilize the mathematical apparatus, are easy to identify as false. This standpoint is completely sterile, though it can certainly be justified on the grounds of the decisions pertaining to the understanding of the term 'truth'.

The attempts which I undertook to rebuild, in a way, the semantic theory of truth are motivated by the facts discussed above. I will aim now at finding such semantic notions that would allow a description of the formal aspects of passing in the research from the real domains being described by the particular empirical sciences, to some idealizations of these domains; I will also search for such semantic notions that would make it possible to account for the way the empirical sciences apply to the real domains, however, through the idealized domains.

The majority of the technical concepts which I am going to make use of, will be explained with the help of the respective definitions. Nonetheless the reader of this paper is expected to show some familiarity with the semantic problems and with the formal apparatus applied in logical semantics (thus, e.g some preliminary knowledge of set theory is here assumed).

1. *Semantics of elementary languages. Arithmetical domains*

Let us consider any relational system of the form $\langle U, R_1, ..., R_m \rangle$, where U is a non-empty set, and $R_1, ..., R_m$ are $p_1, ..., p_m$-ary relations respectively. We do not exclude that some of them are unary relations, i.e. properties. The set U will be called the *range* of the relational system, the sequence $p_1, ..., p_n$ will be called the *similarity type* of this system. Two relational systems of the same similarity type will be said to be *similar*.

Relational systems will also be called *domains* since they are capable of being domains of a language. The domain $\langle U, R_1, ..., R_m \rangle$ will be symbolized by $\mathfrak{M}$. The task that we put forward here is to construct a language L which is conceived to be a language serving for describing the domain $\mathfrak{M}$. The domain $\mathfrak{M}$ will be called the *proper domain* of L. We shall describe the language L and its properties in a sketchy way. Assume that the vocabulary of L consists of:

i. the relational symbols $r_1, ..., r_m$ denoting $R_1, ..., R_m$ respectively.

ii. the individual constants denoting the elements of U; we shall assume that for each v in U there is a constant a_v denoting v.

iii. individual variables $x_1, ..., x_i, ...$ ranging over the set U.

iv. the logical symbols: $\sim$, $\rightarrow$, $\bigwedge$ (the negation symbol, the implication symbol, and the universal quantifier symbol, respectively). If it is necessary we may enrich the vocabulary of L by additional logical symbols defining them by means of those listed so far. Thus, e.g., $a \vee \beta$ is to be treated as an abbreviation for $\sim a \rightarrow \beta$ and $\bigvee xa(x)$ as an abbreviation for $\sim \bigwedge x \sim a(x)$.

v. punctuation signs such as parentheses, comma, dot, etc.

The set of sentences of L is assumed to be defined in the usual manner.

It will be denoted by *Sent* (L). In view of the fact that the vocabulary of L contains the names of all elements of U, the notion of a true sentence of L can be defined in a standard way.[3] Let *Ver* (L) be the set of true sentences of L. Clearly, an atomic sentence

$$r_i\,(a_{v_1}, \ldots, a_{v_k})$$

is a true sentence of L if and only if the relation R_i holds among the elements $v_1, \ldots, v_k$ of U, i.e. $R_i\,(v_1, \ldots, v_k)$. The truth conditions for the sentences of L built by means of logical connectives are to be formulated in a well-known manner. Now let $a \in$ *Sent* (L) be a sentence of the form

$$\bigwedge x\,\beta\,(x)$$

Clearly a is true if and only if for each individual constant a_v the sentence $\beta\,(a_v)$ (i.e. the sentence resulting from the formula $\beta\,(x)$ by substituting for each occurrence of the variable x of the individual constant a_v), is true.

Observe that the adequacy of this condition is due to the fact that for each v in U there is an individual a_v denoting v. However, if we have the notion of a true sentence of L defined, we may easily define the notion of truth for each sublanguage L' of L, even for one which does not contain individual constants at all. By a *sublanguage* of L we clearly mean any language L' such that the set of sentences of L', *Sent* $(L') \subseteq$ *Sent* (L). And obviously we define the set of true sentences of L', *Ver* (L') thus:

$$Ver\,(L') = Ver\,(L) \cap Sent\,(L).$$

It is obvious that the language L which has been thought of as a language for $\mathfrak{M}$, may as well serve as a language for any domain $\mathfrak{M}'$, provided that $\mathfrak{M}'$ is a domain of the same similarity type as $\mathfrak{M}$.

Let $\mathfrak{N} = \langle V, P_1, \ldots, P_m \rangle$. Informally speaking a is true in $\mathfrak{N}$ if it is true assuming that $r_1, \ldots, r_m$ denote $P_1, \ldots, P_r$ respectively, the variables of L range over V, and that the individual constants denote the elements of V in a definite manner. The set of sentences of L which are true in $\mathfrak{N}$ will be denoted by $Ver_{\mathfrak{N}}\,(L)$. Thus $Ver_{\mathfrak{M}}\,(L) = Ver\,(L)$.[4] Note that the description of the domain $\langle U, R_1, \ldots, R_m \rangle$ where U is a set of physical objects of this or another kind, whereas $R_1, \ldots, R_m$ are the relations defined in the set, bears the character of a qualitative description.

We shall however be interested above all in these disciplines whose theorems form a quantitative description of the domain investigated in this branch of science. The languages which make the quantitative description possible must include, apart from the logical terms, also the terms denoting mathematical concepts. At the same time the domains of such languages should differ from the "qualitative" domain $\langle U, R_1, \ldots, R_m \rangle$.

We shall now be concerned with a modification of the concepts previously defined. Consider the system

$$\mathfrak{N} = \langle U, R, F_1, \ldots, F_n \rangle$$

where U is a non-empty set of objects of any kind whatsoever, R is a set of real numbers and $F_1, \ldots, F_n$ are respectively the $q_1, \ldots, q_n$-argument functions with the first argument being the element of R, and their subsequent arguments belonging to U, whereas their values are taken from the set of real numbers R. The functions $F_1, \ldots, F_n$ will be called *quantities* or *arithmetical quantities*.[5] The system $\mathfrak{N}$ will be called an *arithmetical domain*, and the sets U and R, will be called the *first* and the *second range* of the domain, respectively.[6]

Aiming at constructing the language $Ł$ in which it would be possible to give a description of the domain $\mathfrak{N}$, we must face the problem which differs from the one we have been dealing with while constructing a language for the domain $\mathfrak{M}$, in this respect that apart from the individual constants and apart from function symbols we shall have to provide $Ł$ with the names of real numbers and with symbols of certain operations on real numbers (e.g. $+$, $\cdot$, *ln*). At the same time we shall have to introduce two kinds of variables: those which represent the elements of U, and those which represent the elements of R. Correspondingly, the vocabulary of the language $Ł$ will consist of the following symbols:

i. individual constants,
ii. function symbols: $f_1, \ldots, f_n$,
iii. constants symbolizing real numbers,
iv. symbols of mathematical operations $e_1, \ldots, e_l$, assuming that they denote the operations $E_1, \ldots, E_l$, respectively,
v. variables representing the elements of U: $x_1, \ldots, x_i, \ldots$
vi. variables representing the elements of R: $v_1, \ldots, v_i, \ldots$
vii. logical constants: $\sim, \rightarrow, \wedge, =$,
viii. punctuation signs.

The system of mathematical concepts $\langle E_1, ..., E_l \rangle$ will be called *mathematical formalism* of the language $Ł$. We have assumed that $E_1, ..., E_l$ are the operations on numbers; thus they are functions defined on the set R whereas their values are taken from the same set. Let us observe that $Ł$ contains the identity symbol $=$, which did not appear in L. Similarly in case of the language L we should assume that the set of sentences of $Ł$, *Sent* ($Ł$), is defined in a familiar manner. We shall neither define the set of true sentences of $Ł$, *Ver* ($Ł$), nor shall we define the set of sentences of $Ł$ true in a domain $\mathfrak{N}'$ similar to $\mathfrak{N}$. The latter one will be denoted by $Ver_{\mathfrak{N}'}$ ($Ł$). Both concepts may be, however, defined in a quite straightforward manner. We have only to keep in mind that the symbols $e_1, ..., e_l$ are always to be interpreted as $E_1, ..., E_l$, i.e. their meanings do not depend on which arithmetical domain is considered to be the domain described by the language $Ł$. I believe that this remark makes the notion of a sentence true in $\mathfrak{N}'$ clear enough. Note, however, that according to our assumptions, $\mathfrak{N}$ plays a role of the proper model of $Ł$ and hence $Ver(Ł) = Ver_{\mathfrak{N}}(Ł)$.

2. *Physical domains and their idealizations*

We shall pass over now to consideration of the domains which, according to their structure, are similar to the domains investigated in empirical theories utilizing the quantitative description.

Let U be the set of arbitrary material objects and let T be the set of "instantaneous" events. The requirement that the elements of U be the objects of material character and T the events, is impossible to be expressed in the language of set theory; then from a purely formal point of view, U and T are merely two non-empty sets. Let $F_1^*, ..., F_n^*$ be the functions of $q_1, ..., q_n$ arguments respectively, with the first argument always being a member of the set T, and the remaining arguments being elements of U. Let us also assume that $\Omega_1, ..., \Omega_n$ are the sets of values of the respective functions $F_1^*, ..., F_n^*$.

An $n+1$-range system (where $U, T, \Omega_1, ..., \Omega_n$ are the subsequent ranges of this system):

$$\mathfrak{F} = \langle U, T, \Omega_1, ..., \Omega_n, F_1^*, ..., F_n^* \rangle$$

will be called a *physical domain*. We do not decide in any way whatsoever the nature of the objects in Ω_i.

It may however be admitted that with each quantity F_i there is correlated an equivalence relation $\approx_i$ (i.e. a relation which is reflexive, symmetric and transitive) which is defined in the set of the q_i-element systems of the form

$$\langle z, u_1, \ldots, u_{q_i-1} \rangle$$

where $z \in T$, and $u_1, \ldots, u_{q_i-1} \in U$; we may also admit that

$$F_i^* (z', u_1', \ldots, u'_{q_i-1}) = \{\langle z, u_1, \ldots, u_{q_i-1}\rangle : \langle z, u_1, \ldots, u_{q_i-1}\rangle \approx_i \langle z', u_1', \ldots, u'_{q_i-1}\rangle\}.$$

The symbol $\{x: \varphi(x)\}$ denotes the set of these objects x which satisfy the condition $\varphi(x)$. The explanations given above seem to make it clear that instead of the domain $\mathfrak{F}$ the following two-range domain might be taken into consideration

$$\langle U, T, \approx_1, \ldots, \approx_n \rangle.$$

Let us symbolize by $[R]$, the set of the closed intervals of real numbers. Let us then consider some (not necessarily one-one) mappings: $\theta : T \to [R]$, and $\omega_i : \Omega_i \to [R]$, where $i = 1, \ldots, n$. If $z \in T$, then $\theta(z)$ will be called a *time instant* approximately determined by the mapping ω_i. If $u \in \Omega_i$ then $\omega_i(u)$ will be called the *numerical approximation* of the value of u, determined by the mapping ω_i. $\theta(z)$ and $\omega_i(u)$ may be treated as the results of the respective measurements. As long as we do not idealize the measurement operations they give as a result a certain numerical interval but not an isolated number. This is due not only to the fact that each measurement is deformed by an error resulting from imperfect measurement techniques. The kind of objects with which a given magnitude is being measured always determines the upper bound of accuracy with which this magnitude may be determined for these objects. Stepping outside this line the measurement operations become absolutely devoid of any physical meaning. This observation seems trivial; let us consider then only one example. The mass of the pencil that I am writing with may be determined with the accuracy $\pm 10^{-4}$ g, or may be $\pm 10^{-5}$ g; the accuracy may be, perhaps still greater, though obviously not arbitrary. It is subject to constant oscillation in a certain numerical interval if we express the mass in numbers. The dust penetrates the outer surface of the object. Certain atoms of the surface of the pencil evaporate. No object can be perfectly isolated from its surrounding (this would be the condition for a perfect measurement).

We shall define the language $Ł_{\theta, \omega_i}$ aiming at constructing such a language in which the domain $\mathfrak{F}$ might be described in a particular way. Assume first that the vocabulary of the language $Ł_{\theta, \omega_i}$ coincides with the language $Ł$ already defined. Assume also that the set of sentences *Sent* $(Ł_{\theta, \omega_i})$ of the language $Ł_{\theta, \omega_i}$ is identical with the set *Sent* $(Ł)$. However, the set of true sentences *Ver* $(Ł_{\theta\, \omega_i})$ of the language $Ł_{\theta, \omega_i}$ will be defined in quite a different way.

Every system of the foım

$$\mathfrak{N}' = \langle U, R_T, F_1', ..., F_n' \rangle$$

will be called the *idealization* of the domain $\mathfrak{F}$ if and only if the following conditions are satisfied:

(i) $\mathfrak{N}'$ is a system similar to $\mathfrak{N}$ (i.e. $F_1', ..., F_n'$ take, respectively, the same number of arguments as the functions $F_1, ..., F_n$ of the domain $\mathfrak{N}$, and in the same ranges; moreover the values of parameters $F_1', ..., F_n'$ are real numbers).

(ii) There exists a mapping $f: T \rightarrow R_T$ and for an arbitrary $z \in T$, and for arbitrary $u_1, ..., u_{q_i - 1} \in U$ there exists a choice function $\varepsilon^i_{z, u_1, ..., u_{q_i}-1}$ defined on the respective set Ω_i such that

(a) for every $z' \in T, f(z') \in \theta(z')$

(b) for every $w \in \Omega_i, \varepsilon^i_{z, u_1, ..., u_{q_i}-1}, (w) \in \omega_i(w)$

(c) if $F_i^*(z, u_1, ..., u_{q_i - 1}) = w$, then $F_i'(f(z), u_1, ..., u_{q_i - 1}) = \varepsilon^i_{z, u_1, ..., u_{q_i}-1}(w)$.

Note that every idealization $\mathfrak{N}'$ is similar to the domain $\mathfrak{N}$ by virtue of the condition (i) and, at the same time, it was assumed that *Sent* $(Ł_{\theta, \omega_i})$ = *Sent* $(Ł)$. This allows us to admit that the set of sentences of $Ł_{\theta, \omega_i}$ true in $\mathfrak{N}'$ coincides with $Ver_{\mathfrak{N}'}(Ł)$. Now we are in a position to define the set *Ver* $(Ł_{\theta, \omega_i})$ and the set of false sentences of $Ł_{\theta, \omega_i}$. The latter will be denoted by *Fals* $(Ł_{\theta, \omega_i})$.

Df. 1 $a \in$ *Ver* $(Ł_{\theta, \omega_i})$ if and only if $a \in$ *Sent* $(Ł_{\theta, \omega_i})$ and for each domain $\mathfrak{N}'$ if $\mathfrak{N}'$ is an idealization of the domain $\mathfrak{F}$, then a is true in $\mathfrak{N}'$.

Df. 2 $a \in$ *Fals* $(Ł_{\theta, \omega_i})$ if and only if $a \in$ *Sent* $(Ł_{\theta, \omega_i})$ and for each domain $\mathfrak{N}'$ if $\mathfrak{N}'$ is an idealization of the domain $\mathfrak{F}$, then a is false in $\mathfrak{N}'$.

If there are at least two different idealizations $\mathfrak{N}'$, $\mathfrak{N}''$ of $\mathfrak{F}$ then, some sentences of $Ł_{\theta,\,\omega_i}$ are neither true nor false. They will be called *undetermined*; the set of all such sentences will be denoted by $NDet\,(Ł_{\theta,\,\omega_i})$.

Df. 3 $a \in NDet\,(Ł_{\theta,\,\omega_i})$ if and only if $a \in Sent\,(Ł_{\theta,\,\omega_i})$ and $a \in Ver\,(Ł_{\theta,\,\omega_i}) \cup Fals\,(Ł_{\theta,\,\omega_i})$.

The notion of *approximative truth* which we are going to define now will be related to sets of sentences, not to single sentences. Let us denote by $AVer\,(Ł_{\theta,\,\omega_i})$ the class of all sets of sentences of $Ł_{\theta,\,\omega_i}$ which are sets of sentences true in $Ł_{\theta,\,\omega_i}$. The class $AVer\,(Ł_{\theta,\,\omega_i})$ will be defined as follows.

Df. 4 $X \in AVer\,(Ł_{\theta,\,\omega_i})$ if and only if $X \subseteq Sent\,(Ł_{\theta,\,\omega_i})$ and there is such an idealization $\mathfrak{N}'$ of the domain $\mathfrak{N}$ that each sentence $a \in X$ is true in $\mathfrak{N}'$.

Perhaps it is worthwhile noticing that the following assertions are valid.

Assertion I. $a \in Ver\,(Ł_{\theta,\,\omega_i})$ if and only if $\sim a \in Fals\,(Ł_{\theta,\,\omega_i})$.

Assertion II. $a \in NDet\,(Ł_{\theta,\,\omega_i})$ if and only if $\sim a \in NDet\,(Ł_{\theta,\,\omega_i})$.

Let us state also the following important facts.

Assertion III. $Ver\,(Ł_{\theta,\,\omega_i}) \in AVer\,(Ł_{\theta,\,\omega_i})$.

Assertion IV. If $X \in AVer\,(Ł_{\theta,\,\omega_i})$ and $Y \subseteq Ver\,(Ł_{\theta,\,\omega_i})$, then $X \cup Y \in AVer\,(Ł_{\theta,\,\omega_i})$.

We need some semantic notions which are necessary to give a more developed characteristic of the sets $Ver\,(Ł_{\theta,\,\omega_i})$, $Fals\,(Ł_{\theta,\,\omega_i})$, $NDet\,(Ł_{\theta,\,\omega_i})$, $AVer\,(Ł_{\theta,\,\omega_i})$. The notion of tautology and that of entailment are especially important in this respect. We shall define them in a sketchy way. By a *mathematical tautology* of the language $Ł$ we shall understand every sentence a which is true in each domain $\mathfrak{N}'$ similar to the proper domain $\mathfrak{N}$ of $Ł$. I should like to recall here that the meanings of the mathematical symbols $e_1, \ldots, e_i$ of $Ł$ are assumed to be fixed. Now, we shall say that the set of sentences X of $Ł$ *entails* a sentence a if and only if for every domain $\mathfrak{N}'$ similar to $\mathfrak{N}$ if all sentences in X are true in $\mathfrak{N}'$ then a is true in $\mathfrak{N}'$. We shall assume that the notion of mathematical tautology of the language $Ł_{\theta,\,\omega_i}$ coincides with the notion of mathematical tautology of the language $Ł$ and similarly the notion of entailment (or more precisely mathematical entailment) related to $Ł_{\theta,\,\omega_i}$ coincides with that related to $Ł$. With the help of these definitions it may be proved that the following assertions hold.

Assertion V. If a is a mathematical tautology of the language $Ł_{\theta,\omega_i}$ then $a \in Ver(Ł_{\theta,\omega_i})$.

Assertion VI. If $X \subseteq Ver(Ł_{\theta,\omega_i})$, is a sentence of $Ł_{\theta,\omega_i}$ and X entails a then $a \in Ver(Ł_{\theta,\omega_i})$.

Assertion VII. If $X \in AVer(Ł_{\theta,\omega_i})$, a is a sentence of $Ł_{\theta,\omega_i}$ and X entails a then $X \cup \{a\} \in AVer(Ł_{\theta,\omega_i})$.

Assertion VIII. If $X \subseteq Sent(Ł_{\theta,\omega_i})$ and X entails both a and $\sim a$ then $X \in AVer(Ł_{\theta,\omega_i})$.

The intuitive content of all these assertions seems to be quite evident. Perhaps the most important among them is the assertion VII which states among others, that if a set of axioms of a theory is approximately true, then the whole theory, i.e. the set of axioms together with all their consequences, is approximately true as well. Assertion VIII rephrased in a loose way states that none of the inconsistent sets of sentences is approximately true.

3. *Exemplification of the defined concepts*

To make the defined concepts more intuitive we shall produce some examples. Assume that we are given a certain finite (it could be infinite as well) sequence of physical domains

$$\mathfrak{F}_1 = \langle U_1, T_1, \Omega_1^1, \ldots, \Omega_n^1, F_1^*, \ldots, F_n^* \rangle$$

. . . .

. . . .

$$\mathfrak{F}_s = \langle U_s, T_s, \Omega_1^s, \ldots, \Omega_n^s, F_1^*, \ldots, F_n^* \rangle.$$

which are similar to the domain $\mathfrak{F}$ already discussed. These domains differ at most with regard to their ranges; we assume however that the parameters $F_1^*, \ldots, F_n^*$ are identical in all these domains. We might understand it in such a way that these are the same functions $F_1^*, \ldots, F_n^*$ restricted however in each of these domains to the respective terms. Every set of sentences Ξ of the language $Ł$ or of any other fragment of this language containing the symbols $f_r, \ldots, f_n$ will be called an uninterpreted theory of the domains $\mathfrak{F}_1, \ldots, \mathfrak{F}_s$. As a matter of fact, only the set of sentences which describe some particularly important regularities deserves the name 'theory'. An attempt to give account of this remark might lead us into pragmatic considerations which, regarding the character of this paper, should be avoided.

The language $Ł$ already described is unfortunately considerably poor since we had simplicity of our reasoning in mind. Namely, the mathematical formalism of the language $Ł$, consist of some operations on numbers only. Nevertheless, it is suitable for constructng at least some physical theories and to sketch an example of such a theory seems worthwhile taking the trouble.

We are now ready to state axioms for particle mechanics *PM*; they were first given by J. C. C. McKinsey, A. C. Sugar and P. Suppes.[7] The primitive terms of these axioms (or, in other words, the descriptive constants which occur in them) are as follows: P — the *set of particles*, T — an *interval of real numbers* (this set is to represent a certain time interval), s — the *position function* of a particle, m — the *mass* of a particle, f — the *force* exerted by one particle on another, g — the *external force* exerted on a particle. The axioms are as follows:

KINEMATICAL AXIOMS

Axiom P1. The set P is finite and non-empty.

Axiom P2. The set T is an interval of real numbers.

Axiom P3. For every x in P, the function $s(t, x)$ is twice differentiable in T. (Comment: we assume that the variable x represents the elements of the set P while t ranges over the set T).

DYNAMICAL AXIOMS

Axiom P4. For x in P, $m(x)$ is a positive real number.

Axiom P5. For x and y in P and t in T,

$$f(x, y, t) = -f(y, x, t).$$

Axiom P6. For x and y in P and t in T,

$$s(x, t) \times f(x, y, t) = -s(y, t) \times f(y, x, t).$$

Axiom P7. For x in P and t in T,

$$m(x)\,\ddot{s}(t, x) = \sum_{y \in P} f(t, x, y) + g(t, x)$$

(comments: f, g and s symbolize some vector quantities, m is the vector product operation, $\ddot{s}$ is the second derivative of s with respect to the time variable, and Σ is the sum operator).

At the first glance *PM* seems to be far enough from these theories which may be constructed in such language as $Ł$. First, *PM* contains some set theoretical notions: set, interval, set membership relation. Such notions do not appear in $Ł$. Secondly, *PM* contains some vector quantities apart from arithmetical (scalar) ones. Correspondingly, the formalism of *PM* contains some vector operations, e.g. the vector product x. Finally, *PM* contains some operations performable on functions. These are, e.g. derivation and sum operation Σ. Thus the differences between $Ł$ and the language of *PM* seem to be drastic. However, as it was shown in detail by R. Montague in one of his papers,[8] *PM* may be rephrased so as to be capable of being expressed in a language similar to $Ł$. The reader curious to know the details of the procedure is recommended to read the respective paper. We are going to discuss now some problems concerning the interpretation of the empirical theories. Utilizing an empirical theory Ξ in describing the behaviour of certain objects we regard the pertinent fragment of the reality as a domain of the physical theory Ξ. As it is well known *PM* can be used, i.e. for:

1° describing the movements of the bodies of the solar system (in this case these bodies are treated as particles),

2° describing the movement of a missile under the gravitational attraction of Earth,

3° describing the movement of the elementary particles, e.g. electrons in an electromagnetic field.

Each time we apply *PM* to describing some physical objects, and it is worth while remarking that it is the only case we can speak of physical interpretation of *PM*, the description, defined by the laws of *PM*, of the respective system will be approximative as to its character. If the description is adequate, then it is adequate in some interval of accuracy, and, simultaneously, in some interval of time. This obviously pertains not only to *PM* but to an arbitrary empirical theory Ξ. Note also that the counterpart of the concept of time interval in the real world is always some sequence of events. Thus the second range of each of the physical domains is not an interval of time but it consists of a sequence of events.

After these roughly sketched illustrations let us turn back to the general considerations. In what way is the interpretation of a theory carried ont in these systems, how should we describe them? From the formal point of view the question pertains to the passing from the language $\mathcal{L}$ to the language $\mathcal{L}_{\theta, \omega_i}$. As a matter of fact the interpretation of a theory in an arbitrary domain $\mathfrak{F}_i$ is carried on by methods of determining the magnitudes of the quantites F_1^*, ..., F_n^* measured at the elements of U_j in the time moments corresponding to the events of the set T_j on one hand, and by a method or methods of representing particular events in the set T_j by the corresponding real numbers, on the other hand. I have consciously made use of the phrases 'methods of determining the magnitudes', 'methods of representing... by real numbers' instead of the phrase 'measurement methods'. That is because determining the magnitude of some quantity is not necessarily reducible to measurement but it may as well consist of some calculations.

Assume that the numbers characterizing time moments which correspond to the events in the set T_j are determined by methods Λ_1^0, ..., $\Lambda_{S_0}^0$ (we do not reject the possibility of there being not one, but several methods applied). The magnitude of the quantity F_i^* is determined by methods Λ_1^i, ..., $\Lambda_{S_i}^i$. If such methods are available and, moreover, they assign to each $z \in T_j$ exactly one numerical interval and similarly they assign exactly one numerical interval to each $u \in \Omega_i$ (numerical value in some error interval), then some mappings $\theta' : T_j \to [R]$ and $\omega' : \Omega_i \to [R]$ satisfying the conditions put down for in the previous chapter, are set up. Given a theory Ξ we may put a question whether it is approximately true or not in the domain $\mathfrak{F}_j$ with respect to the mappings θ', ω_i' or, formulating the question in a more natural way, whether is it true in the domain $\mathfrak{F}_j$ with the assumption that the magnitudes of the quantities considered in this theory are determined by methods Λ_1^0, ..., $\Lambda_{S_0}^0$, Λ_1^1, ..., $\Lambda_{S_1}^1$, ..., $\Lambda_{S_n}^n$. Note, by the way, that in case the answer to such a question is negative, the theory Ξ, however, need not necessarily be questioned. In some cases the methods of measurement undergo some changes and so do the elaboration of the results of measurement, interpretation of these results, etc.

If we are sometimes apt to evaluate some empirical theories as true or false without relating these evaluations to certain definite methods of determining magnitudes of respective quantities, then — we may suppose — it is due to the fact that in many cases while making such evaluations it is not the methods but the corresponding degrees of precision which play an essential role. Assume that the procedures $\Lambda_1^0, \ldots, \Lambda_{s_0}^0, \ldots, \Lambda_1^n, \ldots, \Lambda_{s_n}^n$ are replaced respectively by $\Lambda_1'^0, \ldots, \Lambda_{s_0}'^0, \ldots, \Lambda_1'^n \ldots, \Lambda_{r_n}'^n$ which define exactly the same mapping θ', ω_i'. In this situation the question of the approximate truth of Ξ in $\mathfrak{F}_j$ will meet the same answer regardless of the system of procedures we relate it to. If however $\Lambda_1'^0, \ldots, \Lambda_{r_n}'^n$ allow for determining the magnitudes in a more precise way, i.e. if they lead to the mappings θ'', ω_i'', which are correlated with the mappings θ', ω' by means of the following conditions:

(a) for any $z \in T_j$, $\theta''(z) \subseteq \theta'(z)$,

(b) for any $u \in \Omega_i^j$ $\omega_i''(u) \subseteq \omega_i'(u)$

then the answer might undergo a change.

Note that the description of the behaviour of a certain system in certain definite time instants given by a theory is approximately true in the intuitive sense if the measurements of the quantities considered in this theory, carried on exactly in these time instants and with some definite accuracy, do not falsify the theory. Note, further, that a theory may lose the character of an approximately true theory of the domain $\mathfrak{F}_j$ not only as a result of the increasing accuracy of measurement of the magnitudes of the quantities considered in this theory. There may also appear a theory which is not approximatively true any longer when we pass from the domain $\mathfrak{F}_j$ to the domain $\mathfrak{F}_j$:

$$\mathfrak{F}_j' = \langle U_j, T_j', \Omega_1^j, \ldots, \Omega_n^j, F_1, \ldots, F_n \rangle$$

where $\mathfrak{F}_j$ differs from $\mathfrak{F}_j'$ only in that $T_j \subseteq T_j'$; so, e.g., if we extend the time interval determined by the events included in T_j, and we pass to considering the time interval determined by the events in T_j'.

The loose observations above suggest the following remark. Given a certain theory Ξ, which is interpreted in any physical domain $\mathfrak{F}_j$ whatsoever, and utilized for the description of this domain, we may aim at improving this theory. There are two possibilities emerging. The first of these possibilities consists in adding to Ξ new theorems, in such a way, however, that the new theory Ξ' thus obtained turns out to be also approximately true. The other possibility consists in modifying Ξ in such a way that the theory preserves its approximative validity in $\mathfrak{F}_j$ even in case when we pass to more precise methods of determining the magnitudes. Following any of these ways of modifying the theory Ξ we shall aim at the same time at keeping the theory Ξ' to be an improvement of Ξ, not only with respect to the domain $\mathfrak{F}_j$, but also in each of the domains $\mathfrak{F}_1, \ldots, \mathfrak{F}_s$, in which these theories may be interpreted. The conceptual apparatus introduced in this paper should allow the precise description and investigation of the mutual relationships of these tendencies which, as it seems, rule the development of science.

Let me state, finally, that I am conscious of the sketchy character of this paper. Many of the problems dicsussed here need far more developement and, above all, closer consideration. I hope, however, that the discussion presented in this paper is concerned with important problems and I believe that the proposals stated above, though imperfect, will turn out to be fertile in further investigations.

Notes

[1] Cf. A. Tarski, 'Der Wahrheitsbegriff in formalisierten Sprachen', *Studia Philosophica*, 1/1935 pp. 261–405. An informal exposition of Tarski's conception of truth, was presented by himself in the article 'The Semantic Conception of Truth', *Philosophy and Phenomenological Research*, 4/1944, pp. 341–375. The latter article may be found also in *Philosophy of Language*, ed. L. Linsky, Urbana: The University of Illinois Press, 1952, pp. 13–47.

[2] The term 'model' will not be used in the present paper although it is usually utilized just in the sense in which we shall here use the term 'domain'. The notion of model is, unfortunately, fairly ambiguous; it is also improper, from our point of view, for another reason. A model – in the common sense – is an object which reconstructs certain properties of the other object; thus it always amounts to some imitation. In this sense, however, we can not say that a fragment of the reality which the given theory describes is a model of that theory (of its language); we might rather say that the theory is a model of the reality and this usage sometimes occurs.

The proposal to employ the term 'domain' in such sense as we design here is due to A. Grzegorczyk. Cf. A. Grzegorczyk, *An Outline of Mathematical Logic*, Warszawa, 1974. Cf. also the same author's 'Zastosowanie logicznej metody wyodrębniania dziedziny rozważań w naukach, technice i gospodarce', *Studia Filozoficzne*, 3–4/1963, pp. 63–75.

[3] Cf. A. Robinson, *Introduction to Model Theory and to the Metamathematics of Algebra*, Amsterdam: North-Holland Publishing Company, 1963.

[4] For more details see M. Przełęcki, *Logic of Empirical Theories*, London: Routledge and Kegan Paul, 1969.

[5] The concept of an arithmetical quantity may be exemplified by any scalar quantity (e.g. mass, temperature, volume).

[6] Consideration of the two-range domains may be somehow reduced to consideration of one-range domains. The same remark pertains to the many-range domains. For this matter see Hao-Wang, 'Logic of many sorted theories', *Journal of Symbolic Logic*, 17, 1952, pp. 105–116.

[7] Cf. J.C. Mc Kinsey, A.C. Sugar, P. Suppes, 'Axiomatic Foundations of Classical Particle Mechanics', *Journal of Rational Mechanics and Analysis*, 2,1953, pp. 253–272. See also P. Suppes, *Introduction to Logic*, D. van Nostrand, 1957.

[8] Cf. R. Montague, 'Deterministic Theories', in: *Decisions, Values and Groups*, Pergamon Press, 1962, pp. 325–369.

THE ATTRIBUTE AND THE CLASS

by

Barbara Stanosz

§ 1. The distributive concept of class, secured in the theory of sets against the antinomies connected with its pre-theoretical stage, plays a very important rôle in the language of modern science. It turned out to be an irreplaceable tool for the formulation of theorems of different branches of mathematics and made possible the development of many new theoretical constructions or even whole new disciplines of great scientific importance. This concept has also applications in the philosophy of language, especially in the so-called theory of referential meaning of expressions, which describes the relation between language and what language can refer to.

As the concept of denotation of the predicate, the set-theoretical concept of class is the direct explicatum of the concept of range (denotation, extension) of general names, as used in traditional logic, of the common 'species' or 'genus' and of the philosophical *universale*. But the question, whether or not this concept is the adequate explicatum of the concept of attribute (property), is controversial. Different opinions in that matter manifest themselves, e.g., in the terminology of semantic works. In some of these works, especially in those devoted to artificial languages, the expression 'denotation of a predicate' is used as exchangeable with the expression 'attribute expressed by a predicate'; others, especially those intended to develop the semantics of natural languages, introduce the distinction between the concept of denotation of a predicate and the concept of the attribute expressed by it.

This distinction is founded upon the supposition that there exist attributes non-identical and co-extensive (i.e. possessed by the same objects); in other words, that the predicates which have the same denotation can express different attributes. If we generalize this supposition in such a way, that it refers to expressions of any syntactical category, it becomes the starting point of the theory of meaning dualism, the fullest version of which is Carnap's theory of extension and intension.[1]

According to the widespread opinion, the forerunner of this distinction is J. S. Mill, the author of the well-known theory of denotation and connotation developed in his *System of Logic* and reproduced later (with different modifications) by the authors of many handbooks of logic. We read in Mill's work: "The name (...) is said to signify the subjects directly, the attributes indirectly; it *denotes* the subjects, and implies, or involves, or indicates, or as we shall say henceforth *connotes*, the attributes."[2] However, the modern version of the distinction between the denotation of a predicate and the attribute it expresses is not closely related to Mill's theory, because the corresponding terms used by Mill differ in their meanings from those used by modern authors. Especially, though Mill had not formulated the condition of identity of attributes connoted by two names, it seems that he would not accept the solution of this problem proposed by modern logicians, because of some aspects of this solution, which we shall analyse later.

The distinction we have in mind appeared as the result of an analysis of modal contexts, in which materially equivalent (i.e. possessing the same denotation) expressions are not exchangeable *salva veritate*. To solve this so-called paradox of intensionality it is usually assumed that the non-exchangeability of such expressions in these contexts is caused by their non-synonymity.[3] In the case of predicates synonymity is identified with the sameness of attributes they express. The condition of exchangeability of expressions in modal contexts is their logical or analytical equivalence; hence two (one-place) predicates Φ_1, Φ_2 express the same attribute if and only if the sentence

$$(*) \qquad \bigwedge x\,[\Phi_1(x) \equiv \Phi_2(x)]$$

is a tautology or a logical consequence of meaning postulates in a given language.

According to this theory (and its different modifications, which can be found in the literature of the subject), these are the linguistic facts which decide whether or not two given predicates express the same attribute: to establish this, it is not necessary (nor sufficient) to refer to extra-linguistic facts, or to experience.

It seems, however, that this way of explicating the term 'attribute' as used in common and philosophical language, is incorrect. Namely, according to the usual meaning of the term (despite its vagueness), two predicates expressing the same attribute may differ in their meanings, being logically non-equivalent. For example, the predicates 'has the colour of a lemon' and 'has yellow colour' express the same attribute though they are not synonyms. The predicates 'has the shape of a sphere' and 'has the shape of a ball' can serve as another illustration.

What ultimately decides whether or not two given predicates express the same attribute is experience. But the decision does not depend only upon the establishment of the material equivalence of these predicates: attributes expressed by the predicates 'is of yellow colour' and 'is of spherical shape' would not be identified even if all yellow objects would be spherical and conversely (i.e. merely on the basis of such coincidence). In addition to the sameness of denotation, the predicates expressing the same attribute satisfy another condition, which we shall analyse later. As the result of the above considerations let us record the following conditions of adequate explication of the common and philosophical concept of attribute:

1. If the predicates Φ_1, Φ_2 are synonymous, then Φ_1, Φ_2 express the same attribute, but the converse does not hold universally.
2. If the predicates Φ_1, Φ_2 express the same attribute, then Φ_1, Φ_2 are materially equivalent, but the converse does not hold universally.

§ 2. The meanings of the predicates used above for illustration include some components which are of importance for our problem; they can be expressed by the phrases 'is such and such with respect to colour' and 'is such and such with respect to shape'. Let us notice that the same components are included in the meanings of some predicates which do not contain the words 'colour' or 'shape': indeed, the predicate 'is of yellow colour' is synonymous with the predicate 'is yellow', and the predicate 'is of spherical shape' — with the predicate 'is spherical'. On the same basis we can assert that the meaning of the predicate 'is sour' includes as its components 'is such and such with respect to taste'. The question whether or not all predicates have such meaning components will be considered later; now let us take into account only those predicates, which are doubtless of this kind.[4]

The predicates: 'has the colour of a lemon', 'is of yellow colour' and 'is yellow' express the same attribute (according to the usual meaning of the term 'attribute'). It is likewise in the case of any set of predicates we choose, if they are materially equivalent and include in their meanings (explicitly or implicitly) the component we describe in the phrase 'is such and such in respect to A'. It justifies acceptance of the following additional condition of adequacy for explication of the concept of attribute:

3*. If the predicates Φ_1, Φ_2 are materially equivalent and both include in their meanings the component *is such and such in respect to A*, then Φ_1, Φ_2 express the same attribute.

To attain a more precise formulation of this condition let us assume tentatively, that 'to be such and such with respect to colour' means 'to belong to the same member of classification of the universe of discourse made by the relation of being equicoloured'. Classification is conceived here in the set-theoretical sense, as the class of classes excluding one another, the union of which is identical with the universe; the relation of being equicoloured is the class of ordered pairs of equicoloured objects. Thus, the identity of two objects in respect to A depends on their belonging to the same class, which is a member of classification of the universe made by the corresponding equivalence relation.[5] Hence, to express the proposition that the meaning of a given predicate includes as its component 'being such and such in respect to this and that', it is enough to say that this predicate is connected by its meaning with this and that equivalence relation defined in the universe (or with this and that classification of the universe). It makes possible the following precisioning of the condition 3*:

3. If the predicates Φ_1, Φ_2 are materially equivalent and connected by their meaning with the same equivalence relation defined in the universe, then Φ_1, Φ_2 express the same attribute.

Let us compare this postulate with the intuitive concept of attribute to establish whether the explication of the phrase 'is such and such in such and such respect' proposed above is adequate. It should be noticed that reference to the common opinions in the question 'what attributes are?' is not necessary for our purposes. Neither is reference to the philosophical literature on 'the nature' of attributes; if it were the case, then our task would be hopeless, because 'definitions' of attributes proposed by philosophers are so different and vague, that it is unikely that we would find any common intuition in them. Our task consists in inquiring into the natural and philosophical concept of attribute, but: (i) at the present stage of analysis we do not seek the definitions of the attribute — we do not ask 'what attributes are?', but only 'what is the condition of identity of attributes?'; (ii) the answer to the first of these questions is — for reasons which we shall consider later — of little importance from the theoretical point of view.

Thus, the problem is whether or not the intuitive concept of attribute is such that the attributes expressed by two predicates are identical whenever the antecedent of (3) is true. Our examples confirm the affirmative answer to this question; the negative answer would require counterexamples. Let us recall that predicates connected by their meanings with the same classification need not be built according to the scheme 'is such and such in such and such respect'; corresponding 'respect' can be included implicitly in their meanings or, being mentioned explicitly, can be termed by different words. Hence, if we accept (3), we should accept, e.g., the following sentence, which is its consequence:

(**) If all yellow objects were spherical, and conversely, and if the classification with respect to colour were identical with the classification with respect to shape (i.e., if each pair of equicoloured objects were a pair of equiform objects, and conversely), then the attribute expressed by the predicate 'is yellow' (or 'has yellow colour') would be identical with the attribute expressed by the predicate 'is spherical' (or 'has spherical shape').

It seems that this consequence of (3) (and other ones of this kind) can be regarded as a reason for rejecting this postulate. It should be noticed, however, that the sentence (**) is a counterfactual conditional, so it has not the force of counterexample: equicoloured and non-equiform objects do exist in reality, like the objects which are equiform and nonequicoloured. Hence, if one wants to use the sentence (**) as a counterexample for the postulate (3), he must substantiate the rejection of (**).

It seems, that the endeavour to do so would follow a line like this: The nature of sense-data, which are for us the basis of assignment of colours to the objects, is different than in the case of assignment of shapes. These differences are given to us in introspection and they decide that we would have distinguished colours from shapes even if the corresponding classifications had been identical.

The reply may be as follows: there exist many cases of identifying attributes, which are predicated about objects on the basis of different sense-data (or even on the basis of data of different senses). As an example let us consider shapes: they are assigned to objects on the basis of eye-data or of touch-data. 'Seeing' the property of being spherical depends on experiencing certain visual sensations, which are clearly distinguished by introspection from corresponding tactile sensations; in spite of that we settle the same problem in both cases: does the given object possess the attribute of being spherical?

Such a way of using the term 'attribute' in natural language argues against the rejection of sentence (**). The non-existence of two predicates which are equivalent and connected by their meanings with the same classification but undoubtedly express different attributes (according to the usual sense of this term) also argues against the negation of (**) and of other similar consequences of the postulate (3).

To develop an explication of the intuitive concept of attribute in a way indicated by the conditions (1)–(3) it should be decided whether or not the condition (3) can be reinforced to the equivalence, i.e., whether or not one can assume that if two predicates express the same attribute, then they are materially equivalent and connected by their meanings with some (the same) classification of the universe.

The answer depends on the settling of the question, whether — according to the natural concept of attribute — each predicate expresses a certain attribute or some predicates (which ones?) express no attributes. Now, it seems that the term 'attribute' is used in common language in such a way that one can distinguish two different concepts of an attribute. According to the first of them, attributes are expressed by: 1) simple predicates, which are not definitional abbreviations of any compound predicates, and 2) compound predicates, which are conjunctions of simple predicates or definitional abbreviations of such conjunctions; according to the second, each predicate — apart from its syntactical structure — expresses some attribute. If we use the term 'attribute' in the first sense, we assume that the sentences '*a* is yellow', '*a* is spherical' and '*a* is yellow and spherical' assign some attributes to the objects *a*, but the sentence '*a* is yellow or spherical' does not: it is only an abbreviation of the sentence '*a* is yellow or *a* is spherical', which asserts that *a* possesses at least one of the two attributes expressed by the predicates 'is yellow', 'is spherical'. If we use the term 'attribute' in the second sense, then we assume that the predicate 'is yellow or spherical', like any other compound predicate, expresses some attribute too.

On the other hand, some predicates which are built from predicates linked by their meanings with certain classifications, are not linked with any classification. For example, the predicate 'is yellow or spherical' is not linked itself in a natural way with any classification a member of which would be the denotation of this predicate: an endeavour to find for it a synonym built according to the scheme 'is such and such in such and such respect' is ineffective. The cause of this fact is the non-transitivity of the relation, which holds between two objects if and only if they are equicoloured or equiform. Being a non-equivalence relation it does not make any classification of the universe. Similarly, the predicate 'is non-yellow' is not linked with any classification, because the relation of being non-equicoloured is not an equivalence relation. On the other hand, the predicate which is a conjunction of predicates linked by their meanings with some classifications, is also linked with some classification, namely, with the classification which is a product of the former ones;[6] the denotation of the predicate 'is yellow and spherical' is a member of the classification of the universe made by the equivalence relation which holds between two objects if and only if they are both equicoloured and equiform.

This parallelism makes it possible to reformulate the postulate (3) in two ways, which correspond to the two natural concepts of attribute. As D_1 we symbolize the strong version of postulate (3), which is related to the first (narrow) concept of attribute, and as D_2, the strong version of (3) related to the second (broad) concept of attribute.

D_1. Predicates Φ_1, Φ_2 express the same attribute if and only if Φ_1, Φ_2 are materially equivalent and connected by their meanings with the same equivalence relation defined in a given universe.

D_2. Predicates Φ_1, Φ_2 express the same attribute if and only if Φ_1,

Φ_2 are materially equivalent and connected by their meanings with the same relation defined in a given universe.

The relation we mention in D_2 is an equivalence relation if Φ_1, Φ_2 are simple predicates and are not definitional abbreviations of any compound predicates; if Φ_1, Φ_2 are compound predicates or definitional abbreviations of some compound predicates, then this relation need not be an equivalence relation: it is a result of corresponding operations (determined by the structure of a given predicate or of its definitional equivalent)[7] performed on the equivalence relations connected with the simple components of these predicates or of their definitional equivalents.

We shall treat D_1 and D_2 as alternative definitions of sameness of attributes expressed by two predicates. The choice of one of them is determined by the decision which of two natural concepts of attribute is to be explicated.

We choose the definition D_2 because it is more general. It will be the basis for our considerations in the next section of this paper.

§ 3. The definition of the phrase 'predicates Φ_1, Φ_2 express the same attributes' can serve as a basis for precisioning of some concepts referring to the classification of attributes, their mutual relations and operations which can be performed on them; for this purpose it is not necessary to decide in advance what kind of entities are attributes. We shall formulate definitions of some such concepts, using the symbols $C(\Phi)$, $D(\Phi)$ and $R(\Phi)$ as abbreviations — respectively — of the phrases: 'attribute expressed by the predicate Φ', 'denotation of the predicate Φ', 'relation with which the predicate Φ is connected by its meaning'.

To begin with, let us distinguish some kinds of attributes with respect to the formal properties and mutual relations of $D(\Phi)$ and $R(\Phi)$.

$C(\Phi)$ is a *proper* attribute $\underset{\text{df}}{\equiv}$ 1°: $R(\Phi)$ is an equivalence relation, 2°; $R(\Phi)$ holds between any two elements of $D(\Phi)$, and 3°: $R(\Phi)$ is not the universal relation.

$C(\Phi)$ is an *improper* attribute $\underset{\text{df}}{\equiv}$ $C(\Phi)$ is not a proper attribute.

$C(\Phi)$ is a *pseudo-proper* attribute	$\underset{\text{df}}{\equiv}$	1°: $R(\Phi)$ is an equivalence relation, and 2°: there are such elements of $D(\Phi)$ that $R(\Phi)$ does not hold between them.

The concept of the proper attribute corresponds to the first of two natural meanings of the term 'attribute' mentioned above; the improper attributes together with the proper ones form the denotation of the term 'attributes' in the second, broader sense. Pseudo-proper attributes (e.g., attributes expressed by the predicate 'is white or yellow') are improper attributes of a special kind; the predicates they are expressed by are connected with the equivalence relations (in contrast to the ones expressing other improper attributes) but the sets of objects they denote are not members of corresponding classifications (in contrast to the predicates expressing proper attributes). Consequently, objects denoted by such a predicate need not be identical in respect to A, though that predicate characterizes them as such and such in respect to A.[8]

To continue the specification of attributes let us assume that 0, U, R_0, R_u symbolize—respectively—the empty set, the universe, the empty relation and the universal relation.

$C(\Phi)$ is an *empty* attribute	$\underset{\text{df}}{\equiv}$	$D(\Phi) = 0$.
$C(\Phi)$ is a *universal* attribute	$\underset{\text{df}}{\equiv}$	$D(\Phi) = U$.
$C(\Phi)$ is a *contradictory* attribute	$\underset{\text{df}}{\equiv}$	$D(\Phi) = 0$ and $R(\Phi) = R_0$.
$C(\Phi)$ is a *tautological* attribute	$\underset{\text{df}}{\equiv}$	$D(\Phi) = U$ and $R(\Phi) = R_u$.

According to this terminology there exist non-identical empty attributes and non-identical universal attributes. This seems to be in agreement with the intuition created by natural language, which demands distinguishing, e.g., the attribute expressed by the predicate 'is a man or a woman' from one expressed by the predicate 'is a white man or a coloured man'.

Let us notice that according to the above definitions contradictory attributes are expressed not only by the predicates the form of which is $\Phi \wedge \sim \Phi$ (or which are logically equivalent to predicates constructed in such a way); similarly, tautological attributes are expressed not only by the predicates the form of which is $\Phi \vee \sim \Phi$ (or which are logically equivalent to predicates constructed in such a way. These consequences of our definitions seem to be also in agreement with the natural concept of attribute, according to which the predicates 'is white and non-white' and 'is white and does not have the colour of snow' express the same, contradictory attribute.

Contradictory attributes and tautological attributes belong to the broad class of attributes, which we call the non-specific ones:

$$C(\Phi) \text{ is a } \textit{non-specific} \text{ attribute } \underset{\mathrm{df}}{\equiv} R(\Phi) = R_0 \text{ or } R(\Phi) = R_u.$$

Hence, the class of non-specific attributes includes also attributes expressed by some non-empty and non-universal predicates, namely by those, which are linked themselves by their meanings with the product or with the union of relations, one of which is a complement of the second, for example, 'is white or non-yellow'. The attribute expressed by such a predicate does not depend upon the kind of relation which is connected with the simple components of that predicate: if we replace those simple components by their material equivalents, we always obtain a predicate expressing the same attribute (the unique condition is the following: any pair of predicates connected with the same relation may be replaced only by pair of predicates connected with the same relation). For example, if 1) all yellow objects were spherical and conversely, 2) all red objects were cylindrical and conversely, then the predicates 'is yellow or non-red' and 'is spherical or non-cylindrical' would express the same attribute, no matter whether or not the classification with respect to colours and the classification with respect to shapes would be identical.

I find it difficult to establish whether the identity of all co-extensive non-specific attributes (which fulfil together the first *resp.* the second member of disjunction in the above definition) is in agreement with common intuition; the problem seems to me undecidable because of the vagueness of the common concept of attribute. Let us remember that extensionalism implies more radical theses in that matter, and intensionalism also identifies, in any case, all contradictory attributes and all tautological ones.

Now we shall define two relations holding between attributes, *inclusion* and *exclusion*, symbolized by $\sqsubset$ and $\sqsupset\sqsubset$ respectively ($\subset$ is used as the symbol of inclusion of classes and $\supset\subset$ as the symbol of exclusion of classes):

$$C(\Phi_1) \sqsubset C(\Phi_2) \overset{\equiv}{_{df}} D(\Phi_1) \subset D(\Phi_2) \text{ and } R(\Phi_1) \subset R(\Phi_2).$$

$$C(\Phi_1) \sqsupset\sqsubset C(\Phi_2) \overset{\equiv}{_{df}} D(\Phi_1) \supset\subset D(\Phi_2) \text{ and } [R(\Phi_1) \subset R(\Phi_2) \text{ or } R(\Phi_2) \subset R(\Phi_1)].$$

We shall also define three operations which can be performed on attributes. Because of their resemblance to the corresponding set-theoretical operations we termed them, respectively, the *complement*, the *product* and the *union* of attributes:[9]

$$\overline{C}(\Phi) \overset{\equiv}{_{df}} C(\sim \Phi).$$

$$C(\Phi_1) \sqcap C(\Phi_2) \overset{\equiv}{_{df}} C(\Phi_1 \wedge \Phi_2).$$

$$C(\Phi_1) \sqcup C(\Phi_2) \overset{\equiv}{_{df}} C(\Phi_1 \vee \Phi_2).$$

§ 4. The definitions we have formulated above illustrate the way of using the definition of identity of attributes to explicate many typical contexts containing the term 'attribute'; this purpose does not require an answer to the question, what is the attribute expressed by a given predicate.

This question (let us call it the representation problem) can be solved on the grounds of our definition in many ways. The relation of expressing the same attribute is reflexive, symmetric and transitive; therefore it makes a certain classification C_0 of the set of predicates. Applying the usual procedure we can define the attribute expressed by the predicate Φ as that member of C_0, which contains Φ, i.e. as the class of predicates expressing the same attribute as the one expressed by Φ. It is also possible, however, to point at the entities of any other kinds as the representations of attributes; the unique condition is that two entities of a given kind must be identical if and only if the corresponding two predicates are materially equivalent and connected by their meanings with the same relation.

The following way of defining attributes seems to be the most natural and convenient:

D_3. The attribute expressed by the predicate Φ is the ordered pair $\langle D(\Phi), R(\Phi)\rangle$.

Since

$$\langle D(\Phi_1), R(\Phi_1)\rangle = \langle D(\Phi_2), R(\Phi_2)\rangle \equiv D(\Phi_1) = D(\Phi_2) \text{ and } R(\Phi_1) = R(\Phi_2)$$

D_3 satisfies the condition mentioned above.

According to the definitions of the complement, the product and the union of attributes introduced in the preceding section, we can characterize these operations as follows:

$$\overline{\langle D(\Phi), R(\Phi) \rangle} = \langle D(\Phi)' \; R(\Phi)' \rangle.$$

$$\langle D(\Phi_1), R(\Phi_1) \rangle \sqcap \langle D(\Phi_2), R(\Phi_2) \rangle = \langle D(\Phi_1) \cap D(\Phi_2), R(\Phi_1) \cap R(\Phi_2) \rangle.$$

$$\langle D(\Phi_1), R(\Phi_1) \rangle \sqcup \langle D(\Phi_2), R(\Phi_2) \rangle = \langle D(\Phi_1) \cup D(\Phi_2), R(\Phi_1) \cup R(\Phi_2) \rangle.$$

So far we have taken into account one-place predicates only. But the field of applications of our ideas is not restricted to them; we can apply the same principles to explicate the concept of attribute expressed by n-place predicate, i.e. the common concept of relation (which differs from the set-theoretical concept of relation in the same way as the common concept of attribute differs from the set-theoretical concept of class).

We shall exemplify that possibility by help of the two-place predicate 'is a friend of'. We suppose that this predicate is connected by its meaning with the classification of the universe (i.e. of the set of ordered pairs of persons) made by the equivalence relation $\boldsymbol{R}^*$ which holds between two pairs $\langle x, y \rangle$, $\langle u, v \rangle$ if and only if x has the same emotional attitude toward y as u toward v.[10] If the denotation of the predicate 'is a friend of' is symbolized by R^*, then the relation of being friends can be represented by the ordered pair

$$\langle R^*, \boldsymbol{R}^* \rangle.$$

Having the relation expressed by the two-place predicate 'Ψ', we can indicate the attribute expressed by the one-place predicate which is the result of binding one of the variables occurring in 'Ψ' by the universal or existential quantifier. For example, the attribute expressed by the one-place predicate 'has a friend' is the pair

$$\langle d(R^*), \boldsymbol{R}^{**} \rangle,$$

where $d(R^*)$ symbolizes the domain of the relation R^* (i.e. the denotation of the predicate 'has a friend') and $\boldsymbol{R}^{**}$ is the relation which holds between y and v if and only if there are such x and u, that $\boldsymbol{R}^*$ holds between $\langle x, y \rangle$ and $\langle u, v \rangle$.[11]

§ 5. Let us make a short summary of our considerations. We have asserted that neither extensionalistic nor intensionalistic interpretation of attributes corresponds to the common and philosophical meaning of the term 'attribute'. Trying to find an adequate explicatum of that term we have established that the attribute expressed by a given predicate depends on: 1) the denotation of the predicate, 2) some special feature of its meaning, which we have recognized as its meaning connection with definite relation holding between elements of the universe of discourse. We have formulated a corresponding definition of identity of attributes expressed by two predicates; this definition has been used to render more precise some contexts including the term 'attribute'. Especially, we have distinguished some kinds of attributes which correspond to those distinguishable by informal philosophical analysis. We have formulated also a certain definition of attribute expressed by a given one-place predicate (simple or obtained by Boolean operations). Finally, we have shown the possibility of analogous explication of the common concept of relation (attribute of ordered tuples) and illustrated the way of construction of attributes expressed by compound predicates, constructed by using non-Boolean operations.

It should be emphasized that—in the author's opinion—all these definitional and classificatory constructions are wholly neutral in respect to the fundamental body of the traditional philosophical problems connected with the concept of attribute. Especially, they do not determine the solution of the problem of objectivity or subjectivity of attributes: the solution of this problem depends on assumptions concerning the nature of the universe of discourse, and we have made no such assumptions. The only theorems we have used here were the theorems of elementary logic and set theory.

However, the presupposition (in the psychological sense of the term) of the ideas presented in this paper is of a philosophical nature. It is the view that classifications are the very foundation of our knowledge because they have been a starting point for the construction of the conceptual apparatus of natural language. The richness of this apparatus in each of its stages ultimately depends on the number of different ways in which we have divided objects of our cognition; the structure of this apparatus is determined by the internal structures of these divisions. The development of our knowledge is accompanied by changes in the conceptual apparatus in these two respects. But its core is permanently connected with the system of some simple divisions, which for this reason can be treated as especially important for retaining of the human species or—in another lingual convention—as ones reflecting the fundamental diversity of the objective world.

Notes

[1] R. Carnap, *Meaning and Necessity*, enlarged edition, Chicago, 1956.

[2] J.St. Mill, *System of Logic ...*, vol. I, sixth edition, London, 1865, p. 32.

[3] See B. Stanosz, 'The Problem of Intensionality', *Studia Filozoficzne*, foreign language edition, 3, 1966.

[4] It seems that at any rate all simple predicates, which are not definitional abbreviations of compound predicates, have such components in their meanings.

[5] An equivalence relation in the set X is any relation, which is reflexive, symmetric and transitive in X. Each equivalence relation in the set X is connected with some classification of X, and conversely: such connection holds between the equivalence relation R in X and the classification $\mathscr{C}$ of X if and only if for every x, y belonging to X, R holds between x and y if and only if x and y both belong to the same member of $\mathscr{C}$. Then we say, that R makes $\mathscr{C}$ of X.

[6] The product of two classifications $\mathscr{C}_1 = \{X_1, ..., X_n\}$, $\mathscr{C}_2 = \{Y_1, ..., Y_m\}$ of a given universe is the set of non-empty products $X_i . Y_j$, where $i = 1, 2, ..., n$ and $j = 1, 2, ..., m$.

[7] If we symbolize as $R(\Phi)$ the relation with which the predicate Φ is connected by its meaning, then we can formulate the simple rules of these operations as follows:

1) $R(\sim \Phi) = R(\Phi)'$; 2) $R(\Phi_1 \wedge \Phi_2) = R(\Phi_1) \cap R(\Phi_2)$; 3) $R(\Phi_1 \vee \Phi_2) = R(\Phi_1) \cup R(\Phi_2)$.

The symbols: $\sim$, $\wedge$, $\vee$ are the negation, conjunction and disjunction (in non-exclusive sense) connectives, respectively; the signs: ', ., + are the symbols of set-theoretical complement, product and union operations, respectively.

[8] Improper attributes can be differentiated further with respect to the differences of the formal properties of corresponding non-equivalence relation. Especially, it seems natural to distinguish the attributes expressed by those predicates which are connected by their meanings with the so-called resemblances, i.e. with the reflexive, symmetric and non-transitive relations.

[9] Let us notice that the operations $-$, $\sqcap$, and $\sqcup$ are governed by Boolean laws. The set of attributes expressed by all predicates of a given language (connected by their meanings with the relations defined in the same universe) is a Boolean algebra; the contradictory attribute plays the role of 0, and the tautological attribute plays the role of 1.

[10] One may doubt whether the predicate 'is a friend of' is really connected by its meaning with that classification (or even whether it is connected with the classification of that universe). Similar doubts arise in the case of other (one- or many-

place) predicates. These doubts show the ambiguity of predicates of the common language.

[11] Of course, the relation R^{**} is not an equivalence relation, so the attribute expressed by the predicate 'has a friend' is improper.

ANALYTICITY AND APRIORITY

by

Adam Nowaczyk

I

It is common knowledge that the definition by which I. Kant introduced the new term 'analytic judgement' into the language of philosophy does not perform its task well.[1] It has been noticed that it does not even render the intentions of its author — intentions which are revealed by other formulations in Kant's writings.[2] They are clear enough to be described by a more or less precise definition. One of the earliest definitions of the term 'analytic statement'[3] was formulated by Gottlob Frege, precisely in order to render Kant's intentions. Frege's definition,[4] which we shall consider a pertinent description of the Kantian concept of analyticity, states that *a statement is analytic in character if it can be proved exclusively by reference to general laws of logic and to definitions.*

It can be shown — as historians of the problem have often done — that the class of statements which Kant terms analytic had already been singled out by philosophers in the 17th century, if not earlier, in the conviction that such statements have specific properties which are important from an epistemological point of view. One of these properties was believed to consist in the fact that the truth of these statements can be established in principle without any reference to empirical data. These statements were also ascribed the highest degree of rational certainty. Hence in pre-Kantian philosophy we may come across the formulations:

(I.1) Analytic, and only analytic, statements are *a priori* in nature.

(I.2) Analytic, and only analytic, statements are certain.

According to the opinions prevailing at that time, *a priori* cognition is implicitly contained in those concepts which are adequate to the essence of objects. These concepts — mainly those occurring in the deductive sciences — were considered by some to be formed through our intellectual contact with the real world, and by others, to be innate. The criterion of truth of *a priori* statements was usually sought in some kind of intuition which enables us unmistakeably to establish appropriate relations among concepts. It is only in the writings of Thomas Hobbes that we find the clearly expressed opinion — one which stood alone at the time — that *a priori* statements are based on linguistic conventions.[5] Since 17th century epistemology included a conviction as to the *a priori* nature of the deductive sciences (called 'demonstrative' at the time), which then included geometry and arithmetic, (I.1) led to the conclusion:

(I.3) The theorems of the deductive sciences (geometry and arithmetic) are contained in the set of analytic statements.

The statements quoted above, in somewhat anachronistic formulations, can be found in Hobbes and Leibnitz, and — although less clearly worded — in other authors as well.

The starting point of Kant's epistemological doctrine is the rejection of (I.3). His decision is justified by the fact that what is claimed in (I.3) is not sufficiently substantiated. Hobbes' view that the axioms of geometry and arithmetic are definitions or consequences of definitions[6] had been adopted as dogma. Leibniz's endeavours to demonstrate that arithmetical theorems can be deduced from the principles of logic and from definitions also failed to substantiate (I.3) sufficiently. Since Kant has important reasons for preserving the validity of the statement concerning the *a priori* nature of the deductive sciences, in rejecting (I.3) he has to reject (I.1) as well and to assume that

(I.4) There are synthetic *a priori* statements.

His further endeavours are intended to explain the basis of synthetic *a priori* cognition. Analytic statements do not particularly interest him. We know that he considered them to be *a priori*, necessary, and universally valid, but he also applied these qualifications to synthetic *a priori* statements. This may give rise to the question: is the concept of analytic statement, in Kant's philosophy, an epistemological category or merely a logical and methodological one? This question will not be answered here. It is posed only because of what it suggests, namely that the concept of analytic statement can occur in various roles: as an epistemological, a methodological, or a semantic category. This fact should be taken into account when we evaluate proposed definitions of that concept from the point of view of their adequacy.

Kant's claim that

(I.5) The basic theorems of geometry and arithmetic are synthetic *a priori* statements

was the subject matter of numerous controversies among philosophers and mathematicians in the 19th and early 20th centuries. The reduction by Frege and Russell of the concepts and theorems of arithmetic to broadly conceived logic, is held to be the strongest argument against those implications of (I.5) which refer to arithmetic. These controversies will not be discussed here, for our task is not to write a history of the problem of analyticity, but only to characterize briefly the theoretical context in which the concept of analyticity occurred.

The picture drawn thus far would be incomplete if we did not consider the theory of cognition advanced by logical empiricism, a theory in which the concept of analytic statement had pride of place. When referring to logical empiricism we do not mean only the philosophy of the Vienna Circle, wherein the problem of analyticity could not for certain reasons even be properly formulated,[7] but also certain viewpoints which can be considered a continuation of that school, and which advocate what can very broadly be called analytic philosophy. To put this rather schematically we may say that we are dealing here with a rejection of the Kantian theorem (I.4) and with a restitution of (I.1) — which restores to the concept of analytic statement the nature of an unmistakably epistemological category — and also of (I.2) and (I.3). These theorems have often been re-worded, but the advances in precision have been insignificant. We must also take note of two facts: (i) the extension of the term 'analytic statement' has been expanded by making it cover definitions by postulates; (ii) the deductive sciences are identified with formal systems, the axioms in such systems being identified with definitions by postulates of the primitive constants of each system respectively. But the most important characteristic of the epistemology of logical empiricism is the rejection not only of the Kantian metaphysics of *a priori* forms, but also of the "myth of the inner eye of reason",[8] which was characteristic of 17th century rationalism, but has had advocates in all periods, our own not excluded. The rejection of that myth is a manifestation of both the anti-metaphysical approach, which is characteristic of logical empiricism, and the standpoint that only those methods of substantiating theorems which are liable to inter-subjective verification are admissible. Logical empiricists do admit, however, that some statements are accompanied by *a priori* assertion. In accordance with (I.1) these are meant to be just analytic statements. This is why they are obliged to explain what makes that assertion valid. The answer usually is:

(I.6) Analytic statements are true (or: hold) by linguistic conventions.

The metaphorical nature of this formulation has resulted in many misunderstandings. In our opinion, it refers to what validates the assertion of an analytic statement, i.e., to the substantiation of analytic statements. This is why we would rather formulate (I.6) thus:

(I.6′) Analytic statements are substantiated by linguistic conventions,

even though this is not free from metaphorical elements, either.

II

The present-day literature of the subject shows a marked polarization of standpoints. One of them, in an extreme version, has found expression in W.V.O. Quine's well-known paper on the two dogmas of empiricism.[9] He points to the difficulties encountered by anyone who tries to formulate a satisfactory definition of the concept of analytic statements, and comes to the conclusion that the very belief that there is a demarcation line between analytic and synthetic statements is groundless. Other authors who represent the same opinion point to the uselessness of the concept of analyticity as applied to natural languages, since those languages lack the criteria which would make it possible to distinguish analytic statements from the others.[10] Those who support the opposite standpoint claim that the concept of analytic statement does, and should, perform an important function in methodological and philosophical reflections on science and its language. They emphasize that the intuitions associated with that concept are clear enough to warrant the claim that some statements are beyond all doubt analytic, and that the vagueness of the concept of analyticity does not justify a rejection of the concept.[11] According to many authors concerned with a logical theory of language the concept of analytic statement must inevitably remain obscure in ordinary language, which is semantically imprecise, but can be rigorously defined with reference to artificial (codified) languages. They also provide many examples of the reconstruction of the concept of analyticity, which are applicable to languages which are properly codified. The first such reconstruction originates with Rudolf Carnap.[12] In recent years many similar proposals have been made by Polish logicians.[13] It would appear that the controversy mentioned above is only to a small extent due to substantial differences, and mainly originates from differences in interpreting the task of constructing the correct definition of the term 'analytic statement'. Certain comments on that task will be made below.

It is usually assumed that the point is to construct a definition which would be formally correct and adequate and which would establish a concept that would be theoretically useful. The requirement of adequacy is most often interpreted as that of extensional or intensional agreement between the explicatum and the explicandum.[14] A selected current or traditional formulation is as a rule made the criterion of adequacy, so understood. Such an approach does not guarantee the scientific usefulness of the explicatum. This is why, in our opinion, the requirement of adequacy should be understood somewhat differently: less importance should be attached to the extensional or intensional agreement between the explicatum and the explicandum (such an agreement is usually problematic in view of the vagueness and obscurity of the explicandum), but a definition should be considered adequate only if it provides an explicatum that can replace the explicandum in a given field in which that explicandum has been used so far.[15] It may easily be noticed that this formulation of the requirement of adequacy leaves much to be desired, and this is why it must be made more specific in each case under consideration. It can, however, generally be said that the adequacy of definitions as understood here does not depend merely on fixing the range of the problems to be discussed, but sometimes on the accepted solutions of such problems as well.

In the preceding section we have outlined the theoretic context in which the traditional concept of analytic statements has been used.

This context determines a certain range of problems, which, as we can easily see, are focussed on the age-old issue of the foundations of human cognition, or more strictly, on that part of human cognition which is supposed not to be based on empirical data. In trying to construct an adequate definition of the concept of analytic statement we should above all take account of this group of problems. We shall henceforth use the *ad hoc* formulation 'epistemological definition of analyticity', if the explicatum provided by that definition is capable of replacing the traditional concept of analytic statement in the range of problems mentioned above. We believe that this condition can be met only if the explicatum denotes a class of statements which have certain specific epistemological characteristics, or to put it more precisely, if it denotes a class of statements which are substantiated in a specific manner which has been singled out in the theory of cognition.

An epistemological definition of analyticity is one which is adequate relative to a specified set of problems. We shall now explain in greater detail what set of problems we have in mind. Among the problems imposed by the theoretical context (outlined above) of the traditional concept of analyticity the first (in logical order) is:

(II.1) Are there statements which can reasonably be accepted without reference to empirical data, i.e., are there *a priori* statements?

It would seem that an answer in the negative should make us abandon the search for an epistemological definition of analyticity (even though it would clear the way for a search for adequate definitions relative to other problems in which the concept of analytic statement is involved, if such a concept exists). For however we might single out the class of analytic statements (on condition that it is a non-empty class), those statements will not stand apart by virtue of epistemological properties, in any case not those which have traditionally been ascribed to them. Moreover, an answer in the negative puts an end to all subsequent problems.

An answer in the affirmative, on the other hand, requires the indication of a statement whose assertion would be both *a priori* and valid. This leads to the question:

(II.2) What makes that assertion valid, i.e., in what does the substantiation of *a priori* statements consist?

The history of philosophy offers many fairly differentiated answers. In the light of the various theses which are answers to (II.2), various definitions of analyticity passing for epistemological definitions are admissible. We shall confine ourselves here to quoting only one of them and to pointing to the issues connected with the version of the epistemological definition of analyticity which that answer entails. We have in mind the thesis of logical empiricism:

(II.3) A statement is *a priori* in nature if and only if it is substantiated by linguistic conventions.

It can be seen that an acceptance of (II.3) narrows down the conditions which an epistemological definition of analyticity must meet. The task then is to single out, as analytic statements, the class of statements, and only those statements, which are substantiated by linguistic conventions. Any other solution results — on condition that (II.3) is assumed — in denying to the analytic statements their specific epistemological nature. That one is in fact endeavouring to preserve the specific epistemological nature of those statements is shown by the broadening of the extension of the concept of analytic statement, which logical empiricists have achieved by making it cover definitions by postulates. Those definitions were held to be true by convention; hence if they were not classed as analytic statements they could not be singled out by the mode of their substantiation (understood as an epistemological category). The same motivation can be ascribed to the reaction to Kazimierz Ajdukiewicz's claim[16] that the substantiation of analytic statements sometimes requires an empirical premiss — a reaction consisting in the claim that Ajdukiewicz's definition is too broad.[17]

The concept of analytic statement in the version suggested by (II.3) must be made relative to a given language. Let us note that (II.3) imposes a certain concept of language, or, to put it more simply, a way of interpreting the term 'language', namely one in which two languages may be considered identical only if the same linguistic conventions hold in both. The set of statements substantiated by linguistic conventions forms the denotation of the term 'analytic statement'. But the description of that set remains extremely vague as long as the conventions in question are not specified. When this description is made precise through an indication of the subject matter and the form of the said conventions, a factual problem immediately arises:

(II.4) Are there statements which are substantiated by linguistic conventions?

That version of the epistemological definition of analyticity which is now under consideration assumes an answer in the affirmative, for if the answer were in the negative the explicatum of the concept of analyticity would be empty and hence theoretically useless.

III

Motives of two kinds may induce us to abandon our endeavours to construct an epistemological definition of analyticity. One such motive may be the devaluation of the above presented issue as meaningless, incomprehensible, or not liable of a precise formulation. The sore point of that issue is that it involves the epistemological concept of a substantiated statement understood as an absolute concept (in the sense of its being relative at most to a given person). The attempts made so far to give precision to that concept have proved unsatisfactory. The other motive may be the rejection of a given thesis, for instance, the thesis of the existence of methods of *a priori* substantiation. Its rejection practically puts an end to the issue which we have associated with the label 'epistemological definition of analyticity'.

Contemporary authors engaged in what is termed the philosophy of science include advocates of two methods of *a priori* substantiation, namely the method of linguistic convention and that of conceptual intuition. The latter provokes familiar objections because — as Ajdukiewicz so rightly says — of its imprecise nature, difficulties in checking its use, and the impossibility of settling controversies between people who refer to the evidence of intuition.[18] Nominalists and moderate realists have additonal objections to this method, of considering concepts as objective entities existing independently of linguistic material. The method of linguistic conventions has long been criticized by the advocates of intuitive cognition and by radical empiricists.

IV

A penetrating criticism of linguistic conventions as a method of substantiation has been carried out by Ajdukiewicz in publications [2] and [3]. His criticism is not intended as an argument in favour of the method of conceptual intuition (which Ajdukiewicz rejects as not being intersubjective), but is in support of radical methodological empiricism, which represents the last stage in the evolution of Ajdukiewicz's views.[19]

In article [2] he singles out two kinds of linguistic (terminological) conventions, namely the semantic and the syntactic. These two kinds of conventions have analogues in two concepts of analytic statement. We shall confine ourselves here to quoting Ajdukiewicz's argumentation in connection with the concept of analytic statement in the semantic sense,[20] which in discussions of the aforementioned article aroused more interest as a rule. Ajdukiewicz's conclusions concerning analytic statements in both senses mentioned above are, incidentally, concurrent with regard to the issues in which we are interested here.

Ajdukiewicz claims that terminological conventions are not a sufficient substantiation of postulates (and hence of their logical consequences, and accordingly of the totality of analytic statements). The term 'substantiation' is interpreted rigorously in this case as a 'guarantee of truth.' It is claimed, therefore, that

(IV.1) The decision that a term λ denotes, in a language L, an object which satisfies the condition $\varphi\ (x/\lambda)$ is not a sufficient condition of the truth of the statement $\varphi\ (\lambda)$ in L.[21]

The arguments in support of (IV.1) are based on the following two premisses:

(IV.2) If there is no object that will satisfy the condition $\varphi\ (x/\lambda)$, then, despite our decision, the term λ will not denote in L an object which will satisfy that condition.

(IV.3) If the term λ does not denote in L an object which satisfies the condition $\varphi\ (x/\lambda)$, then the postulate $\varphi\ (\lambda)$ is not a true statement in L.[22]

These arguments show that, in Ajdukiewicz's opinion, the existence of an object that satisfies $\varphi\ (x/\lambda)$ is a necessary condition of the truth of $\varphi\ (\lambda)$ in the language in which that statement is a postulate.

Of the two premisses on which (IV.1) is based, (IV.2) arouses no objections, but (IV.3), even if it does not give rise to doubts, nevertheless suggests certain questions which, finding no answer in Ajdukiewicz's works, have become a subject of investigation.[23]

In the first place, the problem arises whether in a case when there is no object that will satisfy $\varphi(x/\lambda)$ the term λ denotes in L any object at all? A negative answer is undoubtedly the most natural one, for, in accordance with the decision taken, λ is meant to denote an object which satisfies $\varphi(x/\lambda)$. If that decision remains valid, then λ denotes no object. However, another solution is also possible, and even though it is somewhat artificial, it may prove useful. We can make a general decision concerning all the postulates in L, to the effect that if no object satisfies $\varphi(x/\lambda)$, then λ denotes a certain fixed object.[24] Such an object would not, of course, satisfy $\varphi(x/\lambda)$, and hence $\varphi(\lambda)$ would not be a true statement in L. In that case, certain postulates in L would be false statements in that language.

A second question results from the assumption that the answer to the former question is in the negative. If λ denotes no object because there is no object that will satisfy $\varphi(x/\lambda)$, then is λ a meaningful expression in L? If L has bound variables of the same type as λ, and if such rules of inference as the rule of substitution and the rule of existential generalization do hold in L, then λ cannot be a meaningful expression in L. This is the conclusion arrived at by Marian Przełęcki,[25] who points to the fact that in such languages the statement $\bigvee x\,(x = \lambda)$ is a logical tautology. Hence its metalinguistic analogue $\bigvee x$ (λ denotes x) should be true, which contradicts the assumption that λ denotes no object. Now, if λ is not a meaningful expression in L, then $\varphi(\lambda)$ is not meaningful in L, either. If it is not a meaningful expression in L, $\varphi(\lambda)$ cannot be a true statement in L.

Yet, if L does not satisfy the description given above, then the conclusion that the term λ, which does not denote anything, is not a meaningful expression in L, does not impose itself irresistably. It seems possible to construct a consistent language in which certain terms which have no denotations are nevertheless meaningful expressions. This would seem to apply to everyday language, in which such terms as 'Polyphemus', 'Hamlet', 'the Archangel Gabriel' are held to be meaningful. The problem therefore arises whether the postulates which introduce such terms can be true statements in such a language. For instance, is the statement 'Polyphemus was a Cyclops', which is supposed to be a postulate in everyday language, a 'true' statement in that language even though 'Polyphemus' denotes nothing? Analyses of natural language carried out at an intuitive level suggest that such statements are true. This is clearly at variance with (IV.3), on which Ajdukiewicz's claim (IV.1) is based. But this gives cause for critical reflection: in what sense may we speak about the truth of such statements as 'Polyphemus was a Cyclops' Is the word 'true' used here in its classical sense? The classical concept of truth has a clearly defined sense only if it is related to a specified model of a language.[26] If we mean a language L, interpreted in a specified way by a certain community, we can simply disregard that relativization and say 'a statement which is true in L'. This phrase then denotes those statements in L which are true in the proper model of L, i.e., in the model about which we are speaking when using L. The proper model of L, like any other model of that language, is a two-element system.

(M) $\langle U, C \rangle$,

where U is called the universe of M, and C, the characteristic of M. The characteristic of a model is a sequence of objects which are unambiguously assigned to simple expressions in L. Such objects are called the values of expressions in that model. But the values of simple expressions in the proper model of L are precisely the denotations of those expressions in L. Now if we say that some terms in L denote nothing, then we cannot speak about a proper model of L. Hence the phrase 'a statement which is true in L' is meaningless in such a situation.

This analysis of Ajdukiewicz's arguments in support of (IV.1) leads us to conclude that his claim is based on strong foundations. Accordingly, it is right to conclude that a terminological convention alone, i.e., the decision that a term λ should denote an object which satisfies the condition $\varphi(x/\lambda)$, is not a sufficient condition of the truth of the postulate $\varphi(\lambda)$. Our analysis also confirms the claim that $\varphi(\lambda)$ cannot be a true statement if there is no object that can satisfy $\varphi(x/\lambda)$.

The conclusions drawn by Ajdukiewicz from the theorems we are discussing here go much further. He claims that "(...) terminological conventions alone do not suffice to substantiate postulates."[27] Since analytic statements (in the semantic sense of the term) are postulates and/or their logical consequences, "(...) terminological conventions alone do not suffice to substantiate analytic statements in the semantic sense of the term (...)".[28] He also claims that the substantiation of analytic statements requires, besides a terminological convention, a certain existential premiss. It is the statement $\bigvee x \, \varphi(x/\lambda)$ which is to be the indispensable premiss for the substantiation of the postulate $\varphi(\lambda)$. The last two claims would appear more disputable than those mentioned earlier. This is due to the fact that the concept of substantiation which occurs in them is not clear. This is certainly the epistemological concept of substantiation which occurs in Sec. II of the present paper. Ajdukiewicz himself also considers the explanation of this concept to be extremely difficult.[29] On the other hand, it is precisely this concept which enables him to draw epistemological conclusions from analyses of the conditions of the truth of postulates. He assumes that "(...) the method of substantiating theorems by terminological conventions does not guarantee the truth of those theorems if these are not additionally substantiated by an appropriate existential premiss",[30] and concludes that by abandoning the method of intuition as the method of *a priori* substantiation we have to give up everything that is held to be *a priori* knowledge. In his opinion, analytic statements are not *a priori*, either, for the method of substantiating theorems by terminological conventions is not, in his view, a method of direct substantiation, since it requires a previous substantiation of a premiss. If we are then to avoid a *regressus ad infinitum* when substantiating analytic statements, we shall have to use a method of direct substantiation. But the only method of direct substantiation which Ajdukiewicz was prepared, though not unreservedly, to approve is that of substantiating statements on the basis of sensory observations.

Thus the definition of the concept of analytic statement (in the semantic sense of the term) suggested by Ajdukiewicz is not – at least as he sees it — epistemological in the sense adopted in the present paper. This is so because analytic statements are not distinguished by any specified type of substantiation conceived as an epistemological category. On the contrary, like all other statements they are substantiated empirically (*a posteriori*), even though in their substantiation a certain role is played by linguistic conventions.

V

Can Ajdukiewicz's argumentation convince an advocate of moderate empiricism?[31]

Let us first establish what makes Ajdukiewicz class the method of terminological conventions as a method of indirect substantiation. It would seem that the term 'substantiated', as used in (3), is so interpreted that a statement Z is substantiated (for a given person) if its assertion (by that person) is valid. But, as we know, the assertion of a statement is subject to gradation. Hence the concept of substantiation is subject to gradation, too. But whenever we refer to the substantiation of analytic statements, we always speak about categorical assertion and full substantiation. A statement is fully substantiated if it is subject to categorical assertion and if such assertion is legitimate. Even though the concept of validity of assertion is very unclear, there is no doubt that the validity of the categorical assertion by a person of a certain statement requires that that person should arrive at the assertion of that statement in a manner (according to a rule) which guarantees its truth and hence is fully reliable. This is a necessary — though probably not a sufficient — condition of the validity of categorical assertion. Now the rule that allows us to accept a statement $\varphi(\lambda)$ on the basis of a semantic terminological convention, i.e., on the strength of the decision that λ is to denote an object which satisfies $\varphi(x/\lambda)$, is not fully reliable. It is not, if it is so general that $\varphi(\lambda)$ can have any logical structure (not even excluding the case of self-contradictory expression). But the rule suggested by Ajdukiewicz allows us to accept $\varphi(\lambda)$ on the strength of an appropriate convention and a premiss in the form of $\bigvee x\, \varphi(x/\lambda)$. This rule is fully reliable; it is, however, a *sui generis* rule of inference, and hence a rule of indirect substantiation.

However, it would not be difficult to cite numerous rules which allow us to accept a statement $\varphi(\lambda)$ (which in this case is a postulate) on the strength of a semantic terminological convention — rules which, while being fully reliable, are rules of direct substantiation. Such rules should simply be less general than the one which is criticized by Ajdukiewicz. They allow us to accept $\varphi(\lambda)$, functioning as a postulate, but on condition that such a statement has specified formal or non-formal properties. Such a rule may, for instance, apply only to those statements whose structure is

$$(+) \qquad \bigwedge x\, (\lambda(x) \equiv \ldots x \ldots),$$

where λ is a predicate, and the language in which (+) is a postulate has the logical structure of the first-order predicate calculus. It might also cover other statements, for instance, what are termed bilateral reductive statements. But could it also apply to statements of the type

$$(++) \qquad \wedge x (P(x) \rightarrow \lambda(x)) \wedge \wedge z (R(x) \rightarrow \sim \lambda(x)),$$

which are partial definitions of λ? In such a case, as we know, the denotation of λ, postulated by a semantic terminological convention, may be non-existent. It does not exist if the (previously fixed) denotata of the predicates P and R are not mutually exclusive. This is an empirical issue if the statement

$$(+++) \qquad \wedge x (P(x) \rightarrow \sim R(x)),$$

which is a consequence of $(++)$, is empirical in nature.

It is not difficult, however, to formulate a rule which, while being a fully reliable rule of direct acceptance, will also cover certain statements whose structure is like that of $(++)$. Such a rule could be worded thus:

(R) We are permitted to accept a statement $\varphi(\lambda)$ on the strength of a terminological convention concerning λ, on condition that there is an object which satisfies $\varphi(x/\lambda)$.

As we can easily see, (R) is as fully reliable as the rule suggested by Ajdukiewicz, but unlike the latter it requires no premiss to substantiate $\varphi(\lambda)$. A person applying (R) need not know, i.e., need not have a substantiated conviction, that an object which satisfies $\varphi(x/\lambda)$ exists; it is enough if such an object does in fact exist.

Now (R) is the most general, fully reliable rule for accepting statements solely on the strength of a semantic terminological convention without reference to any premisses whatever. It may, however, be doubted whether, when proceeding in accordance with (R), we are really substantiating $\varphi(\lambda)$; (R) is fully reliable, but in order to substantiate a statement it is certainly not enough to proceed in accordance with a fully reliable rule. If it were, we should have to accept the assertion of any true statement as valid, because anyone who accepts a true statement (regardless of circumstances and accompanying motives) proceeds in accordance with the rule which instructs us to accept a statement Z if that statement is true, a rule which certainly is fully reliable. In refusing to grant (R) the status of a rule of substantiation we are probably bearing in mind that we are not in a position in every case to determine whether this rule has in fact been applied. To establish this requires the knowledge of whether an object that satisfies $\varphi\ (x/\lambda)$ does exist — knowledge which is not always accessible. The example now under consideration suggests that the possibility of deciding whether a given rule has in fact been applied in a given case is also a necessary condition of the validity of that rule. Codification of all the necessary conditions which a rule has to satisfy in order to be granted the status of a rule of substantiation seems difficult even with reference only to those rules which are connected with the issue of the substantiation of analytic statements.

Therefore, can the arguments presented by Ajdukiewicz be considered as decisive in the epistemological controversies over analytic statements? Do they, in particular, decide in the negative the issue of the existence of statements whose substantiation is based exclusively on terminological conventions? We have thus far linked that issue with the existence of valid rules of substantiation which will allow us to accept statements on the strength of terminological conventions alone, without reference to any other premisses. Do such rules exist?

It would appear that this question cannot be answered unambiguously, and that this is the case even regardless of the difficulties mentioned above. When we assess a rule as to its substantiating power we always make use of a system of reference which is the set of accepted (unquestioned) theorems. This is, in the first place, the set of theorems on the basis of which a given rule can be made valid. In the case of the rules of full substantiation making a rule valid means, among other things, demonstrating its full reliability. Such systems of reference, assumed by the various epistemologists, differ from one another, and this results in differences of opinion even when it comes to settling fairly simple issues, such as, for instance, the substantiation of postulates which are equational definitions of predicates. Some writers feel, as Ajdukiewicz does, that to substantiate them we must have, besides a terminological convention, an existential premiss which is a special case of the axiom of set construction. Others claim that such an axiom is not necessary as a premiss on which a postulate is based, but only as a premiss in reasoning intended to demonstrate the validity of the rule which results in the acceptance of that postulate, the rule itself being a rule of direct substantiation by reference to a terminological convention alone. Such controversies clearly point to the need of restricting the concept of substantiation to a specified system of reference. This, however, means abandoning the absolute concept of substantiation as used in traditional epistemology, and thus probably abandoning our aspirations to possession of a knowledge which cannot be rationally questioned. This is so because many systems of theorems to which we wish to restrict the concept of substantiation can be held equally valid at a given stage in the development of science and its methods. It also seems impossible to indicate any criterion of choice between such rival systems except for the significance of the results obtained by procedures based on different systems — the significance of the results being, however, difficult to assess and compare.

In the trend of methodological research which prevails today a special role is assigned to the system of theorems of set theory. This is why the set of such rules of substantiation, whose full reliability can be demonstrated on the basis of set theory enriched with certain general principles of logical semantics, seems to deserve special attention. Such rules would include, besides the classical rules of deduction, certain rules for accepting postulates on the basis of terminological conventions. This would probably enable us to define the extention of the term 'analytic statement' so that we might claim that analytic statements are substantiated by linguistic conventions. Yet, in view of the relative nature of the concept of substantiation involved, nothing could entitle us to ascribe to them such properties as apriority and certainty in the epistemological sense. The issues in which the concept of analyticity so outlined could find applications would therefore differ essentially from the traditional problems of epistemology.

References

[1] K. Ajdukiewicz, *Język i poznanie* (Language and Cognition), Vol. 1 – 1960, Vol. 2 – 1965.

[2] K. Ajdukiewicz, 'Le problème du fondement des propositions analytiques', *Studia Logica*, Vol. VIII, 1958.

[3] K. Ajdukiewicz, 'The Axiomatic Systems from the Methodological Point of View', *Studia Logica*, Vol. IX, 1960.

[4] R. Carnap, *Meaning and Necessity*, Chicago 1947.

[5] G. Frege, *The Foundation of Arithmetic*, Oxford 1950.

[6] H.P. Grice and P.F. Strawson, 'In Defense of Dogma', in: R.R. Ammerman (ed.), *Classics of Analytic Philosophy*, McGraw-Hill, 1965.

[7] Th. Hobbes, *The Elements of Philosophy*, London 1956.

[8] I. Kant, *Critique of Pure Reason* translated by J.M.D. Meiklejohn, London–Toronto 1934.

[9] W. Kneale, 'Are Necessary Truths True by Convention?' *Aristotelian Society Proceedings*, 1947 (Supplement).

[10] W. Kneale and M. Kneale, *The Development of Logic*, Oxford 1962.

[11] M. Kokoszyńska, 'O dwojakim rozumieniu uzasadnienia dedukcyjnego' (Two Concepts of Deductive Justification), *Studia Logica*, Vol. XIII, 1962.

[12] M. Kokoszyńska, 'O dyrektywach inferencji' (On the Rules of Inference), in: *Rozprawy logiczne* (Papers on Logic), 1964.

[13] C.I. Lewis, *An Analysis of Knowledge and Valuation*, La Salle, Illinois 1964.

[14] A. Pap, 'Necessary Propositions and Linguistic Rules', *Archivio di Filosofia*, Rome 1955.

[15] A. Pap, *Semantic and Necessary Truth*, Yale Univ. Press, 1958.

[16] M. Przełęcki, 'O pojęciu zdania analitycznego' (On the Notion of an Analytic Sentence), *Studia Logica*, Vol. XIV, 1963.

[17] W.V.O. Quine, 'Two Dogmas of Empiricism', in: *From a Logical Point of View*, Harvard Univ. Press, 1953.

[18] B. Stanosz, 'Formalne teorie zakresu i treści wyrażeń' (Formal Theories of Extension and Intension of Expressions), *Studia Logica*, Vol. XV, 1964.

[19] R. Suszko, 'Logika formalna a niektóre zagadnienia teorii poznania' (Formal Logic and Selected Issues in Epistemology), *Myśl Filozoficzna*, 1956.

[20] M.G. White, 'The Analytic and the Synthetic: An Untenable Dualism', in: L. Linsky (ed.), *Semantics and the Philosophy of Language*, Univ. of Illinois Press, 1952.

[21] R. Wójcicki, 'Analityczne komponenty definicji arbitralnych' (Analytic Components of Arbitrary Definitions), *Studia Logica*, Vol. XIV, 1963.

Notes

[1] The fragment of Kant's *Critique of Pure Reason* which is usually treated as his definition of the concept of analycity reads as follows: "In all judgments wherein the relation of a subject to the predicate is cogitated (I mention affirmative judgments only here; the application to negative will be very easy), this relation is possible in two different ways. Either the predicate *B* belongs to the subject *A*, as somewhat which is contained (though covertly) in the conception *A*; or the predicate *B* lies completely out of the conception *A*, although it stands in connection with it. In the first instance, I term the judgment analytical, in the second, synthetical." (p. 30). If this is a definition, then, as we can see, it is a partial one, as Kant himself must have realized since he wrote that it had to be extended to cover negative judgements.

[2] See in particular the chapter of the *Critique* entitled 'Of the Supreme Principle of all Analytical Judgments'.

[3] We shall henceforth always speak of analytyc statements, even if texts referred to have analogues to the term *judgement*.

[4] Frege [5].

[5] "... the first Truths were arbitrarily made by those that first of all imposed Names upon Things, or received them from the imposition of others. For it is true (for example) that *Man is a Living Creature*; but it is for this reason, that it pleased men to impose both those Names on the same thing." (Part I, Chapter III, 8)

[6] ... Now *Primary Propositions* are nothing but Definitions, or parts of Definitions, and these only are the principles of Demonstration, being Truths constituted arbitrarily by the Inventors of Speech, and therefore not to be demonstrated. (Part I, Chapter III, 9)

[7] This is linked with the tendency – characteristic of the Vienna Circle – to confine all meaningful problems of philosophy to the logical syntax of language.

[8] This (derogatory) phrase was used to denote the various conceptions of infallible conceptual intuition as an instrument of *a priori* cognition.

[9] Quine [17].

[10] This objection is found, for example, in White [20].

[11] Cf. Grice and Strawson [6].

[12] See Carnap [4].

[13] See Kokoszyńska [11] and [12], Przełęcki [16], Wójcicki [21].

[14] The terms 'explicatum' and 'explicandum' are used here in the sense given them by Carnap (cf. Carnap [4]).

[15] The proposal of such a restriction of the concept of adequacy of definitions is to be found in a paper by B. Stanosz devoted to an explanation of the concepts of the extension and intension of expressions (cf. Stanosz [18]).

[16] Cf. Ajdukiewicz [2].

[17] Cf. Przełęcki [16] and Wójcicki [21].

[18] Cf. Ajdukiewicz [1], Vol. II, p. 343.

[19] For such a criticism see Kneale [9] and [10], Lewis [13], Pap [14] and [15].

[20] This concept is explained by Ajdukiewicz as follows: "... a statement Z is a postulate in a language L if there is a terminological convention in L which says that a term λ which occurs in Z is supposed to denote the object which in place of λ satisfies Z. (...) ... a statement Z is, in a language L, an analytic statement in the semantic sense if it is a postulate in L or a consequence of one or more such postulates." Cf. Ajdukiewicz [2].

[21] The symbol $\varphi(\lambda)$ stands for a statement in L in which λ occurs; the symbol $\varphi(x/\lambda)$ stands for a sentential function obtained by the replacement of λ by a variable x, the latter being of the same type as λ.

[22] Ajdukiewicz has not explicitly formulated such a premiss, but it would seem to render his ideas correctly.

[23] Cf. Kokoszyńska [11], Przełęcki [16], Wójcicki [21].

[24] Such a solution would resemble Frege's interpretation of definite descriptions which do not satisfy the condition of uniqueness.

[25] Cf. Przełęcki [16].

[26] The terms model, proper model, etc., are used here in the sense formulated by Suszko in [19].

[27] Cf. Ajdukiewicz [1], Vol. II, p. 313.

[28] *Ut supra*, p. 314.

[29] *Ut supra*, p. 389.

[30] *Ut supra*, p. 342.

[31] On one occasion K. Ajdukiewicz suggested the term moderate empiricism to denote a standpoint which accords the status of scientific statement not only to statements based on empirical data, but to analytic statements as well. At that time he opposed analytic statements to those substantiated by empirical data (cf. *ut supra*, p. 54).

SOURCES OF THE TEXTS

Kazimierz Twardowski

O jasnym i niejasnym stylu filozoficznym (On Clear and Obscure Styles of Philosophical Writing), first published in *Ruch Filozoficzny*, V, 1919–20. *Symbolomania i pragmatofobia* (Symbolomania and Pragmatophobia), first published in *Ruch Filozoficzny*, VI, 1921.

Zur Lehre vom Inhalt und Gegenstand der Vorstellungen (On the Content and Object of Representations), first published by Hölder Verlag, Vienna 1894.

O czynnościach i wytworach (Actions and Products), first published in *The Festschrift in Commemoration of 1he 250th Anniversary of the Founding of Lwów University, by King Jan Kazimierz* (in Polish), Vol. II, Lwów 1912.

Z logiki przymiotników (Issues in the Logic of Adjectives), first published in the *Proceedings of the 1st Congress of Polish Philosophers,* held in Lwów in 1923 (in Polish).

(The first four papers were reprinted in K. Twardowski's *Selected Philosophical Works* (in Polish), PWN 1965; the fifth was reprinted in *Przegląd Filozoficzny*, XXX, No. 4, 1927).

Tadeusz Kotarbiński

Przegląd problematyki logiczno-semantycznej (A Survey of Logical and Semantic Problems), first published in the *Proceedings of the Łódź Scientific Society* (in Polish), Łódź 1947.

O postawie reistycznej, czyli konkretystycznej (The Reistic or Concretistic Approach), first published in *Myśl Współczesna*, 10/41, 1949.

Uwagi o znaczeniu wyrazów (Comments on the Meaning of Words), first published in the *Papers of the Linguistic Section of the Łódź Scientific Society* (in Polish), VIII, Łódź 1962.

Spór o desygnat (The Controversy Over Designata), first published in *Prace Filozoficzne*, XVIII, PWN 1963.

(The first two papers were published in an English-language version in T. Kotarbiński, *Gnosiology*, Pergamon Press 1966).

Tadeusz Czeżowski

Nazwy okazjonalne oraz imiona własne (Token-Reflexive Words Versus Proper Names), first published in T. Czeżowski, *Odczyty Filozoficzne* (Lectures on Philosophy), Toruń Scientific Society 1958.

Konotacja i denotacja (Connotation and Denotation), first published in T. Czeżowski, *Filozofia na rozdrożu* (Philosophy at the Crossroads), PWN 1965 (written in 1958–63).

Kazimierz Ajdukiewicz

Proposition as the Connotation of Sentence.
Intensional Expressions.
(Both texts were published in *Studia Logica*, XX, 1967, in English-language versions found in K. Ajdukiewicz's papers after his death; these translations had been made privately by O.A. Wojtasiewicz for K. Ajdukiewicz before his visit to the United States in 1959, and were delivered there as lectures; the Polish originals have not been found).

Izydora Dąmbska

W sprawie tzw. nazw pustych (Concerning the So-called Empty Names), first published in *Przegląd Filozoficzny*, XLIV, 1948.

Z filozofii imion własnych (Issues in the Philosophy of Proper Names), first published in *Kwartalnik Filozoficzny*, XVIII, 1949.

Truth and the Concept of Language, first published in a French-language version as *Concept de langue et vérité*, *Actes du XIIe Congrés des Sociétés Philosophiqnes de Langue Française*, Louvain–Paris 1964.

Seweryna Łuszczewska-Romahnowa

Wieloznaczność a język nauki (Ambiguity and the Language of Science), first published in *Kwartalnik Filozoficzny*, XVII, 1948.

Maria Ossowska

Significatio 'per se, i 'per aliud, u Anzelma (*Significatio 'per se,* and *'per aliud,* in Anselm), first published in *Przegląd Filozoficzny*, XXXV, 1932.

Stanisław Ossowski

Analiza pojęcia znaku (An Analysis of the Concept of Sign), first published in *Przegląd Filozoficzny*, XXIX, 1926.

Janina Kotarbińska

Spór o granice stosowalności metod logicznych (The Controversy over the Limits of the Applicability of Logical Methods), first published in *Studia Filozoficzne*, 3/38, 1964.

Kłopoty z istnieniem (Puzzles of Existence), first published in *The Festschrift to Commemorate the 80th Birthday of Tadeusz Kotarbiński* (in Polish), PWN 1967.

Adam Schaff

Wyrazy nieostre i granice ich precyzowania (Vague Words), first published in *Studia Filozoficzne*, 1/16, 1960.

Peter Thomas Geach

Nazwy i orzeczniki (Names and predicables), first published in *Semiotyka Polska.*

Roman Suszko

W sprawie antynomii kłamcy i semantyki języka naturalnego (On the Antinomy of the Liar and the Semantics of Natural Language), first published in R. Suszko, *Zarys elementarnej składni logicznej* (An Outline of Elementary Logical Syntax), PWN 1957.

Zdzisław Kraszewski and Roman Suszko

O klasach normalnych i nienormalnych na terenie języka potocznego (Normal and Non-normal Classes in Current Language), first published in *Studia Logica,* XIX, 1966.

Klasy normalne i nienormalne a teoriomnogościowe i mereologiczne pojęcie klasy (Normal and Non-Normal Classes Versus the Set-Theoretical and the Mereological Concept of Class), first published in *Studia Logica,* XXII, 1968.

Marian Przełęcki

Z semantyki pojęć otwartych (Semantics of Open Concepts), first published in *Studia Logica,* XV, 1964.

Henryk Stonert

Języki i teorie adekwatne z ontologią języka nauki (Languages and Theories Adequate to the Ontology Language of Science), first published in *Studia Logica,* XV, 1964.

Jerzy Pelc

Funkcjonalne podejście do semiotyki logicznej języka naturalnego (A Functional Approach to the Logical Semiotics of Natural Language), first published in Polish in *Studia Filozoficzne,* 2/49, 1967, first published in English in J. Pelc, *Studies in Functional Logical Semiotics of Natural Language,* The Hague–Paris 1971, Mouton, Janua Linguarum, Series Minor 90.

Leon Koj

Zasada przeźroczystości a antynomie semantyczne (The Principle of Transparency and Semantic Antinomies), first published in *Studia Logica,* XIV, 1963.

Witold Marciszewski

O funkcjach semantycznych mowy zależnej (The Semantic Functions of Oblique Speech, first published in *Semiotyka polska.*

Ryszard Wójcicki

Semantyczne pojęcie prawdy w metodologii nauk empirycznych (The Semantic Concept of Truth in the Methodology of Empirical Sciences), first published in Polish in *Studia Filozoficzne*, 3/58, 1969, first published in English in *Dialectics and Humanism*, I, No. 1, 1974.

Barbara Stanosz

Własność i zbiór (The Attribute and the Class), first published in Polish in *Studia Filozoficzne*, 2/49, 1967, first published in English in the fourth English-language issue of *Studia Filozoficzne*, 1970.

Adam Nowaczyk

Analyticity and Apriority (first published in *Semiotyka polska*).

BIOGRAPHICAL AND BIBLIOGRAPHICAL NOTES

AJDUKIEWICZ, Kazimierz (1890–1963). Studied philosophy under K. Twardowski and J. Łukasiewicz at Jan Kazimierz University in Lwów, where he also read mathematics and physics. Ph.D. in 1912 for a thesis on the *a priori* nature of space in Kant's philosophy and the problem of the origin of the spatial nature of ideas. In 1913–4 studied philosophy under E. Husserl and mathematics under D. Hilbert at the University of Göttingen. *Venia legendi* obtained at Warsaw University in 1921 for a thesis on the methodology of the deductive sciences. Professor of Philosophy at the Universities of Warsaw and Lwów, 1925–1939. In 1945–55 held the Chair of the Theory and Methodology of Sciences at Adam Mickiewicz University in Poznań, of which he was Rector in 1948–52. From 1955 to his retirement in 1961 held two Chairs of Logic: one at Warsaw University and one at the Institute of Philosophy and Sociology of the Polish Academy of Sciences; was Vice-director of the Institute. Member of the Polish Academy of Sciences, in Cracow before the World War II and since 1951 in Warsaw. Was Vice-president of the Division of Logic, Methodology and Philosophy of Sciences of the International Union of History and Philosophy of Science; in that capacity he organized an International Colloquium on the Methodology of Science, hold in Warsaw in 1961. Was a member of the International Institute of Philosophy. In 1935–51 he co-edited *Studia Philosophica.* In 1953 he founded *Studia Logica* of which he was editor-in-chief until his death. Was president of the Council of Universities. Held an honorary doctor's degree of the University of Clermont-Ferrand (and died one month before he was to have received an honorary doctor's degree from Poznań University).

Publications on semiotics: O intencji pytania "Co to jest P?" (On the Intention of the Question, "What is P?"), *Ruch Filozoficzny*, VII, 1923; Nazwy i zdania (Names and Sentences), *Ruch Filozoficzny*, IX, 1925; Składniki zdań (Components of Sentences), *Ruch Filozoficzny*, IX, 1925; Analiza semantyczna zdania pytajnego (The Semantic Analysis of Interrogative Sentences) *Ruch Filozoficzny*, X, 1926; O stosowaniu kryterium prawdy (On the Application of the Criterion of Truth), *Przegląd Filozoficzny*, XXX, 1927; *Główne zasady metodologii nauk i logiki formalnej* (The Main Principles of the Methodology of Science and Formal Logic), Warsaw 1928; a review of T. Kotarbiński's *Gnosiology The Scientific Approach to the Theory of Knowledge* (1930: the English-language version appears in the Supplement to T. Kotarbiński, *Gnosiology*, Pergamon Press 1966); O znaczeniu wyrażeń (On the Meaning of Expressions), in *The Festschrift of the Polish Philosophical Society* (in Polish), Lwów 1931; Paradoksy starożytnych (The Paradoxes of the Ancients), *Filomata*, 35/1931; W obronie uniwersaliów (In Defence of Universals), *Ruch Filozoficzny*, XIII, 1932 Logiczne podstawy nauczania (Logical Foundations of Teaching) in *Encyklopedia Wychowania* (Encyclopaedia of Education), Warszawa 1934, II ed. Warszawa–Wilno 1938. Logika (Logic) *Świat i Życie*,

Warszawa 1934; Sprache und Sinn, "*Erkenntnis*", IV, 1934; Das Weltbild und die Begriffsapparatur, *Erkenntnis*, IV, 1934; W sprawie "uniwersaliów" (Concerning Universals), *Przegląd Filozoficzny*, XXXVII, 1934; Über die Anwendbarkeit der reinen Logik auf philosophische Probleme, *Actes du VIII Congrès International de Philosophie*, Prague 1936; Die wissenschaftliche Weltperspektive, *Erkenntnis*, V, 1935; Sinnregeln, Weltperspektive, Welt, *Erkenntnis*, V, 1935; Die syntaktische Konnexität, *Studia Philosophica*, I, 1935; Okres warunkowy w mowie potocznej i w logistyce (The Conditional Sentences in Current Language and in Symbolic Logic), *Ruch Filozoficzny*, XIV, 1936; Die Definition, *Actes du I Congrès International de Philosophie Scientifique*, Paris 1936; Definicja (Definition) in *Język i poznanie. Wybór pism.* vol. I. Warszawa 1960. Problemat transcendentalnego idealizmu w sformułowaniu semantycznym (The Problem of a Transcendental Idealism in Its Semantic Formulation), *Przegląd Filozoficzny*, XL, 1937; *Propedeutyka filozofii dla liceów ogólnokształcących* (Propaedeutics to Philosophy for High Schools) Lwów–Warszawa 1938, II ed. 1947, III ed.1948, IV ed. 1948, V ed. 1950; O tzw. neopozytywizmie (On Logical Positivism), *Myśl Współczesna*, 6–7, 1946; Logika a doświadczenie (Logic and Experience), *Przegląd Filozoficzny*, XLIII (1947); Epistemologia i semiotyka (Epistemology and Semiotics), *Przegląd Filozoficzny* XLIV (1948). Epistemology and Semantics, *Proceedings of the Xth Congress of Philosophy*. Amsterdam 1948; Definicja prawdy a zagadnienie idealizmu (The Definition of Truth and the Problem of Idealism), in *Sprawozdanie Poznańskiego Towarzystwa Przyjaciól Nauk* (The Proceedings of the Poznań Society of the Friends of Science) (in Polish), Poznań 1949; O pojęciu istnienia (On the Notion of Existence) in "Sprawozdania Poznańskiego Towarzystwa Przyjaciół Nauk", Poznań 1949; *Zagadnienia i kierunki Filozofii* (*Teoria Poznania Metafizyka*) (Problems and Trends in Philosophy: Gnosiology and Metaphysics), Warszawa 1949. On the Notion of Existence (Some Remarks Connected with the Problem of Idealism) *Studia Philosophica*, IV, 1949–50; Logic and Experience, *Synthese*, VIII, 6–7, Busum, Netherlands, 1950; W sprawie artykułu prof. A. Schaffa o moich poglądach filozoficznych (Concerning A. Schaff's Article on My Philosophical Views), *Myśl Filozoficzna*, 2(8), 1963; *Abriss der Logik*, Berlin 1958; O definicji (On Definition), *Normalizacja*, 3–4 (1956). Okres warunkowy a implikacja materialna (The Conditional Sentence and Material Implication), *Studia Logica*, IV, 1956; *Three Concepts of Definition*, *Logique et Analyse*, I, 3–4, Brussels 1958; Le problème du fondement des propositions analytiques, *Studia Logica*, VIII 1958; Związki składniowe między członami zdań oznajmujących (Syntactical Relationships Between Parts of Declarative Sentences), *Studia Filozoficzne*, 6 (21), 1960; A Method of Eliminating Intensional Sentences and Sentential Formulae, *Atti del XII Congresso Internationale di Filosofia*, V, Florence 1961; Definicja (Definition) in *Wielka encyklopedia powszechna PWN*, vol. II, Warszawa 1963; Zagadnienie empiryzmu a koncepcja znaczenia (The Problem of Empiricism and the Conception of Meaning), *Studia Filozoficzne* 1 (36), 1964; *Pragmatic Logic*, Reidel 1974 (first published in Polish in 1965); Intensional Expressions, *Studia Logica*, XX, 1967; Proposition as the Connotation of Sentence, *Studia Logica*, XX, 1967.

CZEŻOWSKI, Tadeusz (b. 1889). Studied philosophy, mathematics and phsics at Lwów University under K. Twardowski, J. Łukasiewicz, W. Sierpiński and M. Smoluchowski. *Venia legendi* in philosophy, Lwów University, 1920. Appointed professor of philosophy in Wilno University in 1923; in 1945 took the Chair of Philosophy (later changed into the Chair of Logic) at the newly founded Copernicus University in Toruń. Since 1948 the editor of *Ruch Filozoficzny*. A member of the Editorial Boards of *Studia Logica*, *Library of the Classics of Philosophy* and *Studia Filozoficzne*. Retired in 1960.

Publications on semiotics: Imiona i zdania (Nouns and Sentences), *Przegląd Filozoficzny*, XXI, 1918; Arystotelesa teoria zdań modalnych (Aristotle's Theory of Modal Statements), *Przegląd Filozoficzny*, XXXIX, 1936; *Propedeutyka filozofii* (Propaedeutics to Philosophy), Lwów 1938; *Główne zasady nauk filozoficznych.* (Main Principles of Philosophical Disciplines), Toruń 1946; Prawda a język, (Truth and Language), *Wiedza i Życie* XV, 3 (1945) *Logika, podręcznik dla studiujących nauki filozoficzne* (Logic, a Textbook for Philosophy Students), Warszawa 1949, II ed. 1968; Uwagi o klasycznej definicji prawdy (Comments on the Classical Definition of Truth), in *The Festschrift in Commemoration of the 75th Anniversary of the Scientific Society in Toruń* (in Polish), Toruń 1952; *Odczyty filozoficzne* (Lectures on Philosophy), Toruń 1958 (2nd enlarged edition in 1969); *Filozofia na rozdrożu* (Philosophy at the Crossroads), Warsaw 1965.

DĄMBSKA, Izydora (b. 1904). Studied philosophy at Lwów University under K. Twardowski, 1922–6. Ph. D. in 1927 for a thesis on E. Goblet's theory of judgements, *venia legendi* in 1946 in Warsaw University for a thesis on irrationalism *versus* scientific cognition. In 1930–1 studied in Vienna, Berlin and Paris. In 1946–50 lectured on Philosophy at Warsaw University and Poznań University. Since 1965 has worked at the Institute of Philosophy, and Sociology of the Polish Academy of Sciences. Member of the Institut International de Philosophie; member Polish Semiotic Society. Appointed professor in 1955; held the Chair of History of Philosophy in the Jagellonian University in Cracow from 1957 to 1964; Member of the editorial board of the *Archives Internationales d'Histoire des Idées*; has cooperated with the *Bibliographie de Philosophie*, sponsored by the Institut International de Philosophie in Paris, since 1937.

Publications on semiotics: Z semantyki zdań warunkowych (Semantic Issue of Conditional Sentences), *Przegląd Filozoficzny*, XLI, 1938; W sprawie tzw. nazw pustych (Concerning So-called Empty Terms), *ut supra*, XLIV, 1–3, 1948; Z filozofii imion własnych (Issues in the Philosophy of Proper Names), *Kwartalnik Filozoficzny*, XVIII, 3–4, 1949; Sur le concept de compréhension, *Atti del Congresso Internazionale di Filosofia*, V, Florence 1960; Niektóre pojęcia gramatyki w świetle logiki (Some Concepts of Grammar in the Light of Logic), in: *Philosophical Essays Dedicated to Roman Ingarden* (in Polish), Warsaw 1964; Milczenie jako wyraz i jako wartość (Silence as Expression and Value), *Roczniki Filozoficzne KUL*, 1964; Concept de langue et vérité, *Actes du XIIe Congrès des Sociétés de Langue Française*, Louvain–Paris 1964; Koncepcja języka w filozofii Kazimierza Ajdukiewicza (The Concept of Language in Kazimierz Ajdukiewicz's Philosophy),

Ruch Filozoficzny, XXIV, 1965; Sur certains aspects de philosophie linguistique, *Actes du XIIe Congrès des Sociétés de Language Francaise*, *O narzędziach i przedmiotach poznania* (Instruments and Objects of Cognition), Warsaw 1967; O funkcjach semiotycznych milczenia (The Semiotic Functions of Silence), *Studia Semiotyczne* (Semiotic Studies) ed. J. Pelc, II, 1971; Analiza pojęcia oznaki w semiotyce stoickiej (The Analysis of the Concept of Indexical Sign in Stoic Semiotics), *Studia z historii semiotyki* ed. I. Sulowski, 1971; Aletheia i alethes w dialektyce stoickiej na podstawie "Adversus logicos" Sekstusa Empiryka (Aletheia and Alethes in Stoic Dialectic as seen in the "Adversus Logicos" of Sextus Empiricus), *ut supra*. O konwencjach semiotycznych (Semiotic Conventions), *Studia Semiotyczne* (Semiotics Studies) ed. J. Pelc, IV, 1972; Niektóre zagadnienia semiotyki stoickiej w świetle traktatu Sekstusa Empiryka "Przeciw Logikom" (Some Problems of stoic semiotics in "Adversus logicos" of Sextus Empiricus). *Studia z Historii Semiotyki* (Semiotic Historical Studies) ed. J. Sulowski, II, 1973; *Znaki i myśli. Wybór pism z semiotyki, teorii nauki i historii filozofii* (Signs and Thoughts, Selected Papers in Semiotics, Epistemology and History of Philosophy) Warsaw, 1975; *O konwencjach i konwencjonalizmie* (Conventions and Conventionalism) Wrocław, 1975; Langue et science dans la philosophie de Leibniz. Semiotic Historical Studies ed. J. Sulowski, III, 1976.

GEACH, Peter Thomas (b. 1916). Studied the humanities at Balliol College Oxford, 1934–8. Following World War II, tutor at Cambridge University, where he attended L. Wittgenstein's lectures; cooperated with Von Wright. Connected with Birmingham University since 1950 (reader in logic, 1961; professor of logic in the Department of Philosophy, 1966). Member, the British Academy, since 1966. In 1957–67 lectured as a visiting professor at Oxford, Cornell University of Michigan, and University of Chicago.

Publications on semiotics: *Reference and Generality*, Cornell Univ. Press, 1962 (rev. ed. 1968); articles in *Mind*: On Rigour in Semantics (1949), Subject and Predicate (1950), On Names of Expressions (1950), The Doctrine of Distribution (1956); in *the Aristotelian Society Proceedings*: The Law of Excluded Middle (1956); in *Analysis*: Designation and Truth (1947–8), Russell's Theory of Descriptions (1949–50), On "Insolubilia" (1954–5), On Beliefs about Oneself (1957–8), Imperative and Deontic Logic (1957–8), Russell on Meaning and Denoting (1958–9), Is it right to say "or" is a conjucntion? (1958–9), Ryle on Namely-Riders (1960–1), Namely-Riders Again (1961–2), Referring Expressions Again (1963–4), Dr Kenny on Practical Inference (1965–6); in *The Journal of Philosophy*: On Complex Terms (1965), Complex Terms Again (1965), Intentional Identity (1967); in *Ratio*: Logical Procedures and the Identity of Expressions (1965–6); in the *Philosophical Review*: Assertion (1965); in *Synthese*: Quine's Syntactical Insights (1968).

KOJ, Leon (b. 1929). Studied philosophy at the Roman Catholic University in Lublin, 1948–54. Ph. D. for a thesis on the logical analysis of names of expressions (Maria Skłodowska-Curie University in Lublin, 1962), *venia legendi* in logic for a thesis on the relationships between semantics and pragmatics (Warsaw University, 1971). Since 1956 at the Department of Philosophy, Lublin University, Member. Polish Semiotic Society.

Publications on semiotics: Analiza logiczna nazw wyrażeń (A Logical Analysis of Names of Expressions), *Ruch Filozoficzny*, 1–2, 1960; Zasada przeźroczystości a antynomie semantyczne (The Principle of Transparency and Semantic Antinomies), *Studia Logica*, XIV, 1963; Nazwy cudzysłowowe (Quotational Names), *Studia Logica*, XV, 1964; Dwie koncepcje semantyki (Two Concepts of Semantics), *Studia Filozoficzne*, 4, 1966; O definiowaniu wyrażeń okazjonalnych (On Defining Token-Reflexive Expressions), *Annales UMCS*, Sectio F, Vol. XIX, 2, 1964; O definiowaniu pojęć wielotreściowych (On Defining Concepts with Multiple Content); *Aktualne Problemy Informacji i Dokumentacji*, XIII, 6, 1968; Podstawy denotowania (The Foundations of Denoting), *Studia Semiotyczne* (Semiotic Studies), ed. J. Pelc, I, 1970; *Semantyka a pragmatyka. Stosunek językoznawstwa i psychologii do semantyki* (Semantics Versus Pragmatics. The Relation which Linguistics and Psychology Bear to Semantics), PWN 1971; Analiza pytań. Problem terminów pierwotnych logiki pytań (An Analysis of Questions. The Problem of Primitive Terms in the Logic of Questions), *Studia Semiotyczne*, II, 1971; Analiza pytań II (An Analysis of Questions II), *Studia Semiotyczne* 1972; Jerrolda J. Katza teoria znaczeń (Jerrold J. Katz's Theory of Meanings), *Studia Semiotyczne* 1973; Psychologiczna geneza pojęcia zbioru (The Psychological Origin of the Concept of Set), *Annales U.M.C.S.* Sectio I, vol. 2, 1975.

KOTARBIŃSKA, Janina (b. 1901). Studied philosophy at Warsaw University under T. Kotarbiński, 1921–7. Ph.D. for a thesis on the problem of explanation in natural science in J. S. Mill and E. Meyerson (Warsaw University, 1927), *venia legendi* for a thesis on indeterminism in contemporary physics, biology, the social sciences and humanities (Warsaw University, 1935). Professor of gnosiology, logic and methodology in Łódź University, 1945–51; professor of logic in Warsaw University from 1951 to her retirement in 1972. Dean of the Faculty of Philosophy, Warsaw University, 1958–60. Member: Polish Semiotic Society. Directorial Board of: Polish Philosophical Society, Semiotic Studies and Historical Division of Logic, Methodology and Philosophy of Science of the Polish Committee of the International Union of History and Philosophy of Science.

Publications on semiotics: Definicja (On Definition), *Studia Logica*, II, 1955; Pojęcie znaku (The Concept of Sign), *Studia Logica*, VI, 1957; On Ostensive Definitions, *Philosophy of Science* 1961; Controversy on the Applicability Limits of Logical Methods, *Logique et Analyse*, 1965; Troubles with Existence, *Akten d. XIV Internationalen Kongresses für Philosophie*, Wien 1968; Wyrażenia okazjonalne (Token-reflexive Expressions), *Studia Filozoficzne*, 1/68, 1971.

KOTARBIŃSKI, Tadeusz (b. 1886). Studied mathematics in Cracow and architecture in Lwów and Darmstadt. 1906–7; studied philosophy (under K. Twardowski) and the classics at Lwów University, 1907–12, where in 1912 he took his Ph. D. for a thesis on utilitarianism in the ethical systems of Mill and Spencer. Appointed professor in the newly reopened Warsaw University in 1919 (where in 1928–30 he served as Dean of the Faculty of Arts). In 1939–1945 he lectured at Underground Warsaw University. In 1945–9 he was Rector of the newly founded Łódź University, where he stayed till 1951. Held the Chair of Logic in Warsaw University from 1951 untill his retirement in 1960. Member, Polish Academy of Sciences, since 1951 (President, 1957–62). Member, Institut International de Philosophie (president, 1960–3, honorary president since 1963), Institut de Philosophie Politique, Union des Associations Internationales, Académie de la Philosophie des Sciences, Union Internationale de l'Histoire et de la Philosophie des Sciences. Foreign member of the Soviet, Bulgarian, Mongolian and Serbian Academies of Sciences and of the Societas Scientiarum Fennica, and extraordinary member of the British Academy. Honorary doctor's degrees from the Université Libre in Brussels, Łódź University, the Jagiellonian University in Cracow, the University of Florence, and Oxford University. Member, Polish Semiotic Society.

Publications on semiotics: Sprawa istnienia przedmiotów idealnych (The Problem of the Existence of Ideal Objects), *Przegląd Filozoficzny*, XXIII, 1920; O istocie doświadczenia wewnętrznego (On the Nature of Inner Experience), *Przegląd Filozoficzny*, XXV, 1922; *Gnosiology. The Scientific Approach to the Theory of Knowledge*. Transl. by O. Wojtasiewicz, G. Bidwell and C. Pinder, 1966 (first published in Polish in 1929; the English version also includes a number of papers written after 1945); Le Réalisme radical, *Proceedings of the Seventh International Congress of Philosophy*, Oxford–London 1931; Uwagi na temat reizmu (Comments on Reism), *Ruch Filozoficzny*, XII, 1 (10), 1930–2; Zasady reizmu (The Principles of Reism), *Proceedings of the Poznań Society of Friends of Science* (in Polish), V, 1, 1931; O różnych znaczeniach słowa "materializm" (On the Various Meanings of the Term "Materialism"), *Przegląd Filozoficzny*, XXXIV, 1931; The Fundamental Ideas of Pansomatism, *Mind*, LXIV, 256, 1955; Z dziejów pojęcia teorii adekwatnej (A History of the Concept of Adequate Theory), *Przegląd Filozoficzny*, XXXIX, 1936, and XL, 3, 1937; Humanistyka bez hipotez (The Humanities without Hypostases), *Myśl Filozoficzna*, 1, 1952; Zdania niepsychologiczne (Non-psychological Statements), in *Selected Works* (in Polish), Vol. II, Warszawa 1958; Essai de réduire la connaissance psychologique à l'extraspection, *Atti del XII Congresso Internazionale di Filosofia*, Florence 1960; The Basic Postulate of Concretism, *Zeichen und System der Sprache*, Vol. I, Berlin 1961; *Kurs logiki dla prawników* (A Course of Logic for Lawyers), Warszawa 1961; Uwagi o znaczeniu wyrazów (Comments on the Meaning of Words), *Proceedings of the Linguistic Committee of the Łódź Scientific Society* (in Polish), VIII, 1962; Spór o desygnat (The Controversy over Designata), *Prace Filozoficzne*, XVIII, 1963; List w sprawie pojęcia znaczenia (A Letter Concerning the Concept of Meaning) in: E. Grodziński, *Znaczenie słowa w języku naturalnym* (The Meaning of a Word in Natural Language), Warszawa 1964; Psychological Propositions, in: B. B. Wolman (ed.), *Scientific Psychology*, 1965; Franz Brentano comme réiste, *Revue Internationale de Philosophie*, 78, 4, 1966.

KRASZEWSKI, Zdzisław (b. 1925). Publication on semiotics: Zagadnienie intencjonalności (The Problem of Intensionality), *Studia Filozoficzne*, 3 (12), 1959; Pojęcie i kombinatoryka znaku (The Concepts and the Combinatorics of Sign), *Prace Filozoficzne*, XX, 1969; (with R. Suszko) the two papers included in the present book (previously published in Polish in *Studia Logica*, XIX, 1966, and XXII, 1968).

ŁUSZCZEWSKA-ROMAHNOWA Seweryna (b. 1904). Studied philosophy and mathematics at Lwów University under K. Twardowski, K. Ajdukiewicz, R. Ingarden and H. Steinhaus, 1922–8. Ph.D. for a thesis on token-reflexive words (Lwów University 1932). Connected with Poznań University from 1947 until her retirement in 1974 and in 1957–61 with the Department of Logic Polish Academy of Science; appointed professor of logic in 1962.

Publications on semiotics: see the paper included in the present volume (previously published in Polish in 1938).

MARCISZEWSKI, Witold (b. 1930). Studied philosophy at the Roman Catholic University in Lublin 1953–7, under S. Kamiński, where in 1959 he received his Ph. D. for a thesis on analytic statements in Thomas Aquinas and in analytic philosophy. Connected with Warsaw University since 1963, *venia legendi* in Warsaw University, 1971, for a thesis on the foundations of a logical theory of beliefs. Member of the board, Polish Semiotic Society.

Publications on semiotics: O stosunku eksplikacji do konfirmacji (The Relation between Explanation and Confirmation), *Roczniki Filozoficzne KUL*, 1961; Dzieła sztuki jako znaki (Works of Arts as Signs), *Estetyka*, 1963; Sposoby streszczania i odmiany streszczeń (Methods of Summarizing and Types of Summaries), *Studia Semiotyczne* (Semiotic Studies), ed. J. Pelc, I, Wrocław–Warszawa 1970; On uses of "to believe", *Akten des XIV Internationalen Kongresses für Philosophie*, III, Vienna 1969; Dawida Hume'a empiryczna teoria sądu (David Hume's Empiricist Theory of Judgements), *Studia Semiotyczne*, II, 1971; "Significatio" u Dunsa Szkota ("Significatio" in Duns Scotus' Works), *Studia z historii semiotyki* (Studies in the History of Semiotics), ed. J. Sulowski, Wrocław–Warszawa 1971; *Podstawy logicznej teorii przekonań* (Foundations of a Logical Theory of Beliefs), PWN 1972; Formalna charakterystyka dziedziny rozważań jako podstawa indeksu rzeczowego (Formal Characteristics of a Universe of Discourse as the Basis of a Subject Index), *Studia Semiotyczne*, III, 1972; Problem istnienia przedmiotów intencjonalnych (The Problem of Existence of Intentional Objects), *Studia Semiotyczne*, IV, 1973; Analiza semantyczna pytań jako podstawa reguł heurystycznych (A Semantic Analysis of Questions as a Basis of Heuristic Rules), *Studia Semiotyczne*, V, 1974; Organizacja semantyczna tekstu (The Semantic Structure of Texts), *Studia Semiotyczne*, VI, 1975; A Syntactic Description of Oblique Speech in terms of Categorial Grammar, *Revista Brasileira de Linguistica*, forthcoming; Syntaktische Konnexität und Textkonnexität, *Studia Grammatica*, 1977.

NOWACZYK Adam (b. 1936). Studied philosophy at Warsaw University, 1954–9; Ph. D. for a thesis on the concept of analytic statements as related to selected issues in gnosiology and semantics (Warsaw University, 1967, under M. Przełęcki). 1961–1969, Department of Logic, Łódź University, later the Polish Academy of Sciences.

Publications on semiotics: Lingwistyczna teoria logicznej konieczności w świetle krytyki Arthura Papa (The Linguistic Theory of Logical Necessity as Criticized by Arthur Pap), *Studia Filozoficzne*, 1 (28), 1962; On the Adequate Analysis of the Concept of Synonymity, *Studia Filozoficzne*, 3, 1966; Pojęcie zdania analitycznego w problematyce teoriopoznawczej (The Concept of Analytic Statement in Gnosiological Issues), *Zeszyty Naukowe Uniwersytetu Łódzkiego*, Series I, No 49, 1967; Zaimki zamiast zmiennych i operatorów (Pronouns Instead of Variables and Operators), *Studia Semiotyczne* (Semiotic Studies); ed. J. Pelc, II, Wrocław–Warszawa 1971; W sprawie formalnej definicji języka (On a Formal Definition of Language), *Studia Semiotyczne*, VI, 1975; (with Barbara Stanosz) *Logiczne podstawy języka* (The Logical Foundations of Language), Ossolineum 1975; Numerical Constructs as Theorems of Empirical Theories, *Studia Logica*, 1, XXXV, 1976.

OSSOWSKA, Maria (1896–1974). Admitted to Warsaw University in 1915. Ph. D. for a thesis on Stoic axiology (Warsaw University, 1921 under J. Łukasiewicz). *Venia legendi* for papers on semantics (Warsaw University, 1932). In 1933–5 studied in London, Oxford, and Cambridge. Appointed professor at Łódź University in 1945. Returned to Warsaw University, where she worked until her retirement in 1966. In 1949–52 and 1956–1966, Head, Department of Moral Science. Visiting professor in the United States in 1960 and 1967.

Publications on semiotics: Semantyka Stanisława Szobera (Stanisław Szober's Semantics), *Przegląd Filozoficzny*, XXVIII, 1928; Stosunek logiki i gramatyki (Relationships between Logic and Grammar), *Kwartalnik Filozoficzny*, V, 1929; Słowa i myśli (Words and Thoughts), *Przegląd Filozoficzny*, XXXIV, 1931; Significatio «per se» i «per aliud» u Anzelma (Significatio «per se» and «per aliud» in Anselm), *Przegląd Filozoficzny*, XXXV, 1932; Ocena i opis (Evaluation and Description), *Kwartalnik Filozoficzny*, ..., 1938; O dwóch rodzajach ocen (Two Kinds of Evaluations), *Kwartalnik Filozoficzny*, XVI, 1964; Qu'est-ce qu'un jugement de valeur, *Proceedings of the Xth International Congress of Philosophy*, Amsterdam 1948; Oceny i normy (Evaluations and Norms), *Kwartalnik Filozoficzny*, XVII, 1949.

OSSOWSKI, Stanisław (1897–1963). Studied at the Universities of Warsaw and Wilno and at the Sorbonne. Ph.D. for a thesis on an analysis of the concept of sign (Warsaw University, 1924, under T. Kotarbiński). *Venia legendi* for his book *U podstaw estetyki* (The Foundations of Aesthetics) (Warsaw University, 1933). In 1933–5 studied in London, Oxford and Cambridge. In 1945 appointed professor in Łódź University. Returned to Warsaw Universtiy in 1947, where he worked until his death. Head, Department of Sociology in 1947–1952 and 1952–1963; 1956–1961 head, Department of Sociology, Polish Academy of Sciences. Visiting professor in numerous universities in Europe and America. Founder of the Polish Sociological Society and its president until his death. Co-founder of the International Sociological Society, of which he was elected vice-president in 1959.

Publications on semiotics: See the paper included in the present book.

PELC Jerzy (b. 1924). Studied philosophy under T. Kotarbiński and W. Tatarkiewicz and Polish literature in the underground Warsaw University, and later at Cracow University under R. Ingarden and again at Warsaw University, where in 1951 he obtained his Ph.D. for a thesis on the content and form of literary works. *Venia legendi* for his book *O pojęciu tematu* (On the Concept of Theme), Ossolineum 1961 (Institute of Philosophy and Sociology Polish Academy of Sciences, 1961). In 1945–1950 he worked at the Department of Polish Literature, Warsaw University; in 1951–1971 at the Department of Logic, Warsaw University, appointed Assistent Professor, 1954, since 1972, head Department of Logical Semiotics Warsaw Univ. In 1957–1962, member of the Directorial Board, Institute of Philosophy and Sociology, Polish Academy of Sciences, and at the Department of Logic of the Institute. Appointed professor in 1971. Member, Editorial Board, *Studies in Semiotics* (Paris), 1966–8, *Semiotics* (Paris), sponsored by Association Internationale de Sémiotique, since 1969 *Philosophy and Social Action* (Delhi), since 1975 member, Editorial Board. Co-founder and Secretary General of the Polish Semiotic Society. Member, Philosophical Committee of the Polish Academy of Sciences; member of the board of the Polish Philosophical Society. Member, Comité Directeur Provisoire de l'Association Internationale de Sémiotique, Paris, 1967–8; member of the board of the Association Internationale de Sémiotique, Paris, since 1969; member, Institut International de Philosophie, Paris, since 1971; fellow, Netherlands Institute for Advanced Study in the Humanities and Social Sciences, 1974–75. The editor and founder of *Studia Semiotyczne* (Semiotic Studies), sponsored by the Polish Semiotic Society, since 1970.

Publications on semiotics: *Poglądy Rudolfa Carnapa na kwestie znaczenia i oznaczania, przegląd* (Rudolf Carnap's Views on Meaning and Denoting), Wrocław 1960; Funkcje semantyczne a forma i treść w sztuce (Semantic Functions versus Form and Content in Art), in: *Charisteria* (Philosophical Articles Dedicated to Władysław Tatarkiewicz on His 70th Birthday), Warszawa 1960; O wartości logicznej i charakterze asertywnym zdań w dziele literackim (Truth-value and Modality of Sentences in Literary Works: Sentence Versus Proposition), *Estetyka*, I, 1960, *O pojęciu tematu* (On the Concept of Theme), Wrocław 1961; Semantic Functions as Applied to the Analysis of the Concept of Metaphor, in: *Poetyka-Poetics-Pojetika*, The Hague–Warszawa, Mouton, 1961; Zdanie a sąd w dziele literackim: wartość logiczna i charakter asertywny quasi-sądu (Sentence Versus Proposition in Literary Works: Truth-value and Modality of Quasi-judgements), *Estetyka*, III 1962; Quasi-sądy a dzieło literackie (Quasi-judgements and Literary Work), *Pamiętnik Literacki*, LIV, 3 1963; *Logika i język, studia z semiotyki logicznej* (Logic and Language, Studies in Logical Semiotics, selected, translated, with an Introduction by), Warszawa 1967; Imiona własne w języku naturalnym, prolegomena do teorii (Proper Names in Natural Language: Prolegomena to a Theory), in: *Fragmenty filozoficzne, seria trzecia* (Articles Dedicated to Tadeusz Kotarbiński on His 80th Birthday), Warszawa 1967; Nominal Expressions and Their Real or Fictitious Referents, in: *Akten d. XIV Internation. Kongress. fur Philos.*, Wien 1969; Meaning as an Instrument, *Semiotica*, I, 1 1969; A Functional Approach to the Logical Semiotics of Natural Language, in: *Sign, Language, Culture*, The

Hague–Paris 1970, Mouton, Janua Linguarum, Series Maior I; *Studia semiotyczne* (Semiotic Studies), ed. Wrocław, I (1970), II (1971), III (1972), IV (1973), V (1974), VI (1975), VII (1976); *Studies in Functional Logical Semiotics of Natural Language*, The Hague–Paris 1971, Mouton. Janua Linguarum, Series Minor 90; *O użyciu wyrażeń* (On Use of Expressions), Wrocław 1971; On the Concept of Narration, *Semiotica*, III, 1, 1971; 'W sprawie programu prac z zakresu historii semiotyki' (On a Program of Work in the History of Semiotics), in: *Studia z historii semiotyki*, ed. J. Sulowski, vol. II, Wrocław 1973; Nominal Expressions and Literary Fiction', in: *Aesthetics in Twentieth-Century Poland*, ed. J. G. Harrell, A. Wierzbiańska, Lewisburg 1973, Bucknell Univ. Press; Intensjonalność, język empiryczny i język intencjonalny (Intensionality, Empirical Language, and Intentional Language), in: *Studia semiotyczne*, IV 1975; Kilka uwag na temat związków między semiotyką, logiką a językoznawstwem '(Some Remarks on Relations Between Semiotics, Logic, and Linguistics), in: *Studia semiotyczne*, VII 1976; Semiotics as a Science, a Method and a Program, in Proceedings of Semiotic Symposium at Brown University, Providence, Rhode Island, 1976—in press, Peter de Ridder Press, Holland.

PRZEŁĘCKI, Marian (b. 1923). Studied philosophy at Łódź University, 1945–9. Ph. D. for a thesis on extra-formal criteria of correctness of definitions in the natural sciences (Warsaw University, 1957 under J. Kotarbińska). *Venia legendi* for a thesis on theoretical concepts in empirical theories (Warsaw University, 1961). In 1947–1951 he worked at the Department of Logic, Łódź University, and since 1952 at the Department of Logic, Warsaw University; reader 1961. Appointed professor in 1971. In 1955–1961, at the Department of Logic, Polish Academy of Sciences. Member, Polish Philosophical Society, Polish Semiotic Society; member the Editorial Board of *Studia Logica*, *Studia Filozoficzne*, *Erkenntnis*.

Publications on semiotics: O tzw. definicjach operacyjnych (Operational Definitions), *Studia Logica*, III, 1955; Prawa a definicje (Laws and Definitions), in: *Prawa nauki* (Laws of Science), PWN, 1957; Logiczna analiza rozwoju pojęcia pierwiastka chemicznego (A Logical Analysis of the Development of the Concept of Chemical Element), *Studia Filozoficzne*, 1, 1957; W sprawie terminów nieostrych (Concerning Vague Terms), *Studia Logica*, VIII, 1958; Postulat empiryczności terminów przyrodniczych (The Requirement of Empiricity of Terms in the Natural Sciences), in: *Fragmenty Filozoficzne*, PWN 1959; Operacjonizm (Operationism), *Archiwum Historii Filozofii i Myśli Społecznej*, 5, 1959; Interpretacja systemów aksjomatycznych (The Interpretation of Axiomatic Systems), *Studia Filozoficzne*, 6, 1960; Pojęcia teoretyczne a doświadczenie (Theoretical Concepts and Experience), *Studia Logica*, XI, 1961; The Concept of Genotype, in: *Form and Strategy in Science*, Reidel Co., 1964; O pojęciu zdania analitycznego (The Concept of Analytic Statement), *Studia Logica*, XIV, 1963; O definiowaniu terminów spostrzeżeniowych (Defining Observation Terms), in: *Rozprawy logiczne* (Papers on Logic), PWN 1964; Z semantyki pojęć otwartych (Semantics of Open Concepts), *Studia Logica*, IV, 1964; W sprawie istnienia przedmiotów teoretycznych (On the Existence of Theoretical Objects), in: *Teoria i doświadczenie* (Theory and Experience), PWN 1966; Teorie empiryczne w ujęciu logiki współczesnej (Empirical Theories as Interpreted in Contemporary Logic), in: *Fragmenty Filozoficzne*, PWN 1967; (with R. Wójcicki) The Problem of Analyticity, *Synthese*, 19, 1969; *The Logic of Empirical Theories*, London 1969. Problem interpretacji języka empirycznego w ujęciu teorio-modelowym (A Model Theoretic Approach to the Problem of Interpretation of Empirical Languages), *Studia Filozoficzne*, 1, 1972; O pewnych filozoficznych konsekwencjach semantycznej definicji prawdy (Some Philosophical Consequences of the Semantic Definition of Truth), *Studia Filozoficzne*, 6, 1973; Empirical Meaningfulness of Quantitative Statements, *Synthese*, 26, 1974; W sprawie obserwacyjnej definiowalności terminów teoretycznych (On the Observational Definability of Theoretical Terms), *Acta Universitatis Wratislaviensis*, No. 290, Wrocław 1976; O pojęciu nieistotnego występowania terminów (On the Concept of Vacuous Occurrence of Terms), *Studia Filozoficzne*, 3, 1976; Interpretation of Theoretical Terms: In defence of an Empiricist Dogma, in: *Formal Methods in the Methodology of Empirical Sciences*, Wrocław 1976.

SCHAFF Adam (b. 1913). Studied law and economics at Lwów University and at the École des Sciences Politique et Économiques in Paris, and philosophy in the Soviet Union, where he obtained the equivalents of Ph. D. and *venialegendi* (at the Soviet Academy of Sciences Institute of Philosophy in Moscow, in 1941 and 1945, respectively). Appointed professor in Łódź University in 1945, he moved to Warsaw University in 1948, where he remained until 1968. Member, Polish Academy of Sciences, since 1951. Rector, Institute for the Training of Research Workers (later renamed the Social Science Institute) of the Polish United Workers' Party Central Committee, 1950–6. Director, Institute of Philosophy and Sociology. Polish Academy of Sciences, 1956–1968; member, Polish Semiotic Society. Chairman, Council of Directors of the European Centre for the Coordination of Research and Documentation in the Social Sciences, Vienna, since 1963.

Publications on semiotics: *Pojęcie i słowo* (Concepts and Words), Książka 1946; *Z zagadnień marksistowskiej teorii prawdy* (Issues in the Marxist Theory of Truth), KiW 1951; Wyrazy nieostre i granice ich precyzowania (Vague Terms and the Limits of Their Precision), *Studia Filozoficzne*, 1 (16), 1960; *Introduction to Semantics*, Pergamon Press 1962; *Language and Cognition*, McGraw-Hill 1973 (both previously published in Polish).

STANOSZ Barbara (b. 1935). Studied philosophy in Warsaw University, 1952–7. Ph. D. for a thesis on the semantic functions of expressions as interpreted in formal logic, Warsaw University 1965, under R. Suszko. *Venia legendi* for a thesis on the logical foundations of the semantics of natural language, Warsaw University 1975. Since 1957, at the Department of Logic, Warsaw University; member, Polish Semiotic Society.

Publications on semiotics: Znaczenie i oznaczanie a paradosk intensjonalności (Meaning and Denoting Versus the Paradox of Intensionality), *Studia Filozoficzne*, 2 (41), 1965; Własność i zbiór (The Attribute and the Class), *Studia Filozoficzne*, 2 (49), 1967. The Problem of Intensionality, *Studia Filozoficzne*, 3, 1966; O pojęciu języka prelogicznego (The Concept of Prelogical Language), *Studia Semiotyczne*, (Semiotic Studies), ed. J. Pelc, I, 1970; Formal Theories of Extension and Intension of Expressions, *Semiotica*, II, 1, 1970; Kodeks języka naturalnego (A Code of Natural Language), *Studia Semiotyczne*, II, 1971; Problemy definicji prawdy dla języka naturalnego (Problems of the Definition of Truth for Natural Language), *Studia Filozoficzne*, 5 (72), 1971; Meaning and Interpretation, *Semiotica*, XI, 4, 1974; Status poznawczy semantyki (The Cognitive Status of Semantics), *Studia Semiotyczne*, V, 1974; (with Adam Nowaczyk) *Logiczne podstawy języka* (The Logical Foundations of Language), Wrocław 1976.

STONERT, Henryk (b. 1923). Studied philosophy and mathematics at the Universities of Warsaw and Cracow. Ph. D. for a thesis on definitions in the deductive sciences (Warsaw University 1956 under T. Kotarbiński). *Venia legendi* for a thesis on the logical foundations of practical disciplines (Warsaw University, 1967). Since 1949 at the Department of Logic, Warsaw University, reader 1967; member, Polish Semiotic Society.

Publications on semiotics: *Definicje w naukach dedukcyjnych* (Definitions in the Deductive Sciences), Wrocław 1959; *Język i nauka* (Language and Science), Warszawa 1964; Języki i teorie adekwatne z ontologią języka nauki (Language and Theories Adequate to the Ontology of Scientific Language), *Studia Logica*, XV, 1964; Język i ontologia (Language and Ontology) *Kultura*, 25, 1968; Semantyczne determinacje problematyki naukowej (Semantic Determinants of Research Problems) in: *Studia z dziejów semiotyki* (Studies in the History of Semiotics), ed. J. Sulowski, Ossolineum 1971.

SUSZKO, Roman (b. 1919). Publications on semiotics: in addition to the three papers included in the present book, *Z teorii definicji* (Issues in the Theory of Definitions), Poznań Society of the Friends of Science, VII, 5, 1949; Formalna teoria wartości logicznych (A Formal Theory of Logical Values), *Studia Logica*, VI, 1957; Logika formalna a niektóre zagadnienia teorii poznania (Formal Logic and Selected Issues in Gnosiology), *Myśl Filozoficzna*, 2–3, 1957; *Zarys elementarnej składni logicznej* (An Outline of Elementary Logical Syntax), Warsaw University Publications, 1957; Syntactic Structure and Semantical Reference, *Studia Logica*, VIII and IX, 1958 and 1960; Concerning the Method of Logical Schemes, the Notion of Logical Calculus and the Role of Consequence Relation, *Studia Logica*, XI, 1961; O kategoriach syntaktycznych i denotacjach wyrażeń w językach sformalizowanych (Syntactic Categories and Denotations of Expressions in Formalized Languages), in: *Papers on Logic Dedicated to K. Ajdukiewicz* (in Polish), PWN 1964; Ontology in the Tractatus of L. Wittgenstein, *Notre Dame Journal of Formal Logic*, IX, 1968; Reifikacja sytuacji (Reification of Situations), *Studia Filozoficzne*, 2 (69),1971.

TWARDOWSKI, Kazimierz Jerzy Adolf (1866–1938). Studied philosophy under F. Brentano in Vienna University, 1885–9. Ph. D. for the thesis Idee und Perzeption. Eine erkenntnistheoretische Untersuchung aus Descartes (Vienna University, 1891). *Venia legendi* for the thesis Zur Lehre vom Inhalt und Gegenstand der Vorstellungen (Vienna University, 1894). Appointed professor in Lwów University in 1895, where he remained until his retirement in 1930. since 1930 honorary Professor. Dean, Faculty of Philosophy, Lwów University, in 1900–1 and 1904–5; Rector, Lwów University, 1914–7. Co-founder (1887) of the Vienna Philosophical

Society, and its vice-president, 1887–9. Founder (1904) of the Polish Philosophical Society, of which he was president until his death; in 1911 he founded *Ruch Filozoficzny* and in 1935 another periodical *Studia Philosophica*. Organized the first university seminar on philosophy in Poland (1897) and started the first classes in experimental psychology in Poland (1901). In 1912 he established the Polish section of the international bibliography of philosophy, *Die Philosophie der Gegenwart*, published in Heidelberg. In 1935, he was elected member of the editorial board of *Acta Psychologica*, published in Amsterdam. Honorary doctor's degrees from the Universities of Warsaw and Poznań; member Polish Academy of Sciences.

Publications on semiotics: in addition to the papers included in the present book, *Zur Lehre vom Inhalt und Gegenstand der Vorstellungen*, Wien 1894; *Wyobrażenia i pojęcia* (Ideas and Concepts), Lwów 1898; O tak zwanych prawdach względnych (On What Is Called Relative Truth), in *The Lwów University Festschrift Commemorating the Foundation of the Jagellonian University in Cracow* (in Polish), Lwów 1900; *Ueber begriffliche Vorstellungen*, Leipzig 1903 (expanded Polish version in 1924); O idio- i allogenetycznych teoriach sądu (Idiogenetic and Allogenetic Theories of Judgements), *Przegląd Filozoficzny*, X, 1907; O czynnościach i wytworach. Kilka uwag z pogranicza psychologii, gramatyki i logiki (On activity and products, some remarks from the Border Line of Psychology, Grammar, and Logic) in: *The Festschrift Commemorating the Foundation of Lwów University*; Co znaczy 'doświadczalny' (The Meaning of the Term 'Experimental'), *Ruch Filozoficzny*, II, 1912; Granice puryzmu (The Limits of Purism), *Ruch Filozoficzny*, III, 1913; Co znaczy 'fizyczny' (The Meaning of the Term 'Physical'), *Ruch Filozoficzny*, V, 1919–20; O prawdzie formalnej (On Formal Truth), *Ruch Filozoficzny*, VII, 1922–3. *O istocie pojęć* (On the Nature of Concepts), Lwów 1924.

WÓJCICKI, Ryszard (b. 1931). Studied physics at Wrocław University and philosophy at Warsaw University. Ph. D. for a thesis on arbitrary definitions (Wrocław University, 1961 under M. Lutman-Kokoszyńska). *Venia legendi* for a thesis on the semantic criteria of empirical meaningfulness (Wrocław University, 1966). Since 1968, at the Institute of Philosophy and Sociology, Polish Academy of Sciences, member of the editorial board *Studia Logica*. Since 1970 the Head of the Section of Logic of the Institute, and since 1974 the Editor in Chief of *Studia Logica*. Member of the Editorial Board of *Erkenntnis* and *Poznań Studies in Philosophy of Science and the Humanities*.

Publications on semiotics: Definicje arbitralne i ich analityczne komponenty (Arbitrary Definitions and Their Analytic Components), *Studia Logica*, XIV, 1963; Sensowność terminów teoretycznych. ... (Meaningfulness of Theoretical Terms ...), *Ruch Filozoficzny*, 1964; O warunkach relatywnej sensowności terminów (Conditions of the Relative Meaningfulness of Terms), in: *Teoria i doświadczenie* (Theory and Experimental Data), PWN 1966; Analityczność, syntetyczność, empiryczna sensowność zdań (The Analyticity, Syntheticity, and Empirical Meaningfulness of Statements), *Studia Filozoficzne*, 3, 1966; Semantical Criteria of Empirical Meaningfulness, *Studia Logica*, XIX, 1966; (with M. Przełęcki) The Problem of Analyticity, *Synthese*, 19, 3–4, 1969.

INDEX